Rajendra Chandra Padalia, Dakeshwar Kumar Verma, Charu Arora
and Pramod Kumar Mahish (Eds.)
Essential Oils

Also of interest

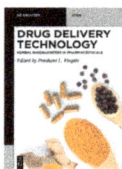

Drug Delivery Technology.
Herbal Bioenhancers in Pharmaceuticals
Prashant L. Pingale (Ed.), 2022
ISBN 978-3-11-074679-2, e-ISBN (PDF) 978-3-11-074680-8

Chemical Sciences in the Focus
Volume 1: Pharmaceutical Applications
Ponnadurai Ramasami, 2021
ISBN 978-3-11-071072-4, e-ISBN (PDF) 978-3-11-071082-3

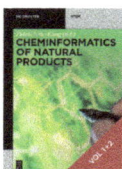

Cheminformatics of Natural Products
Volume 1 Fundamental Concepts
Volume 2 Advanced Concepts and Applications
Fidele Ntie-Kang (Ed.), 2020/2021
Set ISBN 978-3-11-067214-5

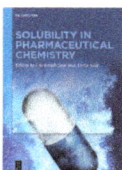

Solubility in Pharmaceutical Chemistry
Christoph Saal und Anita Nair (Eds.), 2020
ISBN 978-3-11-054513-5, e-ISBN (PDF) 978-3-11-055983-5

Essential Oils

Sources, Production and Applications

Edited by
Rajendra Chandra Padalia, Dakeshwar Kumar Verma,
Charu Arora and Pramod Kumar Mahish

DE GRUYTER

Editors

Dr. Rajendra Chandra Padalia
CSIR-Central Institute of Medicinal and
Aromatic Plants (CIMAP)
Pantnagar 263 149
Uttarakhand
India
Email: padaliarc@rediffmail.com

Dr. Charu Arora
Department of Chemistry
Ghasidas University
Bilaspur
Chhattisgarh 495009
India
Email: charuarora77@gmail.com

Dakeshwar Kumar Verma, Ph.D.
Department of Chemistry
Govt. Digvijay Autonomous Postgraduate
College
Rajnandgaon 491441
Chhattisgarh
India
Email: dakeshwarverma@gmail.com

Dr. Pramod Kumar Mahish
Department of Biotechnology
Govt. Digvijay Autonomous Postgraduate
College
Rajnandgaon 491441
Chhattisgarh
India
Email: drpramodkumarmahish@gmail.com

ISBN 978-3-11-079159-4
e-ISBN (PDF) 978-3-11-079160-0
e-ISBN (EPUB) 978-3-11-079161-7

Library of Congress Control Number: 2022947370

Bibliographic information published by the Deutsche Nationalbibliothek
The Deutsche Nationalbibliothek lists this publication in the Deutsche Nationalbibliografie;
detailed bibliographic data are available on the internet at http://dnb.dnb.de.

© 2023 Walter de Gruyter GmbH, Berlin/Boston
Cover image: egal/iStock/Getty Images Plus
Typesetting: Integra Software Services Pvt. Ltd.
Printing and binding: CPI books GmbH, Leck

www.degruyter.com

Contents

Author list

Sushma Kholiya
CSIR-Central Institute of Medicinal and
Aromatic Plants (CIMAP)
Research Centre
Pantnagar 263 149
Uttarakhand
India

Amit Chauhan
CSIR-Central Institute of Medicinal and
Aromatic Plants (CIMAP)
Research Centre
Pantnagar 263149
Uttarakhand
India

Dipender Kumar
CSIR-Central Institute of Medicinal and
Aromatic Plants (CIMAP)
Research Centre
Pantnagar 263 149
Uttarakhand
India

Venkatesha KT
CSIR-Central Institute of Medicinal and
Aromatic Plants (CIMAP)
Research Centre
Pantnagar 263 149
Uttarakhand
India

R. K. Upadhyay
CSIR-Central Institute of Medicinal and
Aromatic Plants (CIMAP)
Research Centre
Pantnagar 263 149
Uttarakhand
India

R. C. Padalia
CSIR-Central Institute of Medicinal and
Aromatic Plants (CIMAP)
Research Centre
Pantnagar 263 149
Uttarakhand
India
and

Academy of Scientific and Innovative
Research (AcSIR)
CSIR-Human Resource Development Centre
(CSIR-HRDC)
Postal Staff College Area
Sector 19, Kamla Nehru Nagar
Ghaziabad 201 002
Uttar Pradesh
India
E-mail: rc.padalia@cimap.res.in

Saeed Mollaei
Phytochemical Laboratory
Department of Chemistry
Faculty of Sciences
Azarbaijan Shahid Madani University
Tabriz
Iran
E-mail: s.mollaei@azaruniv.ac.ir

Poopak Farnia
Mycobacteriology Research Centre (MRC)
National Research Institute of Tuberculosis
and Lung Disease (NRITLD)
Shahid Beheshti University of Medical
Sciences
Tehran
Iran

Saeid Hazrati
Department of Agronomy
Faculty of Agriculture
Azarbaijan Shahid Madani University
Tabriz
Iran
E-mail: saeid.hazrati@azaruniv.ac.ir

Joshua H. Santos
Chemicals and Energy Division
Department of Science and Technology –
Industrial Technology Development Institute
DOST Compound
General Santos Avenue
Bicutan
Taguig, Metro Manila
Philippines
Email: jhsantos@dost.gov.ph

https://doi.org/10.1515/9783110791600-203

Mark Lloyd G. Dapar
Department of Biology
College of Arts and Sciences
and
Center for Biodiversity Research and
Extension in Mindanao
and
Microtechnique and Systematics Laboratory
Natural Science Research Center
Central Mindanao University
University Town
Musuan
Bukidnon 8714
Philippines
Email: f.marklloyd.dapar@cmu.edu.ph

Shekhar Verma
University College of Pharmacy Raipur
Pandit Deendayal Upadhyay Memorial Health
Sciences and Ayush University of
Chhattisgarh
Chhattisgarh 493661
India
Email: shekharpharma@gmail.com

Sonam Soni
University College of Pharmacy Raipur
Pandit Deendayal Upadhyay Memorial Health
Sciences and Ayush University of
Chhattisgarh
Chhattisgarh 493661
India

Nagendra Chandrawanshi
School of Studies in Biotechnology
Pandit Ravishankar Shukla University
Raipur
Chhattisgarh
India

Shivendra Singh Dewhare
School of Studies in Life Science
Pandit Ravishankar Shukla University
Raipur
Chhattisgarh
India

Rushendran R
Department of Pharmacology
SRM College of Pharmacy
SRM Institute of Science and Technology
Kattankulathur 603 203
Chengalpattu
Tamil Nadu
India

Anuragh Singh
Department of Pharmacology
SRM College of Pharmacy
SRM Institute of Science and Technology
Kattankulathur 603 203
Chengalpattu
Tamil Nadu
India

Siva Kumar B
Department of Pharmaceutical Chemistry
SRM College of Pharmacy
SRM Institute of Science and Technology
Kattankulathur 603 203
Chengalpattu
Tamil Nadu
India

Ilango K
Department of Pharmaceutical Quality
Assurance
SRM College of Pharmacy
SRM Institute of Science and Technology
Kattankulathur 603 203
Chengalpattu
Tamil Nadu
India
Email: ilangok1@srmist.edu.in

Shivendra Singh Dewhare
School of Studies in Life Science
Pt. Ravishankar Shukla University
Raipur
Chhattisgarh
India
Email: shivendraprsu@gmail.com

Nagendra Kumar Chandrawanshi
School of Studies in Biotechnology
Pt. Ravishankar Shukla University
Raipur
Chhattisgarh
India

Shekhar Verma
University College of Pharmacy
Pt. Deendayal Upadhyay Memorial Health
Sciences and Ayush University of
Chhattisgarh
Chhattisgarh
India

Sonam Soni
University College of Pharmacy
Pt. Deendayal Upadhyay Memorial Health
Sciences and Ayush University of
Chhattisgarh
Chhattisgarh
India

Pramod Kumar Mahish
Government Digvijay Autonomous Post
Graduate College
Rajnandgaon
Chhattisgarh
India

Nagendra Kumar Chandrawanshi
School of Studies in Biotechnology
Pt. Ravishankar Shukla University
Raipur
Chhattisgarh
India
E mail: chandrawanshi11@gmail.com

Deepali
School of Studies in Biotechnology
Pt. Ravishankar Shukla University
Raipur
Chhattisgarh
India

Anjali Kosre
School of Studies in Biotechnology
Pt. Ravishankar Shukla University
Raipur
Chhattisgarh
India

Shivendra Singh Dewhare
School of Studies in Life Science
Pt. Ravishankar Shukla University
Raipur
Chhattisgarh

Shekhar Verma
University College of Pharmacy
Pandit Deendayal Upadhyay Memorial Health
Sciences and Ayush University of
Chhattisgarh
Chhattisgarh
India

Pramod Kumar Mahish
Government Digvijay Autonomous Post
Graduate College
Rajnandgaon
Chhattisgarh
India

Ashish Kumar
Department of Biotechnology
Sant Gahira Guru University
Sarguja
Ambikapur
Chhattisgarh
India

V. Shanthi
Department of Microbiology and
Biotechnology
St. Thomas College
Bhilai
Chhattisgarh
India

Shubha Diwan
Department of Microbiology and
Biotechnology
St. Thomas College
Bhilai
Chhattisgarh
India
E-mail: shubha2315@gmail.com

Siham Abdulrazzaq Salim
Department of Biology
College of Education
Al-Iraqia University
Baghdad
Iraq

Deepali Koreti
School of Studies in Biotechnology
Pt. Ravishankar Shukla University
Raipur
Chhattisgarh
India
Email: ranukoreti27@gmail.com

Anjali Kosre
School of Studies in Biotechnology
Pt. Ravishankar Shukla University
Raipur
Chhattisgarh
India

Pramod Kumar Mahish
Government V.Y.T. PG Autonomous College
Durg
Chhattisgarh
India

Nagendra Kumar Chandrawanshi
School of Studies in Biotechnology
Pt. Ravishankar Shukla University
Raipur
Chhattisgarh
India

Shri Ram Kunjam
Government Digvijay (Autonomous) Post
Graduate College
Rajnandgaon
Chhattisgarh
India

Vatsala Soni
PEC-Punjab Engineering College
Chandigarh 160012
India

Dipti Bharti
Department of Applied Sciences and
Humanities
Darbhanga College of Engineering
Darbhanga 846005
Bihar
India

Meenakshi Bharadwaj
Department of Chemistry
IEC University
Baddi
Himachal Pradesh
India

Vaishali Soni
PGIMER-Post Graduate Government Institute
of Medical Education and Research
Chandigarh 160012
India

Richa Saxena
Department of Biotechnology
Invertis University
Bareilly 243001
Uttar Pradesh
India

Charu Arora
Department of Chemistry
Guru Ghasidas University
Bilaspur495009
Chhattisgarh
India

Himanshu Pandey
M.B.G.P.G College Haldwani
Kumaun University
Nainital 263139
Uttarakhand
India

Sushma Kholiya
M.B.G.P.G College
Kumaun University
Haldwani
Nainital 263139
Uttarakhand
India
and
CSIR-CIMAP Research Center
Pantnagar 263149
Uttarakhand
India

R. C. Padalia
CSIR-CIMAP Research Center
Pantnagar 263149
Uttarakhand
India

Priyanka Tiwari
M.B.G.P.G College
Kumaun University
Haldwani
Nainital 263139
Uttarakhand
India

Ameeta Tiwari
M.B.G.P.G College
Kumaun University
Haldwani
Nainital 263139
Uttarakhand
India
Email: Tiwari.ameeta@gmail.com

Meenakshi Sharma
IEC University
Baddi 174103
Himachal Pradesh
India

Dipti Bharti
Department of Applied Sciences and
Humanities
Darbhanga College of Engineering
Darbhanga
Bihar 846005
India

Vatsala Soni,
PEC-Punjab Engineering College
Chandigarh 160012
India
Vaishali Soni,
PGIMER-Post Graduate Government Institute
of Medical Education and Research
Chandigarh 160012
India

Charu Arora
Department of Chemistry
Guru Ghasidas University
Bilaspur 495009
Chhattisgarh
India

Meenakshi Sharma, Dipti Bharti, Vatsala Soni, Vaishali Soni and
Charu Arora

Chapter 1
Introduction and general properties
of essential oils

Abstract: This chapter covers the literature data summarizing, on the one hand, the
chemistry of essential oils and, on the other hand, their most important activities.
Essential oils, which are complex mixtures of volatile compounds particularly
abundant in aromatic plants, are mainly composed of terpenes biogenerated by the
mevalonate pathway. These volatile molecules include monoterpenes (hydrocarbon
and oxygenated monoterpenes) and also sesquiterpenes (hydrocarbon and oxygenated sesquiterpenes). Furthermore, they contain phenolic compounds, which are
derived via the shikimate pathway. Thanks to their chemical composition, essential
oils possess numerous biological activities (antioxidant, anti-inflammatory, antimicrobial, etc.) of great interest in food and cosmetic industries, as well as in the
human health field.

Keywords: Introduction of essential oils, chemical composition, general properties

1.1 Introduction

Essential oils are complex mixtures of volatile chemicals found in living organisms.
The attraction of medicinal and aromatic plants is continuously growing due to increasing consumers demand and interest in these plants for culinary, medicinal,
and other anthropogenic applications [1].

The term "essential oil" dates back to the sixteenth century. Essential oils or
"essences" owe their name to their flammability. Numerous authors have attempted
to provide a definition for essential oils [2]. "The essential oil is the product obtained from a vegetable raw material, either by steam distillation or by mechanical
processes from the epicarp of Citrus, or dry distillation."

Meenakshi Sharma, IEC University Baddi, Himachal Pradesh 174103, India
Dipti Bharti, Department of Applied Science and Humanities, Darbhanga College of Engineering,
Darbhanga, Bihar 846005, India
Vatsala Soni, Punjab Engineering College, Chandigarh 160012, India
Vaishali Soni, PGIMER-Post Graduate Government Institute of Medical Education and Research,
Chandigarh 160012, India
Charu Arora, Department of Chemistry, Guru Ghasidas Vishwavidyalaya, Bilaspur-495009, C.G

https://doi.org/10.1515/9783110791600-001

Essential oils are widely used in flavor, food, fragrance, and cosmetic industries for various applications. Contact allergy to them is well known and has been described for 80 essential oils. The relevance of positive patch test reactions often remains unknown. Knowledge of the chemical composition of essential oils among dermatologists is suspected to be limited, as such data are published in journals not read by the dermatological community. Therefore, the authors have fully reviewed and published the literature on contact allergy and on the chemical composition of essential oils.

These volatile oils are generally liquids and are colorless at room temperature. Essentials oils are insoluble in water but soluble in alcohol, ether, and fixed oils. They have a refractive index and a very high optical activity. They have a characteristic odor, are usually liquids at room temperature, and have a density less than unity, with the exception of a few cases (cinnamon, sassafras, and vetiver). These volatile oils present in herbs are responsible for different scents that plants emit. They are widely used in cosmetics industry, perfumery, and also aromatherapy. The latter is intended as a therapeutic technique including massage, inhalations, or baths using these volatile oils [3]. The last key will serve as chemical signals allowing the plant to control or regulate its environment (ecological role), attraction of pollinating insects, repellent to predators, inhibition of seed germination, or communication between plants (emission signals chemically signaling the presence of herbivores, for example). Moreover, aromatic plants are the major source of essential oils which also possess antifungal or insecticide and deterrent activities. These may be found in almost all parts of aromatic plants that may contain essential oils such as leaves, flowers, bark, seeds, fruits, wood, rhizome, root, and root bark.

1. Leaves (eucalyptus, cedar, and laurel)
2. Leafy branches (pine)
3. Herbaceous parts (oregano, mint, and sage)
4. Flowers (rose and jasmine)
5. Dried buds (cloves)
6. Bark (cinnamon and cassia)
7. Wood (sandalwood, cedarwood, and rosewood)
8. Bulb (onion and garlic)
9. Roots (angelica, vetiver, and orris)
10. Rhizomes (ginger and acorus)
11. Fruits (aniseed, fennel, coriander, and cumin)
12. Fruit peel (orange and lemon)
13. Pseudofruit (juniper)
14. Seed (carrot seed, mustard seed, and cardamom)
15. Root bark (sassafras and xylopia)
16. Balsam (storax and Peru balsam)
17. Oleo-gum resin (frankincense, myrrh, and mastic)
18. Oleoresin (turpentine and opopanax)
19. Lichen (oak moss and tree moss)

This chapter describes the mechanism of essential secondary metabolite biosynthesis, essential oil extraction, essential oil chemical profile, and its pharmacological potential against the top list of human killer diseases as presented by the World Health Organization.

1.2 Chemical composition

1.2.1 Biosynthesis

Terpenoid and phenylpropanoid derivatives are the main components found in essential oils. In most plants, their essential oils contain terpenoids at around 80%. But the presence of phenylpropanoid derivatives affords the essential oils' significant flavor, odor, and piquant. Along with substantial improvements in analytical techniques, characteristics such as specific gravity, optical rotation, refractive index, and solubility continue to be important in determining the quality of essential oils and aroma compounds [4]. The secondary metabolites of volatile plants known as essential oils have a molecular mass less than 300 and a nice odor. The essential oils are weakly soluble in water but are soluble in alcohol, nonpolar solvents, waxes, and oils [5, 6]. There are few exceptions to the rule that most essential oils are colorless or pale yellow, such as chamomile (*Matricaria chamomilla*), and chrysanthemum which produces blue color oil, with the exception of sassafras, vetiver, cinnamon, and clove's essential oils, which are denser than water. The changes in characteristics such as solubility in ethanol, relative density, refractive index, and optical purity of a particular essential oil indicate the adulteration and modulation in chemical composition [7].

1.3 Essential oil extraction

Essential oil extraction is one of the critical points that can affect the chemical profile of the essential oil. Sensu stricto, essential oils are volatile odorant complex mixture obtained by distillation. The extraction of essential oils from plants is one of key processes for their end uses and to improve the yield and quality of essential oil. There are various methods employed for extraction of essential oils, which includes both traditional and modern techniques in many parts of the globe. By using several distillation process techniques, essential oils are separated from aromatic plants. Other volatile isolates, however, are also obtained using solvent extraction, cold press, and various other methods. Numerous techniques, including the most popular ones like hydrodistillation, steam distillation, cohobation, cold pressing, maceration, solvent extraction, and simultaneous distillation–extraction techniques, and supercritical fluid techniques, can be employed for this purpose [8]. Despite the fact that these methods have been employed for essential oil extraction for a long time, their use has revealed a wide range of drawbacks, including the loss of some volatile compounds, low extraction efficiency, the degradation of unsaturated or ester compounds through thermal or hydrolytic effects, and potential toxic solvent residues in extracts. The oil extraction companies concentrated on the development of emergent extraction technologies as a result of the rise in energy costs and the advent of the "Green Era." Several new methods are currently available for the extraction of essential oils from plants, including supercritical fluid extraction, pressurized liquid extraction, pressurized hot water extraction, membrane-assisted solvent extraction, solid-phase microextraction, microwave-assisted extraction, and ultrasound-assisted extraction, in order to overcome the drawbacks of traditional methods of extraction. Most of these extraction methods lead most of the time to artifactual products as well as transformed products. To better understand, the following sections will present the most used methods and their principal limits in the way of modification of the original chemical profile of essential oils [9–11].

1.3.1 Distillation method

Distillation methods are a group of methods using steam as a compound vector or transporter. In fact, in distillation method, the plant material may be immersed in water or may not, and after heating to water's boiling point [12], the impression formed in the reactor by steam as well as high temperature will produce the vaporization of these volatile compounds from their stockade cell to the environment of the reactor. The gas is pouched throughout a cooler. The condensation of water and volatilized compounds from their vapor to water phase forms a mixture that can be separated based on their density.

1.3.2 Hydrodistillation

In this method, the plant material is completely soaked in water, indicating that the raw materials come into direct contact with hot water [13]. Depending on the quality of the material treated each time, it may float on the water or may totally submerge. The water is heated to the boiling point using any standard method, such as direct flame, steam jacket, closed steam coil, or, in rare instances, a perforated steam coil. Evaporation, condensation, and separation based on difference in essential oil density are same as in other distillation processes. Before conducting any field distillations for large-scale manufacturing, a small-scale water distillation in glassware (Clevenger apparatus for laboratory purpose), which works on the same principle of hydrodistillation, should be carried out to see if any changes occur during the distillation process [14]. This technique is very simple and inexpensive, and with simple-to-build units it is suitable for on-field extraction of essential oils. Moreover, this technique is recommended to fine powdered materials of plant parts, mainly spices. This distillation process is treated as an art by local distillers, who rarely try to optimize both oil yield and quality; therefore, it requires a great deal of expertise and procedure knowledge. Other disadvantages of this technique are that some of the oil components like esters are sensitive to hydrolysis, and acyclic monoterpenes, hydrocarbons, and aldehydes are susceptible to polymerize during hydrodistillation. Sometimes the loss in yield and quality of essential oils observed is due to solubility of oxygenated components such as phenols and incomplete vaporization of high-boiling oil components [15–17].

1.3.3 Cold pressing

Cold press method, also known as scarification method, is a low-cost technique for extraction of essential oils of certain aromatic plants, especially for extraction of oils from the peels of citrus fruits by applying mechanical pressure in the form of crushing and pressing in low temperature. Different types of cold press machines, namely, gear, rack, hydraulic press, and scrapers used for cold extraction at industrial level. The essential oil extracted from this method also contains certain plant pigments, coumarins, glucosinolate residues, and other volatiles, which are not part of the same essential oil extracted from distillation methods. The cold-pressed essential oils are of higher purity retaining the real aroma of the extraction and fetch premium price in trade [8, 18].

1.3.4 Steam distillation

In the process of steam distillation, steam is generated in a satellite steam generator/boiler and then it is driven through a pipe into a still where the plant material is placed on a perforated tray just as in hydro-cum-steam distillation [17]. A significant benefit of satellite steam generation is the ease with which the volume of steam can be controlled. This method is most frequently used for the industrial manufacturing of essential oils on large batches through one satellite boiler. The major advantages of steam distillation are its controlled distillation as steam can be regulated through boiler, no thermal degradation of essential oil constituents, and consistent quality of the extracted essential oils. The only disadvantages are due to its higher capital expenditure and requirement of trained persons for boiler operation. In terms of essential oil yield, distillation speed, and consistent quality of the essential oil, the steam distillation is superior over hydro-cum-steam distillation and hydrodistillation. The most common method for extracting essential oils on a large scale is steam distillation. However, hydro-cum-steam distillation requires very less investment to establish the distillation setup as compared to steam distillation, and is therefore the most frequently used distillation process for farmers and budding entrepreneurs. In specifically designed field distillation units, commonly known as improved field distillation units, the inclusion of calandria with smoke pipe inside the still acts as in-built small boiler, reduces the fuel consumption, and generates balanced steam for essential oil extraction [15–17].

1.3.5 Solvent extraction

Essential oils can also be extracted using organic solvents such as acetone, hexane, petroleum ether, methanol, or ethanol. In this method, the plant material is submerged or macerated with the organic solvent for a certain period. The essential oils along with pigment and few other components become miscible with the used solvents. After filtration, the essential oil's saturated supernatant was evaporated by spontaneous evaporation or using a rotary evaporator under pressure to the concentrate form known as concrete. The concrete is composed of wax, fragrance, and essential oil. The essential oil is extracted from this concrete by using ethyl alcohol, which is actually known as absolute, wax-free residue; and the essential oils from this absolute are distilled at low temperature under controlled conditions. The main disadvantage of this technique is longer extraction time and quality deterioration due to the presence of solvent trace. This method is useful for delicate flower extraction, which consists of various thermolabile constituents. The concrete and absolutes have their own demands in perfumery industry, subject to meet the regulatory regulation for trace of the solvent [18, 19].

1.3.6 Enfleurage

This is one of the traditional methods to extract the fragrant volatile/essential oils of flowers and delicate plant parts. However, the traditional perfumers in some parts of India, France, Egypt, and Indonesia still use this technique for oriental perfumery material. In this method, the flower petals are spread over grease and cold fat (tallow or lard) for a period of time with specially designed chassis, and the essence is absorbed over the fat till absorption saturation. The essential oils/volatiles/fragrances are then extracted using alcohol. Thereafter, the alcoholic mixture of the essential oil is vacuum distilled to separate the oil from the alcohol. This technique is time-consuming, labor-intensive, and expensive [8].

1.3.7 Microwave-assisted essential oil extraction

Microwave-assisted essential oil extraction is a variant of the distillation method where the heating source has been changed from the normal electric heating cap by the microwave. The advantage here is the hypothetical increase in the extraction yield because the increase in yield is not as spectacular as difficult. It is true that it is better to crush the plant material, but in comparison to the classic distillation method [20], the essential oil yield is systematically the same. The principle of this method is based on the change of the polarity of water by the waves and of course the heating that will play the same role as in classic distillation method. This method has, in addition, the limit of the normal distillation method, the fact that the microwave can lead to chemical stereoswitching from one isomer to another.

1.3.8 Ultrasound-assisted extraction (UAE)

The ultrasound-assisted extraction (UAE) technique has been recognized one of the best extraction techniques for various herbal industries. It is also known as ultrasonic extraction or sonication and utilizes ultrasonic energy for the extraction. UAE is used to isolate volatile components from natural materials at room temperature while using organic solvents, which reduces the amount of solvents used, speeds up the extraction process, and increases the extract yield as compared to traditional procedures. This method is well suited for thermolabile and unstable compounds and is more selective when compared to other extraction techniques [21].

1.4 Method for chemical analysis

There is a myriad of technique and methods for essential oil chemical profiling. All the methods used in organic chemistry can be used here. Due to their volatility nature, the compounds that constitute essential oil are preferably analyzed by gas chromatography (GC). GC alone does not provide enough data for good chemical proofing. Therefore, many other analytical tools have been used such as mass spectrometry (MS), infrared (IR) spectroscopy, and nuclear magnetic resonance. As well, many techniques have been used to make the GC a better tool for chemical profiling, and these include chiral-selective GC and multidimensional GC. Advances in liquid chromatography have highlighted the usefulness of high-pressure liquid chromatography (HPLC) as a tool for essential oil analysis. Many variants are available to date as multidimensional HPLC, HPLC-MS, and HPLC-GC. GC coupled with MS is used for its popularity, and the Fourier-transform IR spectroscopy is used for its simplicity, for being environment-friendly, and for long-term cost-effectiveness [22].

1.5 Chemistry of essential oils

Essential oils are produced by various differentiated structures, especially the number and characteristics of which are highly variable. Essential oils are localized in the cytoplasm of certain plant cell secretions, which lie in one or more organs of the plant, namely, the secretory hairs or trichomes, epidermal cells, internal secretory cells, and the secretory pockets. These oils are complex mixtures that may contain over 300 different compounds [4]. They consist of organic volatile compounds, generally of low molecular weight (<300). Their vapor pressure at atmospheric pressure and at room temperature is sufficiently high so that they are found partly in the vapor state [5, 6]. These volatile compounds belong to various chemical classes: alcohols, ethers or oxides, aldehydes, ketones, esters, amines, amides, phenols, heterocycles, and mainly the terpenes. Alcohols, aldehydes, and ketones offer a wide variety of aromatic notes, such as fruity ((E)-nerolidol), floral (linalool), citrus (limonene), and herbal (γ-selinene).

Furthermore, essential oil components belong mainly to the vast majority of the terpene family. Structure and names of some mono and sesquiterpenoids are reported in Figure 1.1. Many thousands of compounds belonging to the family of terpenes have so far been identified in essential oils [7], such as functionalized derivatives of alcohols (geraniol and α-bisabolol), ketones (menthone and p-vetivone) or aldehydes (citronellal and sinensal), esters (γ-terpinyl acetate and cedryl acetate), and phenols (thymol).

Essential oils also contain non-terpenic compounds biogenerated by the phenylpropanoids pathway, such as eugenol, cinnamaldehyde, and safrole.

Essential oils have a very high variability of their composition, both in qualitative and quantitative terms. Various factors are responsible for this variability and can be grouped into two categories:

- Intrinsic factors related to the plant, and interaction with the environment (soil type and climate) and the maturity of the plant concerned, even at harvest time during the day
- Extrinsic factors related to the extraction method and the environment

The factors that determine the essential oil yield and composition are numerous. In some cases, it is difficult to isolate these factors from each other as they are interrelated and influence each other. These parameters include the seasonal variations, plant organ, degree of maturity of the plant, geographic origin, and genetics [23–25].

Several techniques are used for the trapping of volatiles from aromatic plants. The most often used device is the circulatory distillation apparatus described by Cocking and Middleton [26] introduced in the European Pharmacopoeia and several other pharmacopoeias. This device consists of a heated round-bottom flask into which the chopped plant material and water are placed and is connected to a vertical condenser and a graduated tube, for the volumetric determination of the oil. At

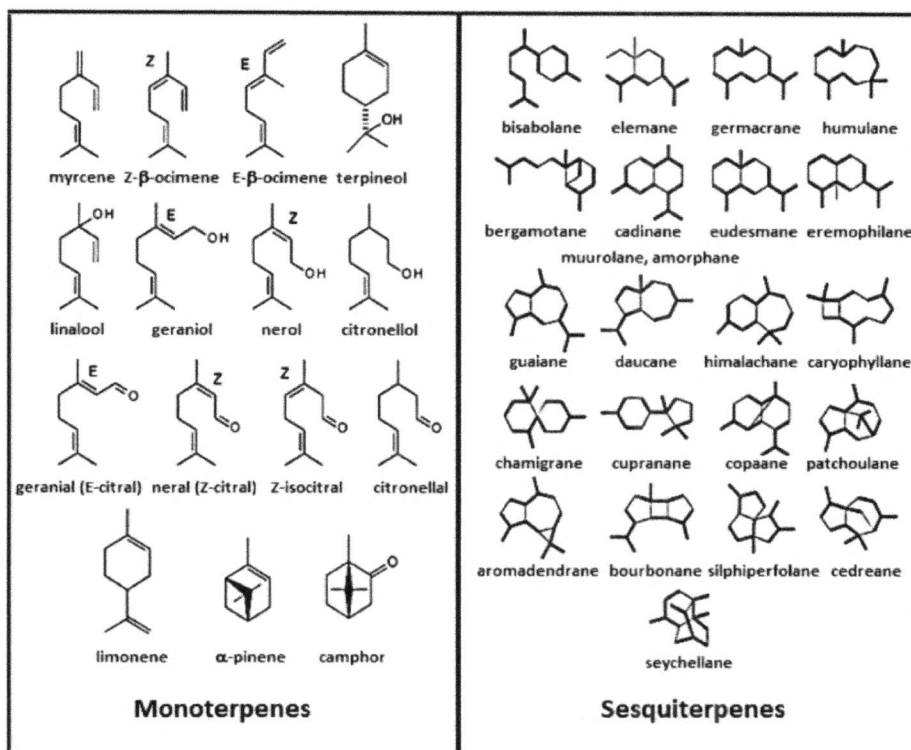

Figure 1.1: Monoterpenes and sesquiterpenes.

the end of the distillation process, the essential oil is separated from the water phase for further investigations. The length of distillation depends on the plant material to be investigated. It is usually fixed to 3–4 h.

A further improvement was the development of a simultaneous distillation–solvent extraction device by Likens and Nickerson in 1964 [27]. The device permits continuous concentration of volatiles during hydrodistillation in one step using a closed-circuit distillation system.

1.6 Application of essential oils

Essential oils are highly concentrated natural extracts from leaves, flowers, and stems of plants. The most common way to use essential oils is to inhale them, both for their amazing scent and their therapeutic properties. But they can also be used in diffusers and humidifiers, as well as diluted with a carrier oil and applied to the skin. Structure and names of some well known constituents of essential oils are reported in Fig. 1.2.

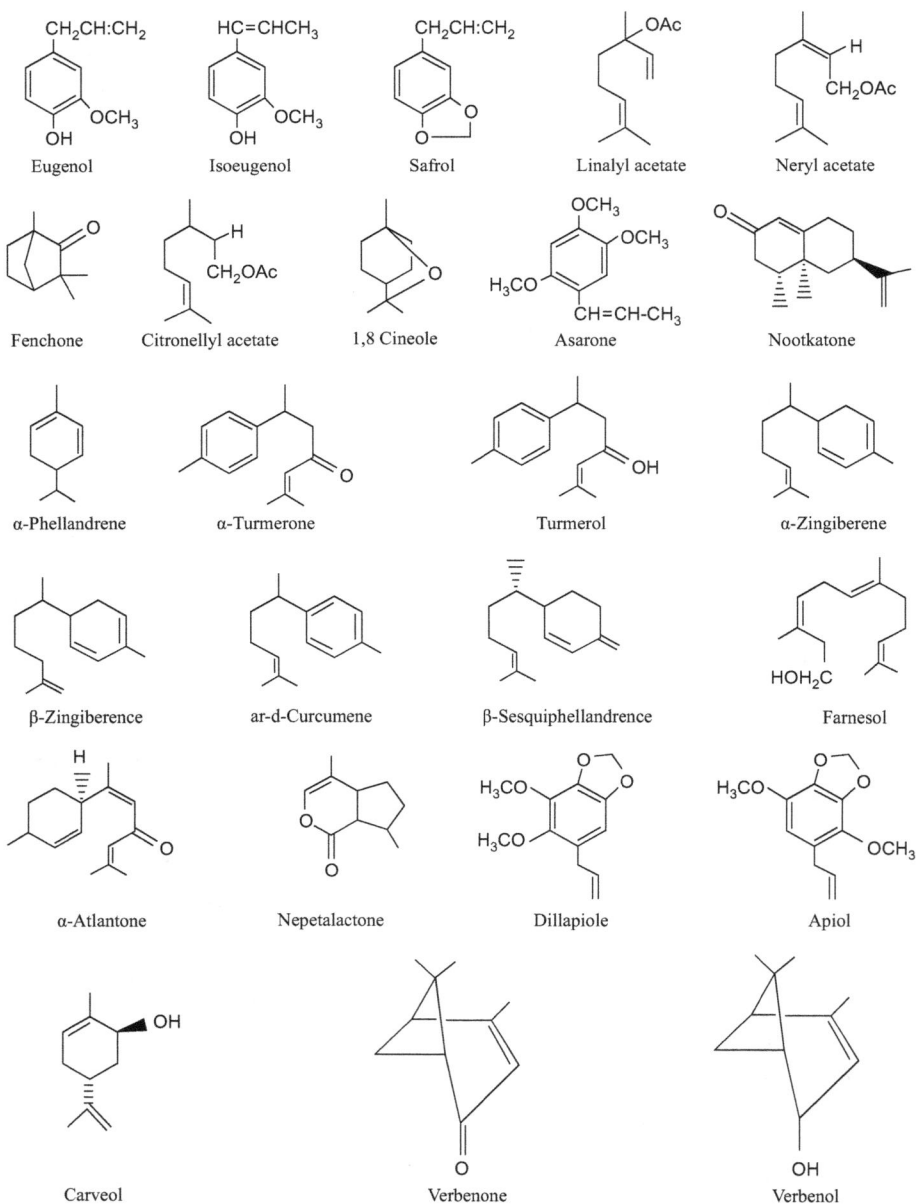

Figure 1.2: Chemical structures of essential oil constituents.

Essential oils have a wide range of medicinal and therapeutic properties. Their antifungal, antibacterial, and antiviral properties make them a useful product in your medicine list. They have also been shown to promote healthy sleep, relieve headaches, and alleviate pain. In addition, essential oils can improve skin conditions, help treat common cold, and encourage healthy digestion.

A diffuser is a device that disperses essential oils into the air. There are several types of essential oil diffusers that you can use for this. For safety reasons, follow the manufacturer's instructions.

Popular types of essential oil diffusers include:
- ceramic
- electric
- candle
- lamp rings
- reed diffuser
- ultrasonic

There is a range of plants that contain potentially active compounds. Manufacturers have turned dozens of these plant oils into essential oil products. Some of the most popular essential oils include:
- lavender
- peppermint
- tea tree
- lemon
- sweet orange
- eucalyptus
- orange
- chamomile
- ylang-ylang

The compounds in these oils may have some health effects on the human body. The following common health conditions may benefit from the use of essential oils:
- headaches
- constipation
- depression
- cold sores
- sinus infections
- sore muscles
- anxiety

1.7 Biological activity of essential oils

1.7.1 Antibacterial activity

The antimicrobial properties of essential oils and of their constituents have been considered [28, 29], and the mechanism of action has been studied in detail [30]. An important feature of essential oils is their hydrophobicity, which allows them to partition into lipids of the cell membrane of bacteria, disrupting the structure, and making them more permeable. This can then cause leakage of ions and other cellular molecules [31–35]. Although a certain amount of leakage of bacterial cells can be tolerated without loss of viability, greater loss of cell contents or critical output of molecules and ions can lead to cell death [36].

Essential oils and/or their constituents can have a single target or multiple targets of their activity. For instance, *trans*-cinnamaldehyde can inhibit the growth of *Escherichia coli* and *Salmonella Typhimurium* without disintegrating the outer membrane (OM) or depleting intracellular ATP. Similar to thymol and carvacrol, *trans*-cinnamaldehyde likely gains access to the periplasm and deeper portions of the cell [37]. Carvone is also ineffective against the OM and does not affect the cellular ATP pool.

It has been reported that essential oils containing mainly aldehydes or phenols, such as cinnamaldehyde, citral, carvacrol, eugenol, or thymol, were characterized by the highest antibacterial activity, followed by essential oils containing terpene alcohols. Other essential oils, containing ketones or esters, such as β-myrcene, α-thujone, or geranyl acetate, had much weaker activity, while volatile oils containing terpene hydrocarbons were usually inactive.

Generally, essential oils characterized by a high level of phenolic compounds, such as carvacrol, eugenol, and thymol, have important antibacterial activities [38–40].

1.7.2 Antioxidant activity

Numerous studies have demonstrated the antioxidant properties of essential oils. The antioxidant potential of an essential oil depends on its composition. It is well established that phenolics and secondary metabolites with conjugated double bonds usually show substantial antioxidative properties.

The essential oils of cinnamon, nutmeg, clove, basil, parsley, oregano, and thyme are characterized by the most important antioxidant properties [41]. Thymol and carvacrol are the most active compounds. Their activity is related to their phenolic structure. These phenolic compounds have redox properties and, thus, play an important role in neutralizing free radicals and also in peroxide decomposition. The antioxidant activity of essential oils is also due to certain alcohols, ethers, ketones, aldehydes, and monoterpenes: linalool, 1,8-cineole, geranial/

neral, citronellal, isomenthone, menthone, and some monoterpenes such as α-terpinene, β-terpinene, and α-terpinolene.

1.7.3 Anti-inflammatory activity

Inflammation is a normal protective response induced by tissue injury or infection and functions to combat invaders in the body (microorganisms and non-self-cells) and to remove dead or damaged host cells. The inflammatory response induces an increase in permeability of endothelial lining cells and influxes of blood leukocytes into the interstitium, oxidative burst, and release of cytokines, such as interleukins and tumor necrosis factor-α. It also stimulates the activity of several enzymes (oxygenases, nitric oxide synthases, peroxidases, etc.), as well as the arachidonic acid metabolism. Recently, essential oils have been used in clinical settings to treat inflammatory diseases, such as rheumatism, allergies, or arthritis. *Melaleuca alternifolia* essential oil was reported to have a considerable anti-inflammatory activity [42–44]. This activity is correlated with its major compound "α-terpineol." The active compounds act by inhibiting the release of histamine or reducing the production of inflammation mediators.

1.7.4 Cancer chemoprotective activity

The varied therapeutic potential of essential oils attracted, in recent years, the attention of researchers for their potential activity against cancer. They and their volatile constituents of the studies target the discovery of new anticancer natural products [41]. Essential oils would act in the prevention of cancer, as well as at its removal. It is well known that certain foods, such as garlic and turmeric, are good sources of anticancer agents. Garlic essential oil is a source of sulfur compounds recognized for their preventive effect against cancer [42–44]. Diallyl sulfide, diallyl disulfide, and diallyl trisulfide are examples.

1.7.5 Cytotoxicity

Due to their complex chemical composition, essential oils have no specific cellular ligands. As lipophilic mixtures, they are able to cross the cell membrane and degrade the layers of polysaccharides, phospholipids, and fatty acids, and permeabilize. This cytotoxicity appears to include such membrane damage. In bacteria, the membrane permeabilization is associated with the loss of ions and the reduction of the membrane potential, the collapse of the proton pump, and the depletion of the ATP pool. Essential oils may coagulate the cytoplasm and damage lipids and proteins. Damage to the wall and the cell membrane can lead to the leakage of macromolecules and lysis.

In addition, essential oils change membrane fluidity, which becomes abnormally permeable, resulting in a leakage of radicals, cytochrome C, Ca^{2+} ions, and proteins, as in the case of oxidative stress. This permeabilization of the outer and inner membranes causes cell death by apoptosis and necrosis. Ultrastructural alteration of the cell can be observed at a plurality of compartments [45–48].

1.8 A survey of oils and plants: allergies by essential oils

In our literature search, we have found 79 essential oils that have caused contact allergy (as demonstrated by positive patch test reactions) or allergic contact dermatitis. For some oils, only one or two such reports are available. For others, however, there is a considerable amount of literature on contact allergy and allergic contact dermatitis, for example, in the case of tea tree oil, ylang-ylang oil, lavender oil, peppermint oil, jasmine absolute, geranium oil, rose oil, turpentine oil, and sandalwood oil. Because of the selection criterion (contact allergy reported), the group of 79 oils is very heterogeneous. Some are high-volume essential oils, such as orange oil (worldwide production 60,000 tons per year) and lemon oil (nearly 9,000 tons), and others are produced and commercialized in very small quantities. The estimated commercial quantity of zdravets oil, for example, which is produced in Bulgaria only, is about 20 kg. Costus root oil is apparently still used in aromatherapy but has been prohibited by IFRA (International Fragrance Association) for use in fragrances and cosmetics since decades because of its sensitizing properties [49]. Others are prohibited or restricted by governmental regulations and laws, as in the case of rosewood oil, guaiac wood oil, and East Indian sandalwood oil, which are produced from endangered species. Table 1.1 provides a list of 91 essential oils and 2 jasmine absolutes with their common name, botanical source, parts of the plant used for obtaining the oil, and if available, the numbers of ISO (International Organization for Standardization, Geneva, Switzerland, www.iso.org) standards. The number of oils/absolutes in Table 1.1 ($n = 93$) exceeds the number of essential oils for which contact allergy has been described ($n = 79$). This is due to the fact that some oils may be obtained from different species (e.g., cedarwood oil, lemongrass oil, thyme oil, and sage oil), from different parts of the same plant (e.g., cinnamon bark oil and cinnamon leaf oil, clove bud oil, clove stem oil, and clove leaf oil), or from various cultivars producing different oils (lavandin abrial oil and lavandin grosso oil). In dermatological literature, however, such data are virtually never provided ("cedarwood oil," "clove oil," "lavandin oil," and "thyme oil"), and therefore, we chose to present the data from all possible source species and plant parts. Full

Table 1.1: List of essential oils and jasmine absolutes with their botanical sources, plant parts used, and ISO numbers.

Serial no.	Common name	Botanical source	Parts of plant used	ISO*
1	*Angelica* fruit oil	*Angelica archangelica* L.	Fruit	
2	*Angelica* root oil	*Angelica archangelica* L.	Rhizome and root	
3	Aniseed oil	*Pimpinella anisum* L.	Fruit	3475
4	Basil oil sweet	*Ocimum basilicum* L.	Flowering aerial top	11043
5	Bay oil	*Pimenta racemosa* (Mill.) J.W. Moore	Leaf	3045
6	Bergamot oil	*Citrus bergamia* (Risso and Poit.)	Pericarp (peel)	3520
7	Black cumin oil	*Nigella sativa* L.	Seed	
8	Black pepper oil	*Piper nigrum* L.	Fruit	3061
9	Cajeput oil	*Melaleuca cajuputi* Powell	Leaf terminal branchlet	
10	Calamus oil	*Acorus calamus* L.	Rhizome	
11	Cananga oil	*Cananga odorata* (Lam.) Hook f. and Thomson, forma macrophylla	Flower	3523
12	Cardamom oil	*Elettaria cardamomum* (L.) Maton	Fruit	4733
13	Carrot seed oil	*Daucus carota* L.	Fruit	
14	Cassia bark oil	*Cinnamomum cassia* (Nees and T. Nees) J. Presl	Bark	3216
15	Cedarwood oil, Atlas	*Cedrus atlantica* (Endl.) G. Manetti ex Carriere	Wood	
16	Cedarwood oil, China	*Cupressus funebris* (Endl.)	Wood	9843
17	Cedarwood oil, Himalaya	*Cedrus deodara* (Roxb. ex D. Don) G. Don	Wood	
18	Cedarwood oil, Texas	*Juniperus ashei* J. Buchholz	Wood	4725
19	Cedarwood oil, Virginia	*Juniperus virginiana* L.	Wood	4724
20	Chamomile oil, German	*Chamomilla recutita* (L.) Rauschert	Flowering tops	19332

Table 1.1 (continued)

Serial no.	Common name	Botanical source	Parts of plant used	ISO*
21	Chamomile oil, Roman	*Chamaemelum nobile* (L.)	Flowering tops	
22	Cinnamon bark oil, Sri Lanka	*Cinnamomum zeylanicum* Blume	Twig and bark of stem	
23	Cinnamon leaf oil, Sri Lanka	*Cinnamomum zeylanicum*	Blume Leaf	
24	Citronella oil, Java	*Cymbopogon winterianus* Jowitt.	Aerial part (leaves)	3848
25	Citronella oil, Sri Lanka	*Cymbopogon nardus* (L.) Rendle	Aerial part (leaves)	3849
26	Clary sage oil	*Salvia sclarea* L.	Flowering top (and leaf)	
27	Clove bud oil	*Syzygium aromaticum* (L.) Merr. and L.M. Perry	Bud	3142
28	Clove leaf oil	*Syzygium aromaticum* (L.) Merr. and L.M. Perry	Leaf	3141
29	Clove stem oil	*Syzygium aromaticum* (L.) Merr. and L.M. Perry	Stem	3143
30	Coriander fruit oil	*Coriandrum sativum* L.	Fruit	3516
31	Costus root oil	*Saussurea costus* (Falc.) Lipsch.	Root	
32	Cypress oil	*Cupressus sempervirens* L.	Twig with leaves	
33	Dwarf pine oil	*Pinus mugo* Turra	Leaf (needle), terminal branchlets	21093
34	Elemi oil	*Canarium luzonicum* (Blume) A. Gray	Wood exudate	10624
35	*Eucalyptus citriodora* oil	*Eucalyptus citriodora* Hook.	Leaf, terminal branch	3044
36	*Eucalyptus globulus* oil	*Eucalyptus globulus* Labill.	Leaf, terminal branch	770
37	Galbanum resin oil	*Ferula gummosa* Boiss.	Root exudate	40716
38	Geranium oil	*Pelargonium* × spp.	Herbaceous part	4731
39	Ginger oil	*Zingiber officinale* Roscoe.	Rhizome	16928

Table 1.1 (continued)

Serial no.	Common name	Botanical source	Parts of plant used	ISO*
40	Grapefruit oil	*Citrus paradisi* Macfad.	Pericarp (peel)	3053
41	Guaiacwood oil	*Bulnesia sarmientoi* Lorentz ex Griseb.	Wood	
42	Hyssop oil	*Hyssopus officinalis* L. ssp. *officinalis*	Flowering top and leaf	9841
43	*Jasminum grandiflorum* absolute	*Jasminum grandiflorum* L.	Flower	
44	*Jasminum sambac* absolute	*Jasminum sambac* (L.)	Aiton. Flower	
45	Juniper berry oil	*Juniperus communis* L.	Fruit, terminal branchlets	8897
46	Laurel leaf oil	*Laurus nobilis* L.	Leaf	
47	Lavandin abrial oil	*Lavandula angustifolia* Mill. × *Lavandula latifolia* Medik. "Abrial"	Flowering top	3054
48	Lavandin grosso oil	*Lavandula angustifolia* Mill. × *Lavandula latifolia* Medik. "Grosso"	Flowering top	8902
49	Lavandin oil	*Lavandula angustifolia* Mill. × *Lavandula latifolia* Medik.	Flowering top	
50	Lavender oil	*Lavandula angustifolia* Mill.	Flowering top	3515
51	Lemon oil	*Citrus limon* (L.) Burm. f.	Pericarp (peel)	855
52	Lemongrass oil, East Indian	*Cymbopogon flexuosus* (Nees ex Steudel) J.F. Watson	Aerial part (leaves)	4718
53	Lemongrass oil, West Indian	*Cymbopogon citratus* (DC) Stapf.	Whole aerial part (leaves)	3217
54	*Litsea cubeba* oil	*Litsea cubeba* (Lour) Pers.	Fruit	3214
55	Lovage oil	*Levisticum officinale* W.D.J.	Koch Root	11019
56	Mandarin oil	*Citrus reticulata* Blanco	Pericarp (peel)	3528
57	Marjoram oil (sweet)	*Origanum majorana* L.	Flowering top	

Table 1.1 (continued)

Serial no.	Common name	Botanical source	Parts of plant used	ISO*
58	Melissa oil (lemon balm oil)	*Melissa officinalis* L.	Aerial parts	
59	Myrrh oil	*Commiphora myrrha* (Nees) Engl. and related *Commiphora* species	Wood exudate	
60	Neem oil	*Azadirachta indica* A. Juss.	Seed	
61	Neroli oil	*Citrus aurantium* L.	Flower	3517
62	Niaouli oil	*Melaleuca quinquenervia* (Cav.) S.T. Blake	Leaves, terminal branchlets	
63	Nutmeg oil	*Myristica fragrans* Houtt.	Seed	3215
64	Olibanum (frankincense) oil	*Boswellia sacra* Flueck.	Wood exudate	
65	Orange oil, bitter	*Citrus aurantium* L.	Pericarp (peel)	9844
66	Orange oil, sweet	*Citrus sinensis* (L.) Osbeck	Pericarp (peel)	3140
67	Palmarosa oil	*Cymbopogon martini* (Roxb.) Will. Watson	Aerial part (leaves)	4727
68	Patchouli oil	*Pogostemon cablin* (Blanco) Benth.	Leaf	3757
69	Peppermint oil	*Mentha × piperita* L.	Aerial parts, leaf	856
70	Petitgrain bigarade oil	*Citrus aurantium* L.	Leaf, twig, little green fruit	8901
71	Pine needle oil (Scots pine oil)	*Pinus sylvestris* L.	Needle, twig	
72	Ravensara oil	*Ravensara aromatica* Sonn.	Twig with leaves	
73	Rosemary oil	*Rosmarinus officinalis* L.	Flowering top, leaf	1342
74	Rose oil	*Rosa × damascena* Mill.	Flower	9842

Table 1.1 (continued)

Serial no.	Common name	Botanical source	Parts of plant used	ISO*
75	Rosewood oil	*Aniba rosaeodora* Ducke, *Aniba parviflora* (Meisn.) Mez.	Wood	3761
76	Sage oil	Dalmatian *Salvia officinalis* L.	Flowering top	9909
77	Sage oil, Spanish	*Salvia lavandulifolia* Vahl	Flowering top	3526
78	Sandalwood oil	*Santalum album* L.	Wood	3518
79	Silver fir oil	*Abies alba* Mill.	Needles	
80	Spearmint oil	*Mentha spicata* L.	Flowering aerial part, leaf	3033
81	Spike lavender oil	*Lavandula latifolia* Medik.	Flowering top	4719
82	Star anise oil	*Illicium verum* Hook. f.	Fruit	11016
83	Tangerine oil	*Citrus tangerina* Hort. ex Tan.	Peel	
84	Tea tree oil	*Melaleuca alternifolia* (Maiden et Betche) Cheel; *Melaleuca linariifolia* Smith; *Melaleuca dissitiflora* F. Muell.	Leaf, terminal branchlet	4730
85	Thuja oil	*Thuja occidentalis* L.	Twig with leaves	
86	Thyme oil	*Thymus vulgaris* L.	Flowering top	
87	Thyme oil, Spanish	*Thymus zygis* L.	Flowering top	14715
88	Turpentine oil, Iberian type	*Pinus pinaster* Aiton	Oleoresin	11020
89	Turpentine oil, Chinese type	*Pinus massoniana* Lamb.	Oleoresin	21389
90	Valerian oil	*Valeriana officinalis* L.	Rhizome, root	
91	Vetiver oil	*Chrysopogon zizanioides* (L.) Roberty	Root	4716
92	Ylang-ylang oil	*Cananga odorata* (Lam.) Hook f. and Thomson, forma genuina	Flower	3063
93	Zdravetz oil	*Geranium macrorrhizum* L.	Aerial part	

literature data on contact allergy to and allergic contact dermatitis from individual essential oils and their chemical composition are presented in the book by de Groot and Schmidt. Not included in this list is hinoki oil obtained from the wood of *Chamaecyparis obtusa* (Siebold and Zucc.) Endl., and allergic contact dermatitis from this essential oil was described recently.

1.9 Side effects

Many people think that because essential oils are natural products, they will not cause side effects. This is not true. The potential side effects of essential oils include:
- **Irritation and burning:** Always dilute oils with a carrier oil before applying it to the skin. Apply a small amount to a small area of skin first to test for any reactions.
- **Asthma attacks:** While essential oils may be safe for most people to inhale, some people with asthma may react to breathing in the fumes.
- **Headaches:** Inhaling essential oils may help some people with their headaches, but inhaling too much may lead to headache in others.

References

[1]   Guenther E. The Essential Oils, D. Van Nostrand Company Inc, New York, NY, USA, 1948, 427.
[2]   Association Française de Normalisation (AFNOR). Huiles Essentielles, Tome 2, Monographies Relatives Aux Huiles Essentielles, 6th edn, AFNOR, Association Française de Normalisation, Paris, France, 2000.
[3]   Carette Delacour AS. Ph.D. Thesis. Université Lille 2; Lille, France: 2000. La Lavande et son Huile Essentielle.
[4]   Sell CS. The Chemistry of Fragrance. From Perfumer to Consumer, 2nd edn, The Royal Society of Chemistry, Cambridge, UK, 2006, 329.
[5]   Vainstein A, Lewinsohn E, Pichersky E, Weiss D. Floral fragrance. New inroads into an old commodity. Plant Physiol 2001, 27, 1383–1389, doi: 10.1104/pp.010706.
[6]   Pophof B, Stange G, Abrell L. Volatile organic compounds as signals in a plant – herbivore system: Electrophysiological responses in olfactory sensilla of the Moth *Cactoblastis cactorum*. Chem Senses 2005, 30, 51–68, doi: 10.1093/chemse/bji001.
[7]   Modzelewska A, Sur S, Kumar KS, Khan SR. Sesquiterpenes: Natural products that decrease cancer growth. Curr Med Chem Anti-Cancer Agents 2005, 54, 477–499, doi: 10.2174/1568011054866973.
[8]   Kubeczka. History and sources of essential oil research. In: Can Baser KH, Buchbauer G (Eds), Handbook of essential oils: Science, technology and applications, CRC Press, Florida, 2010, 3–10.

[9]   Litchenthaler HK. The 1-deoxy-D-xylulose-5-phosphate pathway of isoprenoid biosynthesis in plants. Annu Rev Plant Physiol Plant Mol Biol 1999, 50, 47–65, doi: 10.1146/annurev.arplant.50.1.47.

[10]   Dewick PM. The biosynthesis of C5-C-25 terpenoid components. Nat Prod Rep 2002, 19, 181–222, doi: 10.1039/b002685i.

[11]   Marotti M, Piccaglia R, Giovanelli E. Effects of variety and ontogenic stage on the essential oil composition and biological activity of fennel (*Foeniculum vulgare* Mill.). J Essent Oil Res 1994, 6, 57–62, doi: 10.1080/10412905.1994.9698325.

[12]   Hussain AI, Anwar F, Sherazi STH, Przybylski R. Chemical composition, antioxidant and antimicrobial activities of basil (*Ocimum basilicum*) essential oils depends on seasonal variations. Food Chem 2008, 108, 986–995, doi: 10.1016/j.foodchem.2007.12.010.

[13]   Burt S. Essential oils: Their antibacterial properties and potential applications in foods: A review. Int J Food Microbiol 2004, 94, 223–253.

[14]   Chemat F. Techniques for oil extraction. In: Sawamura M (Ed), Citrus essential oils: Flavor and fragrance, Wiley, New Jersey, 2011, 9–20.

[15]   Holley R, Patel D. Improvement in shelf-life and safety of perishable foods by plant essential oils and smoke antimicrobials. Food Microbial 2005, 22(4), 273–292.

[16]   Ferhat MA, Meklati BY, Chemat F. Comparison of different isolation methods of essential oil from citrus fruits: Cold pressing, hydrodistillation and microwave 'dry' distillation. Flavour Fragr, J2007(22), 494–504.

[17]   Ferhat MA, Boukhatem MN, Hazzit M, Meklati BY, Chemat F. Cold pressing, hydrodistillation and microwave dry distillation of citrus essential oil from Algeria: A comparative study. Elect J Biol, 2016(S1), 30–41.

[18]   Handa SS, Khanuja SPS, Longo G, Rakesh DD. An overview of extraction techniques for medicinal and aromatic plants. In: Extraction technology of Medicinal and aromatic Plants, ICS Unido, International Centre for Science and High Technology Publishing, Trieste, 2008, 21–52.

[19]   Cakaloglu B, Ozyurt VH, Otles S. Cold press in oil extraction: A review. Ukrainian Food J 2018, 7(4), 640–654.

[20]   Guenther E. The Essential Oils. Vol. I. D, Van Nostrand Company, Inc, New York, London, 1949.

[21]   Cox M, Rydberg J. Introduction to solvent extraction. In: Rydberg J, Cox M, Musikas C, Choppin G (Eds), Solvent Extraction Principles and Practice, 2nd edn, Marcel Dekker, New York, 2004, 2–12.

[22]   Seidel V. Initial and bulk extraction. In: Sarker S, Latif Z, Gray A (Eds), Natural Products Isolation, 2nd edn, Humana Press, New Jersey, 2005, 27–35.

[23]   Rakthaworn P, Dilokkunanant U, Sukkatta U, Vajrodaya S, Haruethaitanasan V, Potechaman Pitpiangchan P, Punjee P. Extraction methods for tuberose oil and their chemical components. Nat Sci 2009, 43, 204–211.

[24]   Tatke P, Jaiswal Y. An overview of microwave assisted extraction and its application in herbal drug research. Res J Med Plant 2011, 5, 21–31.

[25]   Toma M, Vinatoru M, Paniwnyk L, Manson T. Investigation of the effects of ultrasound on vegetal tissues during solvent extraction. Ultrason Sonochem 2001, 8, 137–142.

[26]   Anwar F, Hussain AI, Sherazi STH, Bhanger MI. Changes in composition and antioxidant and antimicrobial activities of essential oil of fennel (*Foeniculum vulgare* Mill.) fruit at different stages of maturity. J Herbs Spices Med Plants 2009, 15, 1–16, doi: 10.1080/10496470903139488.

[27]   Cocking TT, Middleton G. Improved method for the estimation of the essential oil content of drugs. Q J Pharm Pharmacol 1935, 8, 435–442.

[28] Nickerson G, Likens S. Gas chromatographic evidence for the occurrence of hop oil components in beer. J Chromatogr 1996, 21, 1–5, doi: 10.1016/S0021-9673(01)91252-X.

[29] Shelef LA. Antimicrobial effects of spices. J Food Saf 1983, 6, 29–44, doi: 10.1111/j.1745-4565.1984.tb00477.x.

[30] Nychas GJE. Natural antimicrobials from plants. In: Gould GW (Ed), New Methods of Food Preservation, 1st edn, Blackie Academic & Professional, London, UK, 1995, 58–89.

[31] Lambert RJW, Skandamis PN, Coote P, Nychas GJE. A study of the minimum inhibitory concentration and mode of action of oregano essential oil, thymol and carvacrol. J Appl Microbiol 2001, 91, 453–462, doi: 10.1046/j.1365-2672.2001.01428.x.

[32] Sikkema J, de Bont JAM, Poolman B. Interactions of cyclic hydrocarbons with biological membranes. J Biol Chem 1994, 269, 8022–8028.

[33] Gustafson JE, Liew YC, Chew S, Markham JL, Bell HC, Wyllie SG, Warmington JR. Effects of tea tree oil on *Escherichia coli*. Lett Appl Microbiol 1998, 26, 194–198, doi: 10.1046/j.1472-765X.1998.00317.x.

[34] Cox SD, Mann CM, Markham JL, Bell HC, Gustafson JE, Warmington JR, Wyllie SG. The mode of antimicrobial action of essential oil of *Melaleuca alternifolia* (tea tree oil). J Appl Microbiol 2000, 88, 170–175, doi: 10.1046/j.1365-2672.2000.00943.x.

[35] Carson CF, Riley TV. Antimicrobial activity of the major components of the essential oil of *Melaleuca alternifolia*. J Appl Bacteriol 1995, 78, 264–269, doi: 10.1111/j.1365-2672.1995.tb05025.x.

[36] Ultee A, Bennink MHJ, Moezelaar R. The phenolic hydroxyl group of carvacrol is essential for action against the food-borne pathogen *Bacillus cereus*. Appl Environ Microbiol 2002, 68, 1561–1568, doi: 10.1128/AEM.68.4.1561-1568.2002.

[37] Denyer SP, Hugo WB. Biocide-induced damage to the bacterial cytoplasmic membrane. In: Denyer SP, Hugo WB (Eds), Mechanisms of Action of Chemical Biocides, the Society for Applied Bacteriology, Technical Series No 27, Oxford Blackwell Scientific Publication, Oxford, UK, 1991, 171–188.

[38] Farag RS, Daw ZY, Hewedi FM, El-Baroty GSA. Antimicrobial activity of some Egyptian spice essential oils. J Food Prot 1989, 52, 665–667.

[39] Cosentino S, Tuberoso CIG, Pisano B, Satta M, Mascia V, Arzedi E, Palmas F. In vitro antimicrobial activity and chemical composition of Sardinian Thymus essential oils. Lett Appl Microbiol 2002, 29, 130–135, doi: 10.1046/j.1472-765X.1999.00605.x].

[40] Dorman HJD, Deans SG. Antimicrobial agents from plants: Antibacterial activity of plant volatile oils. J Appl Microbiol 2000, 88, 308–316, doi: 10.1046/j.1365-2672.2000.00969.x.

[41] Davidson PM. Chemical preservatives and natural antimicrobial compounds. In: Doyle MP, Beuchat LR, Montville TJ (Eds), Food Microbiology: Fundamentals and Frontiers, ASM Press, Washington, DC, USA, 1997, 520–556.

[42] Knobloch K, Weigand H, Weis N, Schwarm HM, Vigenschow H Action of terpenoids on energy metabolism. In: Brunke EJ, (Ed). *Progress in Essential Oil Research: 16th International Symposium on Essential Oils*. De Walter de Gruyter; Berlin, Germany, 1986, 429–445.

[43] Pauli A. Antimicrobial properties of essential oil constituents. Int J Aromather 2001, 11, 126–133, doi: 10.1016/S0962-4562(01)80048-5.

[44] Fabian D, Sabol M, Domaracké K, Bujnékovâ D. Essential oils, their antimicrobial activity against *Escherichia coli* and effect on intestinal cell viability. Toxicol in Vitro 2006, 20, 1435–1445.

[45] Marino M, Bersani C, Comi G. Antimicrobial activity of the essential oils of *Thymus vulgaris* L. measured using a bioimpedometric method. J Food Prot 1999, 62, 1017–1023.

[46] Senatore F, Napolitano F, Ozcan M. Composition and antibacterial activity of the essential oil from *Crithmum maritimum* L. (Apiaceae) growing wild in Turkey. Flav Frag J 2000, 15, 186–189, doi: 10.1002/1099-1026(200005/06)15:3<186::AID-FFJ889>3.0.CO;2-I.

[47] Canillac N, Mourey A. Antibacterial activity of the essential oil of *Picea excelsa* on *Listeria*, *Staphylococcus aureus* and coliform bacteria. Food Microbiol 2001, 18, 261–268, doi: 10.1006/fmic.2000.0397.

[48] Cimanga K, Kambu K, Tona L, Apers S, de Bruyne T, Hermans N, Totté J, Pieters L, Vlietinck AJ. Correlation between chemical composition and antibacterial activity of essential oils of some aromatic medicinal plants growing in the Democratic Republic of Congo. J Ethnopharmacol 2002, 79, 213–220, doi: 10.1016/S0378-8741(01)00384-1.

[49] Ratledge C, Wilkinson SG. An overview of microbial lipids. In: Ratledge C, Wilkinson SG (Eds), *Microbial Lipids*. Volume 1, Academic Press Limited, London, UK, 1988, 3–22.

Sushma Kholiya, Amit Chauhan, Dipender Kumar, Venkatesha KT,
R. K. Upadhyay and R. C. Padalia*

Chapter 2
Essential oils, applications, and different extraction methods

Abstract: The essential oils, synthesized and stored in complex secretary structures, are volatile plant secondary metabolites, also commonly known as ethereal oil or aromatic oil. Essential oils are extensively used in perfumes, in cosmetics, in beverages, in food flavors, and as potential therapeutic agents in aromatherapy and phytomedicines. There are various aromatic plant cultivars which are grown on large scale in India for essential oil production. The quality and quantity of extracted essential oils from aromatic plants are influenced by various intrinsic (genetic) and extrinsic factors such as geographical and agroclimatic conditions, cultivation practices, and harvesting technique along with extraction methods. The traditional extraction techniques such as cold pressing, enfleurage, and bhapka techniques are of significance and still in use in some countries. However, distillation methods (hydrodistillation, steam distillation, and hydrosteam distillation) are mostly used for extraction of essential oil for industrial use. The other sophisticated techniques such as solvent extraction, supercritical fluid extraction, solid-phase microextraction, ultrasound-assisted extraction, and microwave-assisted extraction are used in small scale but they have their own advantages or disadvantages. The choice of a particular extraction process depends upon various factors such as the type of planting material, cultivation practices, nature of planting material to be distilled, as well as the cost-effectiveness of the extraction technique. This chapter elaborates the term "essential oil," its ancient and modern extraction techniques, and its applications along with commitment of CSIR-CIMAP toward essential oil production in India.

Keywords: Essential oil, extraction methods, distillation, terpenes

*Corresponding author: R. C. Padalia, CSIR-Central Institute of Medicinal and Aromatic Plants (CIMAP), Research Centre, Pantnagar 263 149, Uttarakhand, India; Academy of Scientific and Innovative Research (AcSIR), CSIR-Human Resource Development Centre (CSIR-HRDC), Postal Staff College Area, Sector 19, Kamla Nehru Nagar, Ghaziabad 201 002, Uttar Pradesh, India, e-mail: rc.padalia@cimap.res.in
Sushma Kholiya, Amit Chauhan, Dipender Kumar, Venkatesha KT, R. K. Upadhyay, CSIR-Central Institute of Medicinal and Aromatic Plants (CIMAP), Research Centre, Pantnagar 263 149, Uttarakhand, India

https://doi.org/10.1515/9783110791600-002

2.1 Essential oils

Herbs, shrubs, and trees which synthesize and retain odorous substances are known as aromatic plants. Since these odorous constituents are volatile by nature, we can easily detect their scent by simply touching a plant. In his book *Quinta Essential* written in the sixteenth century, Paracelsus von Hohenheim introduced the term "essential oil" for the very first time [1]. The word "essential oil" or "volatile oil" or "ethereal oil" refers to the volatile odorous portion of the plant that is extracted by physical methods, such as hydrodistillation, steam distillation, and hydro-cum-steam distillation. Terpenes and other non-terpene groups form essential oil, which are natural, intricate, and multicomponent in nature. In superficial secretory systems found in various plant sections, the essential oils are biosynthesized and stored [2, 3]. The essential oils are one unique natural product that has long been prized for their therapeutic benefits. Records from the prehistoric era (ancient Egypt, China, India, Mesopotamia, and Persia) reveal that essential oils were employed for a variety of purposes. The essential oils were captured by the ancient Egyptians using an infusion technique. However, Greeks and Romans developed the distillation method for obtaining fragrant oil, which added to the industry's worth. Furthermore, under Islamic culture, the process of extracting oil was enhanced. Europeans finally investigated the nature and chemical makeup of essential oils as science and technology advanced [4–7]. The first half of the nineteenth century saw a considerable advance in the industrialization of oil production due to the rise in demand for fragrance and flavoring components. Hoffmann, Semmler, Guenther, Gildemeister, and Finnemore carried out substantial scientific research on chemical and biological makeup of essential oils and aroma compounds as well as their industrial applications in the subsequent years [8]. Systematic scientific investigations of industrial goods have therefore proven to be essential. The second edition of Parry's monograph, which contains roughly 90 essential oils, was published there in the nineteenth century [8]. Essential oils are "produced from pure, recognised raw materials of plant origin, obtained by hydro-distillation and steam-distillation, mechanical procedures (e.g., essential oil from citrus), or by dry-distillation for some woods," according to AFNOR and the European Pharmacopoeia [9]. Other aromatic compounds obtained from methods like solvent extraction (concretes and absolutes), supercritical fluid extraction (SFE), and microwave-assisted extraction (MAE) are previously not included in the definition of an essential oil [9] but these are modern methods of extracting essential oil from plant materials. The physical and chemical characteristics of essential oils are also different from those of the fatty oils. While essential oils are a combination of volatile chemicals, fatty oils are glycerides of fatty acids. The latter, in contrast to the former, evaporates fast and does not leave any stains on the filter paper [9]. Both scientific and commercial interests in essential oils are rising at the moment.

2.2 Composition of essential oils

The secondary metabolites that make up the essential oils are complex combinations of terpenes and phenylpropenes with low boiling points. The essential oil of each aromatic plant has its own complex chemical profile responsible for its odor, taste, and ultimately its applicability. Terpenes (C_{10}, C_{15}, and occasionally C_{20}) and their oxygenated derivatives make up the majority of various molecules that make up these active compounds chemically, which are an aggregate of many molecules. All chemical components work synergistically for effectiveness and application of essential oils. Additionally, phenylpropanoids, fatty acids and their esters, aliphatics, and phenolics are components of essential oils [2, 3]. There are various types of aromatic plants and spices around the world, and each region's culture and customs will dictate the usage of each one. Given that they have been linked to a variety of pharmacological effects, essential oils are attracting interest from both the academic and industrial worlds. In order to create effective industrial processes, mathematical models are now being developed for both classic and emerging methods of essential oil extraction. Despite the fact that the majority of essential oils are GRAS, their use as preservatives in food is frequently constrained due to flavor considerations because effective antimicrobial dosages may go above sensory acceptable levels. Each essential oil's antimicrobial activity mostly depends on the quality and quantity of its constituent parts, which are influenced by a variety of variables including the plant's growing environment, the extraction process, and other factors. Diffusion and dilution or vapor phase procedures are the most often used techniques for assessing the in vitro antibacterial activity of essential oils. Complex combinations of volatile compounds known as essential oils are normally found in low concentrations in a variety of plant components, such as flowers, buds, bark, herbs, wood, fruits, roots, seeds, leaves, and branches. These substances are made of quite low-molecular-weight organic molecules comprising carbon, hydrogen, oxygen, and occasionally nitrogen and sulfur. Less frequently, chlorine and bromine may also be present, especially in volatiles found in seaweed. The fact that many of essential oil's chemicals were isolated and researched prior to the development of standardized chemical nomenclature makes it difficult to categorize and name them. As a result, the majority of them go by their nonsystematic or popular names. Eucalyptol, limonene, and thymol are few examples of these that are sometimes but not always depend on their source [10]. Terpenoids and non-terpenoids are among the many different types of molecules comprising essential oil, depending on the plant and the technique of extraction, among other things. The short-chain alcohols, phenylpropanoids, acids, ketones, esters, and aldehydes that are generated by the metabolic conversion or degradation of phospholipids and fatty acids are known as nonterpenoid hydrocarbons. Many essential oils also include significant amounts of numerous nitrogen- and sulfur-containing molecules, which contribute distinctive sensory qualities. Here are some examples of these substances as well as plants that are rich in them. The greatest family of naturally occurring volatile compounds present in plants are terpenes, commonly known as isoprenoids. They are formed when two or more molecules of isopropene condense head to tail. Hemiterpenes also include derivatives that contain oxygen, such as phenol and isovaleric acid. Hemiterpenes (5-carbon terpenoids) and triterpenes (30-carbon terpenoids) are some of the terpenoids found in essential oil. The second most frequent constituent of an essential oil is sesquiterpene. Three isopropene units are combined to create them. These substances, which all derived from farnesyl pyrophosphate through distinct cyclization mechanisms, are structurally varied and frequently follow skeletal rearrangement. Due to genetic and environmental factors that affect the expression of genes, different essential oils have different compositions [11]. The makeup of the oil in plants that produce essential oils varies from species to species; some may collect active principles that are dispersed throughout the plant, while others may concentrate them only in specific organs [12].

2.3 Physical and chemical signatures of essential oils

Along with substantial improvements in analytical techniques, characteristics such as specific gravity, optical rotation, refractive index, and solubility continue to be important in determining the quality of essential oils and aroma compounds [13]. The secondary metabolites of volatile plants known as essential oils have a molecular mass less than 300 and a nice odor. The essential oils are weakly soluble in water but are soluble in alcohol, nonpolar solvents, waxes, and oils. There are few exceptions to the rule that most essential oils are colorless or pale yellow, such as chamomile (*Matricaria chamomilla*), chrysanthemum which produces blue color oil, with the exception of sassafras, vetiver, cinnamon, and clove's essential oils, which are denser than water. The changes in characteristics such as solubility in ethanol, relative density, refractive index, and optical purity of a particular essential oil indicate the adulteration and modulation in chemical composition. Essential oil comprises aromatic compounds, which are made up of quite low-molecular-weight organic molecules comprising carbon, hydrogen, oxygen, and occasionally nitrogen and sulfur. Chlorine and bromine may also be present, especially in volatiles found in seaweed. Chemically, essential oils are a combination of aromatic compounds made up mostly of terpenes (C_{10}, C_{15}, and occasionally C_{20}) and their oxygenated derivatives. These aromatic compounds are an aggregation of numerous molecules. Other ingredients found in essential oils include phenylpropanoids, fatty acids and their esters, aliphatics, and phenolics [2, 3]. Some common classes of compounds are tabulated in Table 2.1, and structures of monoterpenoids and sesquiterpenoids are represented in Figures 2.1–2.3. Numerous internal and external factors influence the synthesis of secondary metabolites in

Table 2.1: Common classes of compounds of essential oils.

Class	Example
Terpenes	
Monoterpenes (C_{10}): acyclic, monocyclic, and bicyclic	
Hydrocarbon, alcohol, aldehyde, ketone, oxide, lactone, and ether	α-Pinene, sabinene, myrcene, limonene, geraniol, menthol, borneol, thymol, geranial, nerol, citronellal, bornyl acetate, geranyl acetate, geranyl formate, 1,8-cineole, ocimene oxide, linalool oxide, carvacrol, camphor, carvone, piperitenone, nepetalactone, fenchone, and carvacryl methyl ether
Sesquiterpenes (C_{15}): acyclic, monocyclic, bicyclic, and tricyclic	
Hydrocarbon, alcohol, aldehyde, ketone, oxide, and lactone	Bicyclogermacrene, β-selinene, (*E*)-caryophyllene, α-muurolene, dehydroaromadendrene, α-copaene, germacrene A, germacrene D, (*E*)-nerolidol, α-cadinol, elemol, farnesal, dihydroactinidiolide, cyperotundone, α-humulene oxide, caryophyllene oxide, nootkatone, alantolactone, santamarin, and dehydrocostus lactone
Diterpenes (C_{20}): acyclic, monocyclic, bicyclic, tricyclic, and tetracyclic	
Hydrocarbon, alcohol, and oxide	Sclarene, sclareol, and sclareol oxide
Aliphatic compounds	
Alcohols, aldehydes, esters, ketones, and acids	Hexanol, 3-methyl-3-buten-1-ol, (*E*)-2-hexenal, hexanal, methyl 2-methylbutyrate, (*Z*)-3-hexnyl acetate, 3-octanone, *cis*-jasmone, octanoic acid, and butanoic acid
Aromatic compounds	
Phenol, allyl phenols, phenylpropanoids, phenolic ether, and aromatic aldehydes	Thymol, eugenol, carvacrol, methyl eugenol, methyl cinnamate, methyl salicylate, benzaldehyde, anisaldehyde, *trans*-cinnamaldehyde, and cuminaldehyde
Coumarins and furanocoumarins	Bergapten

the plant system. A plant's genetic buildup has a big impact on the chemical makeup of its essential oils. Because of these genetic and epigenetic characteristics, a plant species may generate essential oils with a variety of chemical profiles, and as a result, the medicinal efficacy of those oils may likewise differ [3]. The geographical and growing factors (such as soil type and soil condition, meteorological conditions, altitude, and cultivation procedures), harvesting seasons, and the time of day when harvesting is carried out all affect the chemical

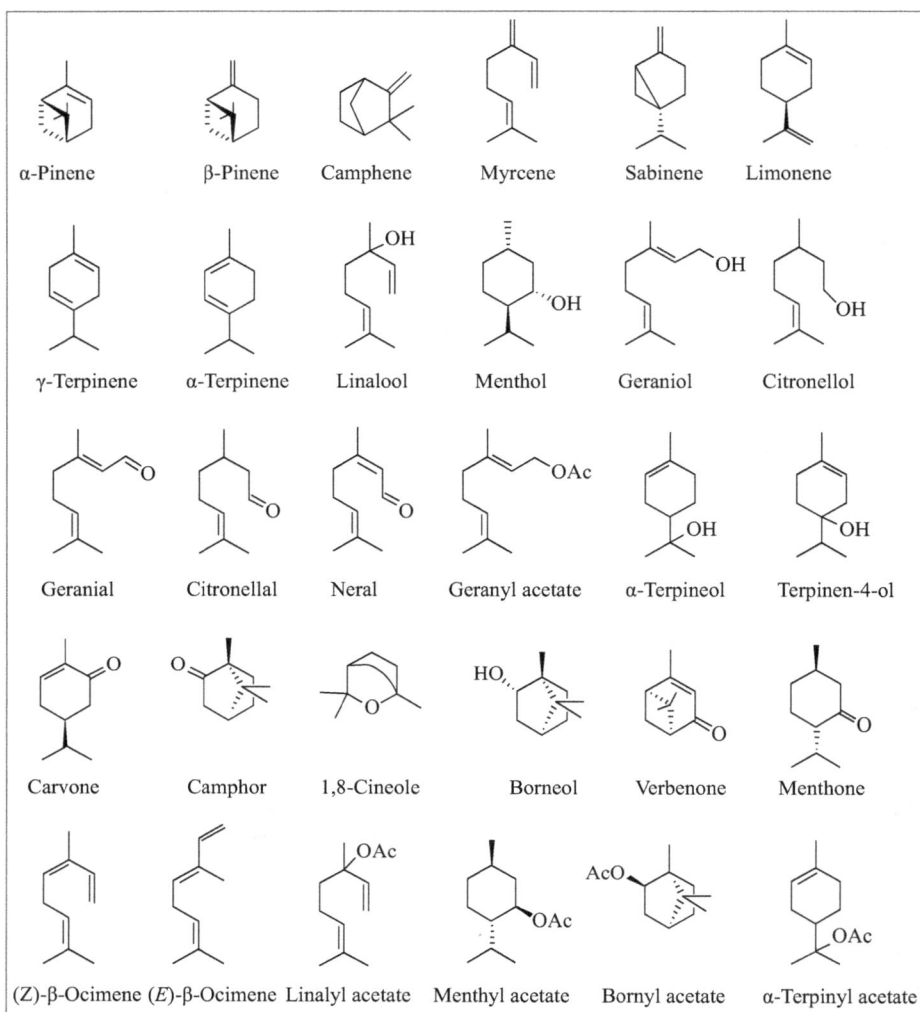

Figure 2.1: Structures of some naturally occurring monoterpenoids.

makeup of the essential oils [10, 14–16]. The concept of chemotypes was developed as a result of these kinds of changes in the chemical profile [3]. As a result of the various extraction methods used, it has also been noted that the chemical makeup of essential oils varies in terms of not only quantitative composition but also qualitative composition. It may have a significant impact on the essential oil's quality metrics. As a result, the extraction technique must be determined by the requirements and nature of the biomass to be processed [17]. Chemical composition of some of the aromatic plant species is tabulated in Table 2.2.

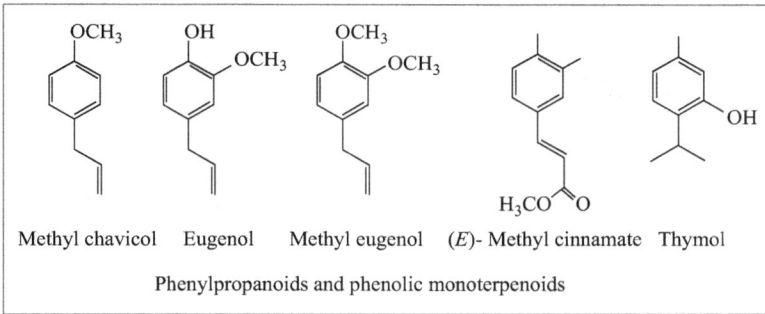

Figure 2.2: Structures of some phenylpropanoids and phenolic monoterpenoids.

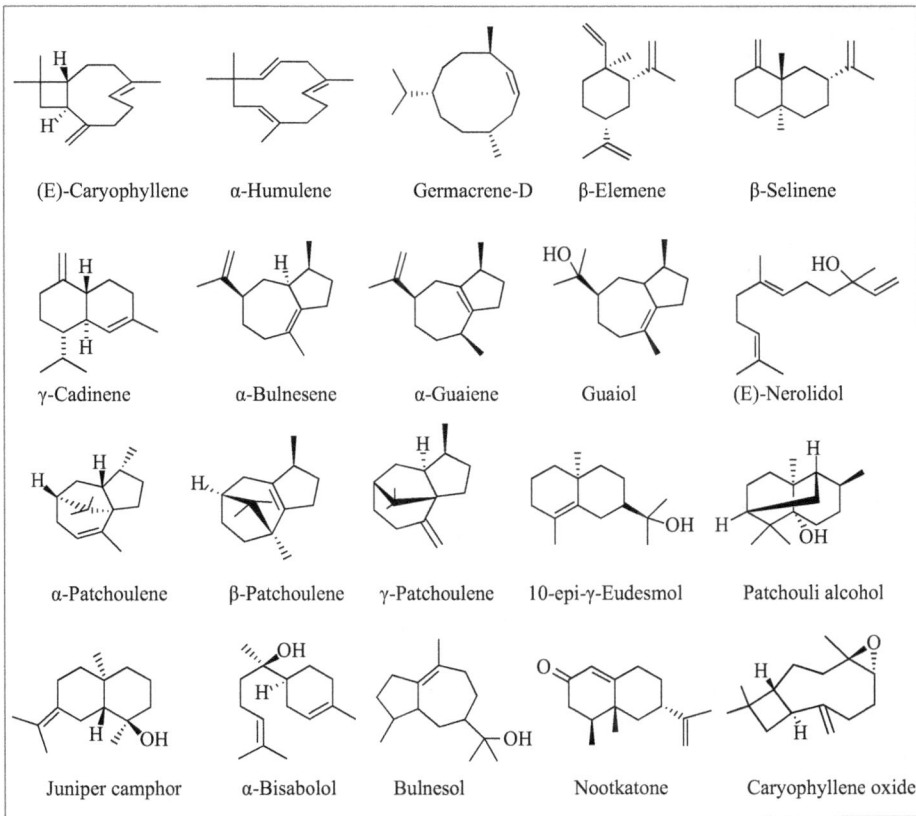

Figure 2.3: Structures of some naturally occurring sesquiterpenoids.

Table 2.2: Chemical markersof some of the traded Indian essential oils.

Plant species	Common name	Marker aroma components
Mentha arvensis	Mentholmint	Menthol and menthyl acetate
Mentha spicata	Spearmint	Carvone
Mentha citrata	Bergamot mint	Linalool and linalyl acetate
Mentha piperita	Peppermint	Menthol, menthone, and menthofuran
Rosa damascena	Damask rose	Geraniol and geranyl acetate
Cymbopogon winterianus	Citronella	Citronellal and citronellol
Cymbopogon flexuous	Lemongrass	Citral
Cymbopogon martini	Palmarosa	Geraniol and geranyl acetate
Ocimum basilicum	Basil	Linalool and methyl chavicol
Ocimum sanctum	Holy basil	Eugenol
Pogostemon cablin	Patchouli	Patchouli alcohol
Pelargonium graveolens	Geranium	Geraniol, citronellal, and citral
Citrus spp.	Lemon/lime/tangerine	Citral and limonene
Elettaria cardamomum	Elaichii	1,8-Cineole and terpinyl acetate
Cinnamomum verum	Cinnamon	Cinnamic aldehyde and eugenol
Syzygium aromaticum	Laung	Eugenol and eugenyl acetate
Lavandula angustifolia	Lavender	Linalool, cineole, and linalyl acetate
Eucalyptus citriodora	Nilgiri	Citronellal
Eucalyptus globulus	Eucalyptus	1,8-Cineole
Salvia rosmarinus	Rosemary	1,8-Cineole and camphor
Santalum album	Sandalwood	
Matricaria chamomilla	Chamomile	Bisabolol and oxides, chamazulene
Melaleuca alternifolia	Tea tree	Terpineol and terpinyl acetate
Salvia sclarea	Clary sage	1,8-Cineole, chamazulene, and scalereol
Valeriana jatamansi	Valerian	Patchouli alcohol and acetate
Chrsopogon zizanioides	Vetiver	Vetivones, khusimol, khusal, and so on

2.4 Essential oil extraction

Essential oils are highly concentrated, volatile, odorous chemicals found in aromatic plants. These come from a variety of plant organs, including flowers, buds, seeds, leaves, twigs, bark, wood, fruits, and roots. The demands of essential oils and their related products increase on a global level. The pharmaceutical, sanitary, cosmetic, food, and agricultural industries use essential oils obtained from plants extensively for bactericidal, fungicidal, antiviral, antiparasitic, insecticidal, medicinal, or cosmetic uses [18]. However, essential oils must first be extracted from the plant matrix in order to be used or studied. The extraction of essential oils from plants is one of the key processes for their end uses and also to improve the yield and quality of essential oils. There are various methods employed for extraction of essential oils, which includes both traditional and modern techniques in many parts of the world. By using several distillation process techniques, essential oils are separated from aromatic plants. Other volatile isolates, however, are also obtained using solvent extraction, cold press, and various other methods. Numerous techniques, including the most popular ones like hydrodistillation, steam distillation, cohobation, cold pressing, maceration, solvent extraction, simultaneous distillation–extraction techniques, and supercritical fluid techniques, can be employed for this purpose [19]. Despite the fact that these methods have been employed for essential oil extraction for a long time, their use has revealed a wide range of drawbacks, including the loss of some volatile compounds, low extraction efficiency, the degradation of unsaturated or ester compounds through thermal or hydrolytic effects, and potential toxic solvent residues in extracts. The oil extraction companies concentrated on the development of emergent extraction technologies as a result of the rise in energy costs and in the advent of the "Green Era." Several new methods are currently available for the extraction of essential oils from plants, including SFE, pressurized liquid extraction, pressurized hot water extraction, membrane-assisted solvent extraction, solid-phase microextraction, MAE, and ultrasound-assisted extraction (UAE), in order to overcome the drawbacks of traditional methods of extraction. By using small amount of solvents, fossil fuel, and harmful compounds, these alternatives to traditional extraction techniques may improve industrial efficiency and contribute to environmental protection [19]. The chemical components and makeup of essential oils are greatly impacted by the extraction processes [20]. The most representative classic and cutting-edge techniques for essential oil's extraction are provided in the following sections. The most suitable and practical technique is carefully chosen to concentrate on the desired biologically active ingredient in the essential oil of plant material used.

2.4.1 Distillation

Distillation is one of the best and most adopted techniques for essential oil extraction. Extraction of essential oils based on distillation is still considered the most common technique in extracting essential oils. The traditional approach of separating volatile components like essential oil from plant materials is distillation using water (hydrodistillation) as a medium. The hydrodistillation is of three types: water distillation, water and steam distillation, and direct steam distillation depending upon the handling of plant materials and use of water as a heating medium to release essential oils from the plant matrix. All these methods are subject to the same theoretical considerations which deal with distillation of two-phase systems. The aromatic plant material is packed in a still and is boiled with water/generated steam and live steam through boilers. Due to the influence of high temperature of boiling water/steam, the essential oil is released from oil glands of plants by evaporation. The vaporized moisture of water and essential oils is condensed through a condenser by indirect cooling with water, and then the oil was separated from water in specially designed separators [24]. The defenses in handling of plant material and water/steam in different distillation methods and the advantages and disadvantages are given in further sections.

In this method, the plant material is completely soaked in water, indicating that the raw materials come into direct contact with the hot water [25]. Depending on the quality of the material treated each time, it may float on the water or may totally submerge. The water is heated to the boiling point using any standard method, such as a direct flame, steam jacket, closed steam coil, or, in rare instances, a perforated steam coil. The evaporation, condensation, and separation based on difference in essential oil density are same as in other distillation processes. Before performing any field distillations for large-scale manufacturing, a small-scale water distillation in glassware (Clevenger apparatus for laboratory purpose), which works on the same principle of hydrodistillation, should be carried out to see if any changes occur during the distillation process [26]. This technique is very simple and inexpensive, and with simple-to-build units, it is suitable for on-field extraction of essential oils. Moreover, this technique is recommended to fine powdered materials of plant parts, mainly spices. This distillation process is treated as an art by local distillers, who rarely try to optimize both oil yield and quality; therefore, it requires a great deal of expertise and procedure knowledge. Other disadvantages of this technique are that some of the oil components like esters are sensitive to hydrolysis; and acyclic monoterpenes, hydrocarbons, and aldehydes are susceptible to polymerize during hydrodistillation. Sometimes the loss in yield and quality of essential oils is observed due to solubility of oxygenated components such as phenols and incomplete vaporization of high-boiling oil components [24–26].

2.4.1.1 Hydro-cum-steam distillation

In this second popular distillation method, the plant material is held above the boiling water on a perforated grid and is not in direct contact with the water. Any of the aforementioned techniques can be used to warm the water. This technique's standout quality is its use of steam that is totally saturated, moist, and never overheated [26]. This distillation technique is widely used in rural areas for extraction of essential oils of cultivated aromatic crops. It is a cost-effective technique with only inclusion of perforated grid to hold the planting material and significantly reduces the disadvantages of hydrodistillation. The increase in essential oil yield, minimal loss of polarmolecules, and lesser requirement of fuel and sustainability in essential oil quality are major advantages of this technique. However, longer distillation time for high-boiling-range essential oils and poor channeling of the steam resulting in slowed distillation are two main disadvantages of this technique. The field distillation units (FDU) used by farmers work on the principle of this technique [24].

2.4.1.2 Steam distillation

In the process of steam distillation, steam is generated in a satellite steam generator/boiler and then it is driven through a pipe into a still where the plant material is placed on a perforated tray just as in hydro-cum-steam distillation [26]. A significant benefit of satellite steam generation is the ease with which the volume of steam can be controlled. This method is most frequently used for the industrial manufacturing of essential oils on large batches through one satellite boiler. The major advantages of steam distillation are its controlled distillation as steam can be regulated through boiler, no thermal degradation of essential oil constituents, and consistent quality of the extracted essential oils. The disadvantages are due to its higher capital expenditure and requirement of trained persons for boiler operation. In terms of essential oil yield, distillation speed, and consistent quality of the essential oil, the steam distillation is superior over hydro-cum-steam distillation and hydrodistillation. The most common method for extracting essential oils on a large scale is steam distillation. However, hydro-cum-steam distillation requires very less investment to establish the distillation setup as compared to steam distillation, and is therefore the most frequently used distillation process for farmers and budding entrepreneurs. In specifically designed FDUs, commonly known as improved FDUs, the inclusion of calandria with smoke pipe inside the still acts as inbuilt small boiler, reduces the fuel consumption, and generates balanced steam for essential oil extraction [24, 26].

2.4.2 Microwave-assisted extraction (MAE)

A lot of attention has recently been paid to the creation of new separation processes for the chemical, food, and pharmaceutical industries. The MAE uses microwave radiation as a source of heating required for essential oil vaporization. The breaking of weak hydrogen bonds, which is facilitated by the molecules' dipole rotation, is one benefit of microwave heating. Because the extraction happens as a result of changes in the cell structure brought on by electromagnetic waves, MAE extractive processes are distinct from those of conventional approaches. This technique is especially well suited for the extraction of thermolabile chemicals because the temperature is kept low throughout the extraction process [30].

2.4.3 Supercritical fluid extraction (SFE)

It is an effective method for extracting plant components, particularly flavors and essential oils, for industrial use [32, 33]. When compared to conventional procedures, this emergent extraction process is typically quicker, more selective toward the substances to be extracted, and more ecologically benign [34]. The foundation of SFE is the use of solvents in their supercritical form, which involves pushing them beyond their critical points in terms of pressure and temperature. Supercritical fluids have peculiar qualities that fall in between those of a gas and a liquid, and these features depend on the fluid's composition, temperature, and pressure. These substances are effective and selective solvents because they are heavy like liquids with gas-like penetrating strength. CO_2 is the preferred supercritical solvent for the extraction of plant components because it is nontoxic, nonreactive, nonflammable, noncorrosive, and high diffusivity, and enables supercritical operation at temperatures and pressures (critical points 31.1 °C and 73 atm) [35, 36]. The benefits of employing SCFs include environmental benefits from the decrease in the use of conventional organic solvents that harm the environment, health and safety benefits, and chemical benefits [37].

2.4.4 Microwave accelerated distillation (MAD)

Dry distillation at atmospheric pressure combined with microwave heating is known as distillation or microwave accelerated distillation (MAD). MAD was designed for use in industrial or laboratory processes for the extraction of essential oils from various aromatic plants. This technique, which is based on a rather straightforward idea, entails adding "fresh" plant materials to a microwave reactor without any additional water or solvent [38].

The choice of a particular extraction process depends on various factors such as the type of planting material, geographical location of cultivation of aromatic crops, nature of planting materials to be distilled, as well as the cost-effectiveness of the extraction technique (Figure 2.4). After distillation of essential oils, the rectification and storage are very important to maintain the quality of the extracted essential oils. The traces of water from essential oils must be removed by adding sodium sulfate and the essential oils may be stored in glass bottles, galvanized iron drums, aluminum containers, and high-density polyethylene drums as per the quantity of distilled essential oils. The distillation and storage of essential oils in CSIR-CIMAP at laboratory and large scales are shown in Figures 2.5 and 2.6.

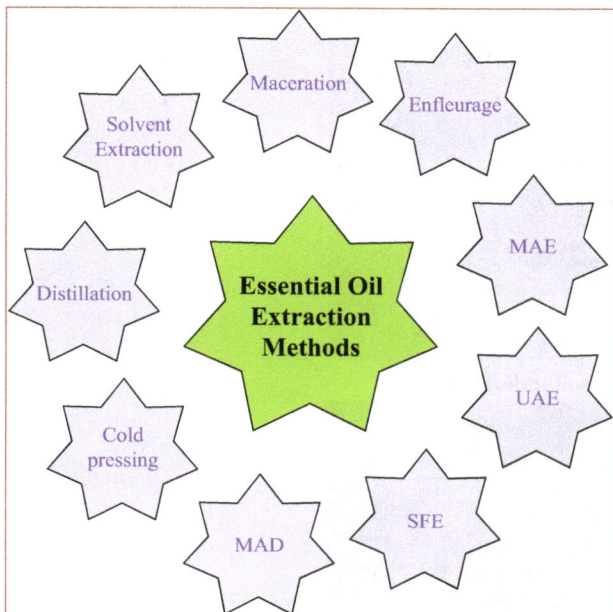

Figure 2.4: Different methods used for essential oil extraction.

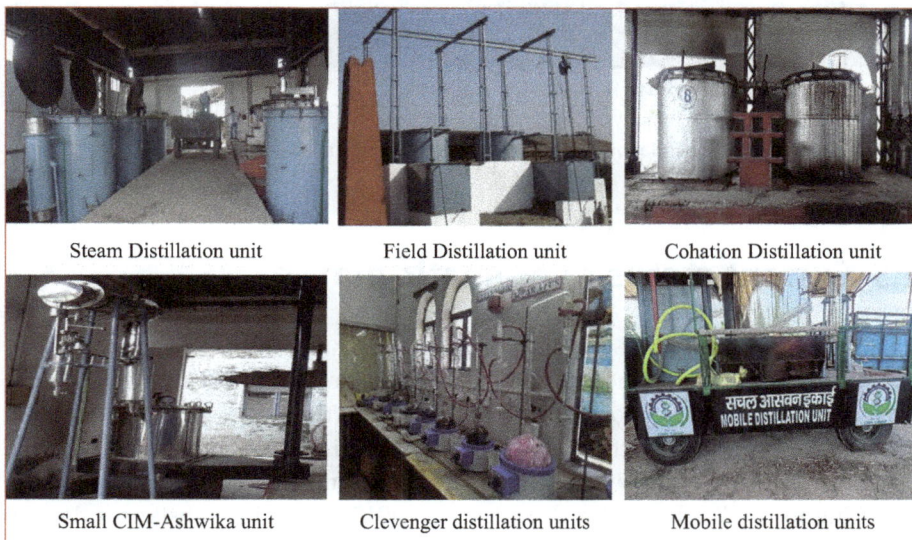

Figure 2.5: Distillation techniques used for essential oil extraction at CSIR-CIMAP.

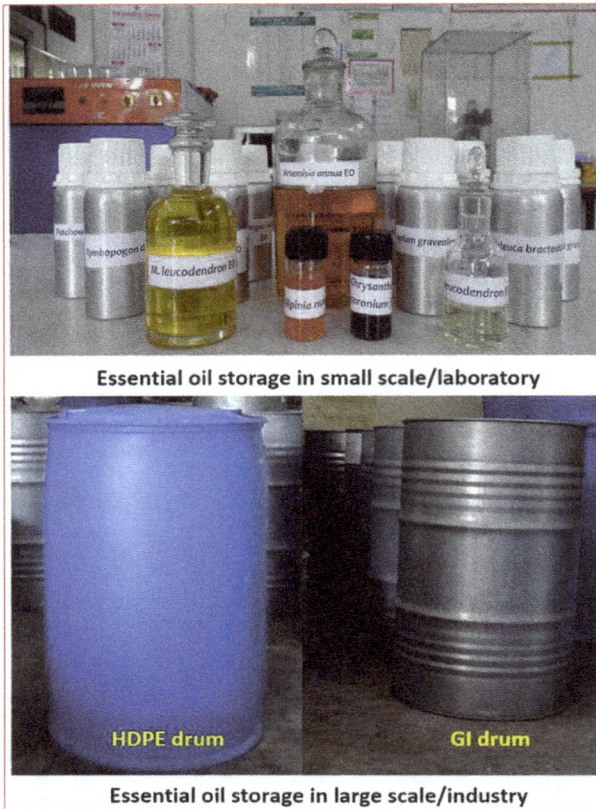

Essential oil storage in small scale/laboratory

HDPE drum   GI drum

Essential oil storage in large scale/industry

Figure 2.6: Storage of essential oils at laboratory and commercial scales.

2.5 Current status of aromatic plants

The tremendous untapped economic potential of aromatic and medicinal plants, notably in the application of herbal remedies, is attracting a lot of attention nowadays throughout the world. It is one of the industries with the fastest growth rates. India has a strong growth rate in this industry (22%), which is significantly greater than the average growth rate of the world (7%). About 30% of medications supplied globally, according to the World Health Organization (WHO), contains substances derived from the actual plant matter. The WHO predicts that the global trade in medical plants might reach US $5 trillion by 2050 [39]. The essential oils are valuable natural products, highly recognized for their medicinal value since ancient times. The Indian states, namely, Gujarat, Uttar Pradesh, Rajasthan, Haryana, Tamil Nadu, Andhra Pradesh, and the states of Himalayan region are major producers of high-quality essential

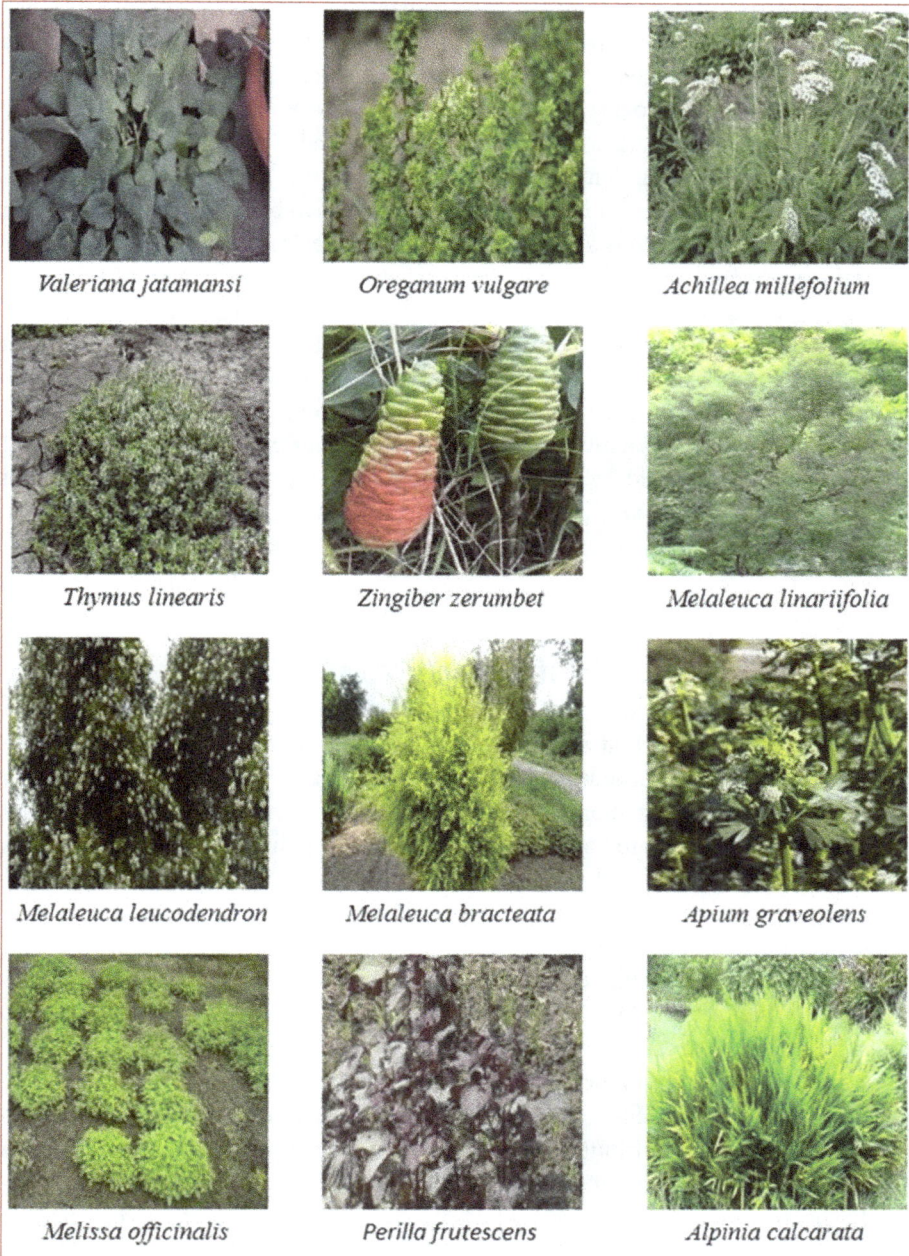

Valeriana jatamansi   *Oreganum vulgare*   *Achillea millefolium*

Thymus linearis   *Zingiber zerumbet*   *Melaleuca linariifolia*

Melaleuca leucodendron   *Melaleuca bracteata*   *Apium graveolens*

Melissa officinalis   *Perilla frutescens*   *Alpinia calcarata*

Figure 2.7: Some potential Himalayan aromatic plants for essential oils.

oils. Fragrant essential oils are used in the development of perfumes and flavors for foods and beverages. The essential oils are significantly used in preparing various products, such as perfumes, cosmetics, soaps, shampoos, or cleaning gels, and this has been scientifically recognized. In addition to this, the essential oils are also widely used as potential therapeutic agents in aromatherapy and phytomedicines. Moreover, in recent times, due to the demand of natural essential oils and their isolates in food flavor, food preservation is on rise because of their relative biosafety value and global acceptance. There are various aromatic plants that are having commercial cultivars for their large-scale cultivation in India. Additional study and development of their suitable varieties and agrotechnologies for extensive cultivation and essential oil extractions are needed for a number of different aromatic plants that have the potential for a variety of fragrance and flavor in terms of essential oils. The Himalayan region and the Indian states of Gujarat, Rajasthan, Haryana, Tamil Nadu, and Andhra Pradesh are the main producers of premium herbal products. Some indigenous and potential aromatic plants that could be cultivated on large scale for essential oil production are mentioned in Figure 2.7.

2.6 Contribution of CSIR-CIMAP in India for aromatic plant cultivation

Central Institute of Medicinal and Aromatic Plants, popularly known as CIMAP, is a frontier plant research laboratory of Council of Scientific and Industrial Research (CSIR). CSIR-CIMAP also explores and develops the use of herbs, plants, flowers, and fruits for medicinal and aromatic purposes in India by way of improving extraction techniques, modern processes, and herbal products. CSIR-CIMAP through various society-oriented projects is popularizing superior varieties and agrotechnologies, and assessment of their suitability for specific climatic regions without causing harm to the environment and creating opportunities for income generation of marginal farmers living in rural areas. CSIR-CIMAP developed and promoted high essential oil and its constituents and specific varieties of various aromatic plants for commercial cultivation and essential oil production in India. CIMAP has given many improved varieties of several aromatic plant species, and some of which are tabulated in Table 2.3 [40]. CSIR-CIMAP has contributed to develop improved varieties of aromatic plants to enhance quality and quantity of essential oil, which results in significant contribution to the economy of farmer as well as of country. Mint Revolution governed by CSIR-CIMAP has extraordinarily yet remarkable impact on essential oil industry, which enlisted India as the largest exporter of mint oil worldwide. Some commercial aromatic plants are shown in Figure 2.8.

Table 2.3: Some improved varieties developed by CSIR-CIMAP.

Plant species with varieties	Improved varieties by CSIR-CIMAP	Percentage composition
Ocimum basilicum	CIM-Saumya	Methyl chavicol (62%) and linalool (24%)
Ocimum africanum	CIM-Jyoti	Citral (68–75%)
Ocimum sanctum	Angana Ayu	Eugenol (40%) and germacrene D (16.6%) Eugenol (83%) and β-elemene (7.4%)
Ocimum gratissimum	Akshay	Thymol (45–55%)
Cymbopogon winterianus	Bio G-1	Citronellal (35–38%), citronellol + geranyl acetate (14–18%), and geraniol (18–24%)
Cymbopogon flexuosus	Krishna	Citral (86%)
Cymbopogon martini	Harsh PRC-1	Geraniol (94%) Geraniol (75–80%)
Mentha arvensis	Kosi CIM-Unnati	Menthol content (75–80%) Menthol content (75–78%)
Mentha spicata	MSS-5	Carvone (68%)
Mentha citrata	Kiran	Linalool (48%) and linalyl acetate (38%)
Mentha piperita	Kukrail	Menthol (34.5%) and menthone (27.9%),
Rosa damascena	Rani Sahiba Noor Jahan	Geraniol (30–35%) and *trans*-rose oxide (10.1%) Citronellol (20.8%) and geraniol (25.3%)
Chamomilla recutita	CIM-Sammohak	Chamazulene content (12.98%)
Salvia sclarea	CIM-Chandni	Linalool (26.2%) and linalyl acetate (52.4%)
Rosmarinus officinalis	CIM-Hariyali	Camphor (26.8%) and 1,8-cineole (20.1%)
Origanum vulgare	CIM-Sudeeksha	Carvacrol (52.3–60%)
Geranium	Cim-BIO 171	Geraniol content (21%)
Tagetes minuta	Vanphool	Dihydrotagetone (32%) and (Z)-tagetone (16.7%)

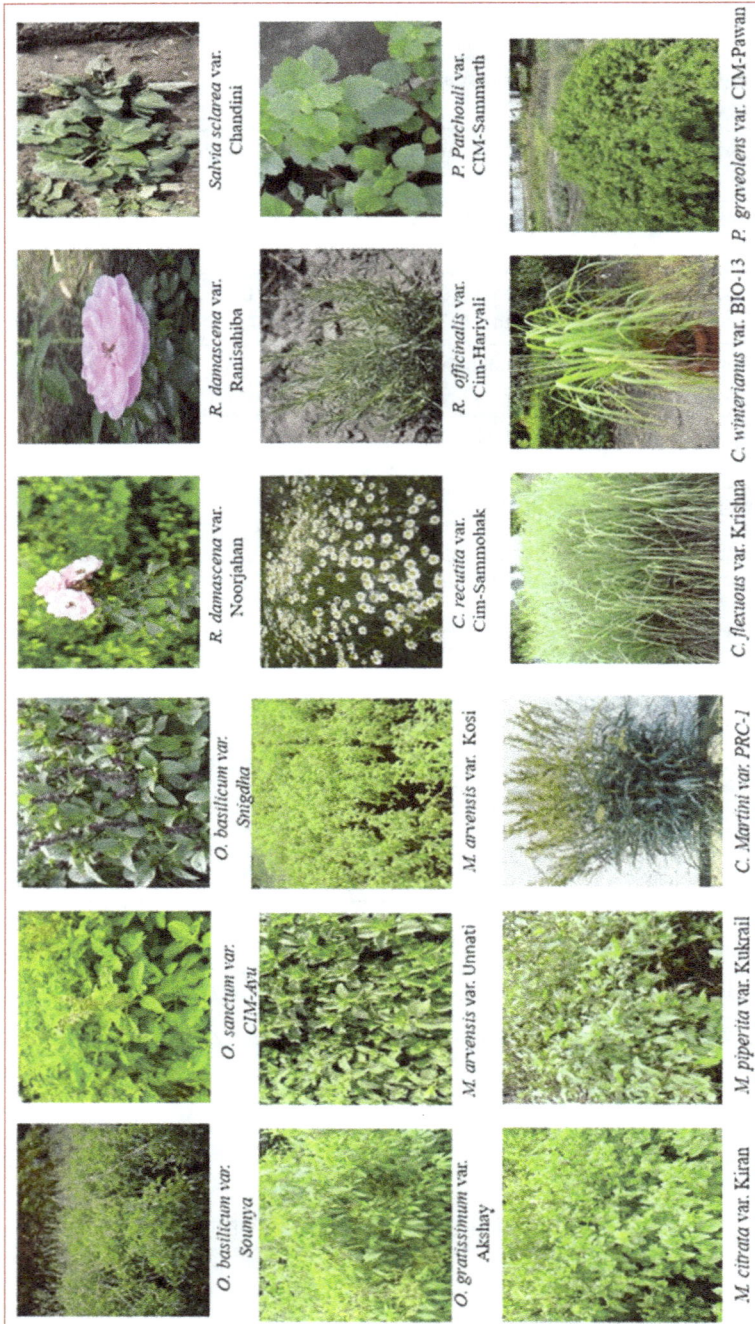

Figure 2.8: Some of the improved cultivars of aromatic plants developed by CSIR-CIMAP for commercial cultivation in India.

2.7 Conclusions

The market for essential oils is rising globally at present due to consumers' growing interest in ingredients from natural sources and their worry over possibly dangerous synthetic chemicals. Essential oils are still among the top-priority natural goods for researchers due to their promising bioactivity. The creation of perfumes, food and beverages, and cosmetics all employ fragrant essential oils. Since ancient times, essential oils have been employed in holistic remedies because they provide therapeutic comforts and improve emotional well-being. It has been established scientifically that essential oils are used extensively in various products such as soaps, cosmetics, fragrance, shampoos, and cleaning gels. The essential oils were found to possess various pharmacological properties like antimicrobial, antioxidant, anti-inflammatory, antiseptic, and anticarcinogenic, and are used in various formulations of health care, personal care, food flavor and pharmaceutical products, as well as in aromatherapy. There are various aromatic plant cultivars for essential oil production which are grown on large scale in India. The quantity and quality of essential oils extracted from aromatic plants are of interest for fetching premium prices and economic gain. The quality (composition of essential oils) depends upon various factors such as geographical and agroclimatic conditions, cultivation practices, harvesting technique, postharvest management, as well as the extraction methods employed for essential oil production. Extraction methods are of importance as they affect both the yield and quality of essential oils. The choice of a particular extraction process depends upon various factors such as the type of planting material, cultivation practices, nature of planting material to be distilled, as well as the cost-effectiveness of the extraction technique. When it comes to essential oils, the extraction procedure changes both the external and interior composition. Using nonstandard extraction techniques can lead to variations in the chemical components of therapeutic plant extracts. To accomplish this, a number of institutions, notably CSIR-CIMAP, are making great efforts to move the essential oil industry to new heights. The cultivation of suitable aromatic plants makes an important contribution to the income and livelihood of relatively poor rural population in general.

References

[1] Guenther E. The Essential Oils, vol. IV, D. Van Nostrand Company, Inc, New York, London, 1950.
[2] Baser KHC. Handbook of Essential Oils: Science, Technology, and Applications, CRC Press, Taylor & Francis Group, Boca Raton London, New York, 2010.
[3] Djilani A, Dicko A. The Therapeutic Benefits of Essential Oils, Nutrition, Well-Being and Health. Bouayed J (Ed.), InTechOpen, London, 2012, ISBN: 978-953-51-0125-3.

[4]   Burt S. Essential oils: Their antibacterial properties and potential applications in foods. Int J Food Microbiol 2004, 94, 223–253.

[5]   Peeyush K, Sapna M, Anushree M, Santosh S. Insecticidal properties of Mentha species. Ind Crops Prod 2011, 34, 802–817.

[6]   Steven BK. Traditional Medicine: A global perspective. Steven BK (ed), Pharmaceutical Press, London, 2010.

[7]   Luqman S, Dwivedi GR, Darokar MP, Kalra A, Khanuja SP. Potential of rosemary oil to be used in drug-resistant infection. Altern Ther Health Med 2007, 13, 54–59.

[8]   Zellner BA, Dugo P, Dugo G, Mondello L. Analysis of essential oils. In: Baser KHC, Buchbauer G Eds., Handbook of Essential Oils. Science, Technology and Applications, CRC Press, Boca Raton, 2010, 151–184.

[9]   Zuzarte M, Salgueiro L. Essential oils chemistry. In: de Sousa DP (ed.), Bioactive Essential Oils and Cancer. Springer International Publishing, Switzerland, 2015, 19–28.

[10] Carson C, Hammer K. Chemistry and bioactivity of essential oils. In: Thormar H Ed, Lipids and Essential Oils as Antimicrobial Agents, Wiley, Chichester, 2011, 203–3.

[11] Charles D, Simon J. Comparison of extraction methods for the rapid determination of essential oil content and composition of basil. J Am Soc Hortic Sci 1990, 115(3), 458–462.

[12] Lawrence B. Natural products and essential oils. In: Swift KA (ed), Advances in Flavours and Fragrances: From the Sensation to the Synthesis. R Soc Chem Cambridge, 2002, 57–64.

[13] Rios JL. Essential oils: What they are and how the terms are used and defined. In: Preedy VR (ed), Essential Oils in Food Preservation, Flavor and Safety. Academic Press, Elsevier, London, 2015, 3–10.

[14] Deans SG, Svoboda KP, Gundidza M, Brechany EY. Essential oil profiles of several temperate and tropical aromatic plants: Their antimicrobial and antioxidant activities. Acta Hortic 1992, 306, 229–232.

[15] Pengelly A. The Constituents of Medicinal Plants: An Introduction to the Chemistry and Therapeutics of Herbal Medicine, Allen & Unwin, 83 Alexander Street Crows Nest NSW 2065 Australia, 2004.

[16] Sangwan NS, Farooqi AHA, Shabih F, Sangwan RS. Regulation of essential oil production in plants. Plant Growth Regul 2001, 34, 3–21.

[17] Stratakos AC, Koidis A. Methods for extracting essential oils. In: Preedy VR (ed), Essential Oils in Food Preservation, Flavor and Safety, Academic Press, Elsevier, London, 2015, 31–38.

[18] Burt S. Essential oils: Their antibacterial properties and potential applications in foods: A review. Int J Food Microbiol 2004, 94, 223–253.

[19] Chemat F. Techniques for oil extraction. In: Sawamura M Ed, Citrus Essential Oils: Flavor and Fragrance, Wiley, New Jersey, 2011, 9–20.

[20] Holley R, Patel D. Improvement in shelf-life and safety of perishable foods by plant essential oils and smoke antimicrobials. Food Microbial 2005, 22(4), 273–292.

[21] Ferhat MA, Meklati BY, Chemat F. Comparison of different isolation methods of essential oil from citrus fruits: Cold pressing, hydrodistillation and microwave 'dry' distillation. Flavour Fragrance J 2007, 22, 494–504.

[22] Ferhat MA, Boukhatem MN, Hazzit M, Meklati BY, Chemat F. Cold pressing, hydrodistillation and microwave dry distillation of citrus essential oil from Algeria: A comparative study. Elect J Biol 2016, S1, 30–41.

[23] Cakaloglu B, Ozyurt VH, Otles S. Cold press in oil extraction: A review. Ukrainian Food J 2018, 7(4), 640–654.

[24] Handa SS. An overview of extraction techniques for medicinal and aromatic plants. In: Handa SS, Khanuja SPS, Longo G, Rakesh DD (ed), Extraction Technology of Medicinal and aromatic

Plants. ICS Unido, International Centre for Science and High Technology Publishing, Trieste, 2008, 21–52.

[25] Kubeczka. History and sources of essential oil research. In: Can Başer KH, Buchbauer G Eds, Handbook of Essential Oils: Science, Technology and Applications, CRC Press, Florida, 2010, 3–10.

[26] Guenther E. The Essential Oils, vol. I, D. Van Nostrand Company, Inc, New York, London, 1949.

[27] Cox M, Rydberg J. Introduction to solvent extraction. In: Rydberg J, Cox M, Musikas C, Choppin G (eds), Solvent Extraction Principles and Practice. 2nd edn Marcel Dekker, New York, 2004, 2–12.

[28] Seidel V. Initial and bulk extraction. In: Sarker S, Latif Z, Gray A (Eds), Natural Products Isolation. 2nd edn Humana Press, New Jersey, 2005, 27–35.

[29] Rakthaworn P, Dilokkunanant U, Sukkatta U, Vajrodaya S, Haruethaitanasan V, Potechaman Pitpiangchan P, Punjee P. Extraction methods for tuberose oil and their chemical components. Nat Sci 2009, 43, 204–211.

[30] Tatke P, Jaiswal Y. An overview of microwave assisted extraction and its application in herbal drug research. Res J Med Plant 2011, 5, 21–31.

[31] Toma M, Vinatoru M, Paniwnyk L, Manson T. Investigation of the effects of ultrasound on vegetal tissues during solvent extraction. Ultrason Sonochem 2001, 8, 137–142.

[32] Reverchon E. Mathematical modeling of supercritical extraction of sage oil. AIChEJ 1996, 42(6), 1765–1771.

[33] Wang L. Energy Efficiency and Management in Food Processing Facilities, CRC Press, Boca Raton, 2008, 351–359.

[34] Mendiola J, Herrero M, Castro-Puyana M, Ibañez E. Supercritical fluid extraction. In: Rostango M, Prado J, Kraus G (Eds), Natural Product Extraction: Principles and Applications. R Soc Chem, Publishing: Cambridge, 2013, 196–201.

[35] Zizovic I, Stamenic M, Orlovic A, Skala D. Supercritical carbon-dioxide extraction of essential oils and mathematical modelling on the micro-scale. In: Berton LP (Ed), Chemical Engineering Research Trends. Publisher: Nova Science Publishers, USA, 2007.

[36] Camel V. Recent extraction techniques for solid matrices super critical fluid extraction, pressurized fluid extraction and micro wave assisted extraction: Their potential and pitfalls. Analyst 2001, 126, 1182–1193.

[37] Shi J, Kassama L, Kakuda Y. Supercritical fluids technology for extraction of bioactive components. In: Shi P Ed, Functional Food Ingredients and Nutraceuticals: Processing Technologies, CRC Press, Boca Raton, 2006, 5–30.

[38] Tigrine-Kordjani N, Meklati BY, Chemat F. Microwave 'dry' distillation as an useful tool for extraction of edible essential oils. Int J Aroma 2006, 16, 141–147.

[39] Bhowmik D, Kumar KPS, Tripathi P, Chiranjib B. Traditional herbal medicines: An overview. Arch Appl Sci Res 2009, 1(2), 165–177.

[40] Improved Varieties of Medicinal and Aromatic Plants CSIR-CIMAP's Contribution, year. (Accessed July 15, 2022, at https://14.139.57.217/cimvariety/)

Shekhar Verma*, Sonam Soni, Nagendra Chandrawanshi
and Shivendra Singh Dewhare

Chapter 3
Sources and raw materials of essential oils

Abstract: Essential oils are the extracted products obtained from various parts of the plant using different extraction methods (distillation, enfleurage, and expression methods). It is a special ingredient in aromatherapy, which is a part of treatment. It contains specific or characteristic aroma of plants from which it is extracted. Because of this unique character, it has several pharmacological values in today's life. It can be removed from the dry and fresh parts of a plant using a suitable solvent or method of extraction. They have various therapeutic effects like mood enhancers, stress reducers, and treatment of headaches, but in higher concentration, they may show some side effects. Hence, it is mandatory to dilute it before use. This study suggests that more well-designed clinical trials are required to ascertain these plant products' efficacy and safety.

Keywords: extraction, essential oils, aromatherapy, distillation

3.1 Introduction

Essential oils are the extracted products obtained from various parts of plants using different extraction methods. Ethereal oil, volatile oil, or aromatic oils are synonyms of essential oil. It is a key ingredient in aromatherapy, which is a part of treatment. It contains a specific or characteristic aroma of plants from which it is extracted. It is a hydrophobic liquid containing a volatile compound with the nature of evaporation at room temperature. Essential oils are top-rated for use in medicinal and cosmetic fields [1].

*Corresponding author: Shekhar Verma, University College of Pharmacy Raipur, Pt. Deendayal
Upadhyay Memorial Health Sciences and Ayush University of Chhattisgarh, Chhattisgarh 493661,
India, e-mail: shekharpharma@gmail.com
Sonam Soni, University College of Pharmacy Raipur, Pt. Deendayal Upadhyay Memorial Health
Sciences and Ayush University of Chhattisgarh, Chhattisgarh 493661, India
Nagendra Chandrawanshi, School of Studies in Biotechnology, Pt. Ravishankar Shukla
University, Raipur, Chhattisgarh, India
Shivendra Singh Dewhare, School of Studies in Life Science, Pt. Ravishankar Shukla University,
Raipur, Chhattisgarh, India

https://doi.org/10.1515/9783110791600-003

3.1.1 Source of essential oils

Plants and their parts are the primary sources of essential oils (Table 3.1). These oils are obtained from them by using various extraction methods [1, 2].

Table 3.1: Essential oils from plant sources.

S. no.	Parts used	Examples
1	Bark	Cassia, cinnamon, sassafras
2	Berries	Juniper
3	Leaves	Basil, bay leaf, cinnamon, guava, lemon grass, peppermint, pine, rosemary, spearmint, tea tree, thyme, wintergreen
4	Flowers	Cannabis, chamomile, Clary sage, clove, jasmine, lavender, orange, rose
5	Peels	Bergamot, grapefruit, lemon, lime, orange
6	Resin	Benzoin, frankincense, labdanum, myrrh
7	Rhizome	Galangal, ginger
8	Roots	Valerian
9	Seeds	Anise, buchu, cumin, celery, flax, nutmeg oil
10	Woods	Agar wood, camphor, cedar, rosewood, sandalwood

3.1.2 Extraction of essential oil

Essential oils or volatile oils are extracted from the raw (*dried* or *fresh*) plant material (stem, barks, leaves, and fruits). They are also obtained from citrus peels by pressing the peels. When plants are mixed with the solvent, chemical constituents get dissolved in the solvent, and this extract is obtained by using various extraction methods [2, 3]. Figure 3.1 shows the extraction procedure of essential oils.

Figure 3.1: Extraction procedure of essential oils.

3.2 Types of extraction methods [4]

When a plant sample is mixed with a suitable solvent, the components (a chemical constituent) from the sample plant get dissolved in the solvent, resulting in herbal extract; at last, after this process, the solvent will be available in the liquid or semi-solid form which gets converted into solid by evaporation. The remaining solvent can be used as preservatives or agents that can help plant cells to break down and release their contents. The essential oil can be extracted by the following methods:
1) Distillation
   a. Water distillation
   b. Steam distillation
   c. Water and steam distillation
2) Extraction
   a. Solvent extraction
   b. CO_2 extraction
   c. Maceration
3) Enfleurage method
4) Expression method

3.3 Distillation

3.3.1 Water distillation

Delicate flowers like roses and orange blossoms would get clumped when exposed to steam in the distillation method. Therefore, the best extraction technique in this case is to submerge fragile stuff in the pure boiling water instead. The water protects the extracted essential oil from heating. The condensed liquids relax and cut loose one another. The remaining water, typically sweet-scented, is cited by many names together with hydrolase, hydrosol, flavored water, essential water, floral water, or flavored distillation.

3.3.2 Water and steam distillation

This technique uses leaves of plants, and the stuff is immersed in water during a still to which heat is applied. Steam is fed into the still the most from outside.

3.3.3 Steam distillation

Steam distillation is the most common and popular technique for extraction of essential oils. This is an appropriate technique for extracting oils from still dry leaves of plants. With the help of steam, oil gets vaporized from the plant; once it passes through the condensation process, then it is converted again into a liquid form. This happens once the moisture evaporates the plant material's volatile compounds, which eventually takes a condensation and assortment method. Steam distillation is the superior generally enforced technique for oil extraction. Heat of steam is responsible for the removal of volatile compounds from plants in the form of vapors. The vaporized compounds rise and enter a cooling chamber, referred to as a condenser unit. Because the vapors cool, they are condensed from steam into a liquid. After being reduced, the essential oil and water from the soluble elements of the plant – known as hydrosols or floral water – are collected into a receiver with two separate outflows. As a result, oil and water do not combine the oil floats higher than the water. The lower outflow can postpone the water so that the higher discharge can delay the oil. Part of the reason that steam distillation is such a popular method of extracting essential oils is that the volatile compounds are also distilled at temperatures that are not up to their boiling points. As such, the natural qualities of the fabric are less doubtless to become altered or diminished throughout the tactic. For this reason, oil distillation is typically thought to be the only accurate extraction technique that permits purity [5–7].

3.3.3.1 Method

1. Associate oversize instrument which is usually made up of stainless steel, containing the things has steam acquired to that.
2. Through the associated body of water, steam is injected through the fabric containing the required oils, plant's aromatic molecules and turning them into vapor.
3. The vaporized plant compounds travel through the condensation flask or the condenser. Here, two separate pipes make it attainable for tight water to exit and cold water from entering the condenser. This makes the vapor cool back to the liquid form.
4. The aromatic liquid by-product drops from the condenser and collects within a receptacle beneath it, known as a centrifuge. As water and oil do not combine, the oil floats on the prime of water. From here, it is siphoned off [6].

3.3.4 Solvent extraction

This methodology extracts essential oil using solvents (especially food-grade solvents like dissolvent and ethanol). It is appropriate for plants containing a low quantity of essentials that are mostly tarry or delicate aromatics unable to resist the pressure and distress of steam distillation. This methodology conjointly produces a more delicate fragrance than any distillation methodology. This method extracts nonvolatile materials, like waxes and pigments, and generally removes them through different processes. When the plant material is treated with the solvent, it produces a waxy hydrocarbon referred to as "concrete." Once this concrete substance is mixed with alcohol, the oil particles are free. The mentioned chemicals employed in the method stay within the oil; therefore, the essential oil is used in fragrances by the perfume business or for aromatherapy functions. Another standard oil extraction methodology is solvent extraction. This contemporary methodology implements food-grade solvents like grain alcohol, benzene, dimethyl, or dissolvent to isolate the essential oils. It is usually used for extracting essential oils from delicate aromatics, like shrubs of rose, that are typically unable to resist the pressure of steam distillation. Moreover, solvent extraction generally produces a more delicate fragrance than most different extraction methods, which adds to its attractiveness surely as a shooting application. During oil extraction, the plant materials are lined by a solvent and dissolved into it. Once the solvent absorbs the oil, the following extract is gaseous; therefore, solely the plant oil is left behind. Technically, the remaining oil is understood as associate absolute, and not a necessary oil. Associate absolute may be a highly focused aromatic substance resembling the plant's natural aroma. In addition, it is a lot more vivid color than a necessary essential oil. As such, solvent extraction is usually done to produce extracts for fragrance or cosmetic applications. The biggest drawback to the current sort of extraction methodology may be that trace amounts of the solvent might not get gaseous. As such, tiny quantities of harsh chemicals might stay within the final absolute that may cause irritation once used [8, 9].

Solvent extraction encompasses the following methods: *hypercritical CO_2 (carbon dioxide)*, *maceration*, and *enfleurage*.

3.3.5 CO_2 extraction

Essential oils derived from the critical carbonic acid gas extraction of herbs unit, like the oils made through distillation, will be utilized in aromatherapy and natural perfumery. Oils derived from steam distillation vary in their qualities, reckoning on the temperatures, pressures, and length of your time applying for the method. The carbonic acid gas extraction method may turn out higher quality oils that have not been altered by the application of high heat, not like the steam distillation method [9].

In carbonic acid gas extraction, none of the constituents of the oil unit is broken by heat. Thus, the distinction between ancient distillation and critical extraction is that carbonic acid gas is employed as a solvent within the latter methodology rather than heated water or steam. The required extraction method operates at temperatures between 95 and 100 °F, whereas steam distillation operates at temperatures between 140 and 212 ° F. In steam distillation, the molecular composition of each plant matter and, therefore, the volatile oil unit is modified because of the temperature applied.

On the other hand, a carbonic acid gas extract is close to the chemical composition of the initial plant from that it is derived because it contains a wider variety of plant's constituents. For example, carbonic acid gas extraction of sweet false chamomile flowers yields an inexperienced extract. The absence of warmth means that it had not been altered from its wild or "denatured" form. The following section is much more similar in composition to the initial flower than the distilled essential oils. Carbonic acid gas extracts contain a lot of plant constituents than the quantity extracted from identical plant exploitation (i.e., steam distillation) [9, 10].

3.3.5.1 Method

- Pressurized dioxide becomes liquid whereas remaining during a gassy state, which implies it is currently "supercritical" during this state, it is pumped up into a plant material-packed chamber.
- Carbonic acid gas has properties like liquid; that is why it acts as a solvent on the natural plant matter, propulsion the oils and different substances like pigment and organic compounds from the plant matter. The volatile oil content then dissolves into the liquid carbonic acid gas.
- The carbonic acid gas is brought back to natural pressure and evaporates back to its gassy state, whereas what is left is the ensuing oil.
- CO_2 extraction may be a kind of volatile oil extraction that uses dioxide because of the solvent. Unlike ancient solvent extraction, no residue is left behind, typically making the ensuing oils purer and safer to use.
- The carbonic acid gas extraction method starts by pressuring the carbonic acid gas till it liquefies. The liquid carbonic acid gas is employed because the solvent extracts the essential oils from the stuff.

Once the carbonic acid gas has absorbed the volatile oil, the extract appears at the normal pressure that causes the carbonic acid gas to revert to gas. As a result, there isn't any residual solvent that gets left behind within the extracted oil. Carbonic acid gas is colorless, odorless, flavorless, and nontoxic; it would not impact the ensuing oil. However, as a result of carbonic acid gas extraction during a fully sealed chamber, it recovers the entire oil from the plant material – including any chemical residue [11].

3.3.6 Maceration

Macerated oils are mentioned as infused oils. Once carrier oils are used as solvents to extract therapeutic properties from stuff. The good thing about macerated oil on top of distilled oil is that adding a plant's essence is captured within the oil, which results in capturing heavier, larger plant molecules than those obtained in the distillation method. This keeps the merchandise nearer to holding additional of the plant's valuable offerings. The ideal stuff to be infused is harvested so it is as dry as doable, as any plant wetness can cause the oil to become rancid and encourage microorganism growth. Adding five hitter of tocopherol or aliment oil (high in nutriment E) can forestall rancidity.

3.3.6.1 Method

1. Stuff is finely cut, crushed, or ground into a moderately coarse powder.
2. Stuff is placed in a very closed vessel.
3. Solvent (menstruum) is added.
4. The mixture is allowed to settle for 1 week and is agitated often.
5. The liquid is strained.
6. Solid residue (marc) is ironed to recover any remaining liquid.
7. Strained and expressed liquids are mixed.
8. Finally, the macerated product is filtered to separate the liquid and plant materials. Then the liquid is transferred into an air-tight container and stored in a cool place for 12 months [11, 12].

3.3.7 Enfleurage

Enfleurage is not customarily used nowadays. However, it is one of the oldest volatile oil extraction strategies that implement fat utilization. By the top of this method, vegetable fat or animal material becomes infused with the flower's fragrance compounds. The fats that the area unit used are odorless and solid at room temperature. The enfleurage method is carried out either "hot" or "cold." In each instance, the fat saturated with fragrance is named "enfleurage pomade."

3.3.8 Cold enfleurage

1. Extremely pure and odorless vegetable or animal material, typically lard or animal oil, is detached over glass plates in an exceeding frame referred to as a chassis and is allowed to line.

2. Contemporary flower petals or whole contemporary flowers are then placed on the prime layer of fat and ironed in. They are then allowed to settle for 1–3 days or a handful of weeks based on the used flowers. Throughout this point, their scent seeps into the fat.
3. The depleted petals are replaced, and the method is repeated until the fat reaches the specified saturation.
4. The ultimate product is the enfleurage pomade: the fat and odoriferous oil. This is often washed with alcohol to separate the floral extract from the remaining fat used to create soap. Once the alcohol evaporates from this mixture, the "absolute" is left over [12].

3.3.9 Hot enfleurage

The sole distinction in this method is that the fats are heated.

3.3.10 Cold-press extraction

This methodology is also referred to as expression or scarification and is employed especially for citrus peels.
1. The whole citrus fruit is placed in an cold-press extraction that automatically pierces into it to rupture the oil sacs that square measure set on the side of the rind. The oil and pigments run down into the device's assortment space.
2. The whole fruit is ironed to squeeze out the juice and oil.
3. The oil and juice that square measure created still contain solids from fruits, like the peel, and should be centrifuged to filter the solids from the liquids.

Finally, the oil gets separated from the juice layer and is then collected in another container.

3.4 Expression (cold-press distillation)

Expression, or cold-press extraction, is often used for uninflected oils from citrus peels. This ancient oil extraction method initially concerned soaking citrus peels in heated water and then hand-squeezing them with a sponge till the oil glands burst. Once the oil was discharged, it was collected within the sponge and squeezed into a group instrumentality wherever the juice and oils would separate over time. Today, the method of expression may be a bit a lot of technical and involves the employment of machinery. The modern expression method consists of the work of a tool that punctures the citrus peels; therefore, the oil sacs on the face of the rind rupture. Once this happens, the oil runs into the device's assortment space. Then, the peels are automatically ironed to squeeze out the oils and juices. Now, ensuing oils and juices can still contain the peel and alternative solid parts of the fruit that should be centrifuged to separate the liquids from the solids. The oil can break free of the juice layer to form the ultimate oil product. If you are fascinated by oil distillation, the USA Science Lab instrumentation features an extensive range of kit choices to assist you to begin. Our intensive inventory includes distillation systems, solvents, centrifuges, and laboratory refrigerators to store your merchandise. For the past decade, we have worked exhaustively to produce fine quality, new, and used laboratory instrumentation at competitive costs. In addition, we provide exceptional client service. Contact the USA Science Lab to find out how our instrumentation will assist you to begin, expand, or enhance your oil business [12].

3.5 Essential oil recovery

The entire liquid (aqueous section and essential oil) is extracted with a volatile solvent. Therefore, the solvent is gaseous without drying over Na sulfate. In this case, the crude product cannot be referred to as a necessary oil; however, the solvent is extracted from the distillation water. This definition fits the *Likens–Nickerson* technique for uninflected volatile organic compounds below either region or reduced pressure. Preferred solvents ought to be volatile (diethyl ether, chloride, pentane, etc.) and not mixable with water. The oil is recovered by decantation; therefore, the half that continues to be on the walls of the glasswork (Clevenger or oil receiver) is retrieved by laundry with a minimum quantity of a volatile solvent. The solvent should be eliminated, so the following product will still be outlined as a necessary oil. The residual solvent should be checked in all told cases as not prodigious. The authors should indicate whether or not the oil has been dried, the character of the desiccant, and, therefore, the yield of this step. If the oil is to be employed in food application, the residual solvent should adjust to the IOFI Code of Practice 1. The distillation yield ought to be expressed differently (weight of oil/weight of dry biomass weight).

3.5.1 Uses

It is obligatory to dilute volatile oil before use because it is not good to be used in focused type:
- In aromatherapy
- In food

- As pesticides
- As straightforward antimicrobial smells like lavender, chamomile, and rosewater could facilitate keeping you calm. You will be able to inspire or rub diluted versions of those oils on your skin. Scientists suppose that they work by causing chemical messages to elements of the brain that affect mood and feeling. Though these scents alone would not take all your stress away, the aroma could assist you in relaxing.
- Cumin oil used in your food will cause blisters if you place it on your skin.
- Citrus oils that are safe in your food could also be dangerous for your skin, mainly if you are exposed into the sun. And also, the opposite is true. Eucalyptus or sage oil could soothe you if you rub it on your skin or breathe it in. however, swallowing them might cause severe complications and seizures. Undiluted oils are too robust to use directly. You will have to be compelled to dilute them, typically with vegetable oils, creams, or bathtub gels, to an answer that solely incorporates a bit 1 Chronicles to five of the volatile oils. Precisely what proportion will vary? The upper the share, the more doubtless you will own a reaction. Thus, it is vital to combine them properly. Confusion or inflamed skin can absorb additional oil and cause unwanted skin reactions. Undiluted oils that you should not use in any respect will be downright dangerous on the broken skin.
- Massaging your scalp frequently with a combination of lavender oil and alternative herb's essential oils could facilitate slow hair loss from alopecia. This autoimmune disorder, which may run in families, makes your body erroneously attack your hair follicles. The analysis goes on to envision if lavender oil may facilitate hair growth, too. In one study, mice were treated with lavender oil up garment worker coats.
- Culinary lavender adds a clean sweetness to simply concern any dish. It also packs ursolic acid, a nutrient that will fight cancer and burn additional calories. Mix your salt-free herbs in the French region seasoning with lavender, rosemary, thyme, fennel, chives, and alternative herbs.

The compounds in these oils could have some side effects on the soma. The subsequent articles gives additional data on common health conditions could like the utilization of essential oils:
- headaches
- constipation
- depression
- cold sores
- sinus infections
- sore muscles
- anxiety
- mood booster
- stress reliever

- improve sleep
- reduce inflammation
- kill bacterium, plants, and viruses

3.5.2 Side effects

Many people suppose that due to essential oils' natural merchandise, they will not cause facet effects. This is often not true. The potential facet effects of essential oils include:

1. Essential oils typically have a far more pungent smell than the plants they are available from and contain higher levels of active ingredients. This must be the quantity of plant matter needed to form volatile oils [13].
2. Headaches: eupnoeic essential oils could facilitate some folks with their headaches; however, an excessive amount of eupnoeic could result in a headache.
3. Irritation and burning invariably dilute oils with a carrier oil before applying it on the skin. Apply low quantity to common space of the skin first to check for any reactions.
4. Asthma attacks: whereas essential oils could also be safe for many folks to inhale, some folks with asthma attacks could react to inhaling the fumes.

3.5.3 Essential oils and their benefits

1. **Lavender oil**
   Cleansing agent
   Added to water to make a room spray or body spritzer
   Side effect
   It can potentially disrupt hormones in young boys.

2. **Tea tree oil**
   Antiseptic
   Antimicrobial
   Antifungal
   Acne
   Athlete's foot and ringworms
   Side effect
   Neurotoxic

3. **Frankincense oil (king of oils)**
   Inflammation
   Mood and sleep

Relieves asthma
Prevents gum disease
Aromatherapy
Present in skin creams

4. **Peppermint oil**
   Anti-inflammatory
   Antifungal
   Antimicrobial
   Fights fatigue
   Lifts mood
   Supports digestion

5. **Eucalyptus oil**
   It soothes a stuffed-up nose by opening your nasal passage so you can breathe easier.
   Relieves pain
   Fights against herpes simplex virus
   Antimicrobial
   Anti-inflammatory
   Side effect
   It should not be ingested and can have dangerous effect on children and pets.

6. **Lemon oil**
   Reduces anxiety and depression
   Reduces pain
   Eases nausea
   Kills bacteria
   Aromatherapy of this oil might improve the cognitive function of people with Alzheimer's disease.
   Side effects
   It can make the skin more sensitive to sunlight and increase the risk of sunburn.

7. **Lemongrass oil**
   Relieves stress
   Anxiety
   Depression
   Kills bacteria
   Prevents growth of fungus in athlete's foot, ringworm
   In type 2 diabetes, it reduces the blood sugar level.

8. **Orange oil**
   Natural cleanser
   Kills bacteria
   Reduces anxiety
   Reduces pain
   Side effect
   Makes the skin more sensitive to sunlight

9. **Rosemary oil**
   Flavoring agent
   Improves the brain function
   Promotes hair growth
   Reduces pain and stress
   Lifts mood
   Reduces joint inflammation

Note: *In case of pregnancy or high blood pressure, it is advisable not to use rosemary oil.*

10. **Cedarwood oil**
    An ingredient in insect repellent, shampoo, and deodorant
    Antioxidant
    Antibacterial
    Aromatherapy

3.5.3.1 How to use essential oils

- By diffusing
- By rolling on the skin
- By ingesting

In addition, Table 3.2 lists out some of the common plant names with their family and also the parts of plants that are used for extraction of essential oils.

3.6 Conclusion

Overall, it can be concluded that plant materials such as leaves, bark, fruits, flowers, root, herbs, and seeds are the main sources of essential oils. These oils can be extracted from all parts of plants by applying various extraction processes.

Table 3.2: Some of the common plant names with their family and description of plant parts used for extraction of essential oils [11, 13].

S. no.	Common name	Biological source	Family	Parts used
1	Amyris	*Amyris balsamifera* L.	Rutaceae	Wood
2	Angelica root	*Angelica archangelica* L.	Apiaceae	Root
3	Anise seed	*Pimpinella anisum* L.	Apiaceae	Fruit
4	Armoise	*Artemisia herba-alba* Asso.	Asteraceae	Herb
5	Asafoetida	*Ferula assa-foetida* L.	Apiaceae	Resin
6	Basil	*Ocimum basilicum* L.	Lamiaceae	Herb
7	Bay	*Pimenta racemosa* Moore	Myrtaceae	Leaf
8	Bergamot	*Citrus aurantium* L.	Rutaceae	Fruit peel
9	Buchu leaf	*Agathosma betulina* (Bergius)	Rutaceae	Leaf
10	Camphor	*Cinnamomum camphora* L. (Sieb.)	Lauraceae	Wood
11	Caraway	*Carum carvi* L.	Apiaceae	Seed
12	Cardamom	*Elettaria cardamomum* (L.) Maton	Zingiberaceae	Seed
13	Carrot seed	*Daucus carota* L.	Apiaceae	Seed
14	Cascarilla	*Croton eluteria* (L.) W. Wright	Euphorbiaceae	Bark
15	Cedarwood	*Cupressus funebris* Endl.	Cupressaceae	Wood
16	Cedarwood	*Juniperus mexicana* Schiede	Cupressaceae	Wood
17	Cedarwood	*Juniperus virginiana* L.	Cupressaceae	Wood
18	Celery seed	*Apium graveolens* L.	Apiaceae	Seed
19	Chamomile	*Matricaria recutita* L.	Asteraceae	Flower
20	Chamomile	*Anthemis nobilis* L.	Asteraceae	Flower
21	Chenopodium	*Chenopodium ambrosioides* (L.) Gray	Chenopodiaceae	Seed
22	Cinnamon bark	*Cinnamomum zeylanicum* Nees	Lauraceae	Bark
23	Cinnamon bark	*Cinnamomum cassia* Blume	Lauraceae	Bark

Table 3.2 (continued)

S. no.	Common name	Biological source	Family	Parts used
24	Cinnamon leaf	*Cinnamomum zeylanicum* Nees	Lauraceae	Leaf
25	Citronella	*Cymbopogon nardus* (L.) W. Wats.	Poaceae	Leaf
26	Citronella	*Cymbopogon winterianus* Jowitt.	Poaceae	Leaf
27	Clary sage	*Salvia sclarea* L.	Lamiaceae	Flowering herb
28	Clove buds	*Syzygium aromaticum* (L.)	Myrtaceae	Leaf/bud
29	Clove leaf	*Syzygium aromaticum* (L.)	Myrtaceae	Leaf
30	Coriander	*Coriandrum sativum* L.	Apiaceae	Fruit
31	Cornmint	*Mentha canadensis* L.	Lamiaceae	Leaf
32	Cumin	*Cuminum cyminum* L.	Apiaceae	Fruit
33	Cypress	*Cupressus sempervirens* L.	Cupressaceae	Leaf/twig
34	Davana	*Artemisia pallens* Wall.	Asteraceae	Flowering herb
35	Dill	*Anethum graveolens* L.	Apiaceae	Herb/fruit
36	Dill	*Anethum sowa* Roxb.	Apiaceae	Fruit
37	Elemi	*Canarium luzonicum* Miq.	Burseraceae	Resin
38	Eucalyptus	*Eucalyptus globulus* Labill.	Myrtaceae	Leaf
39	Eucalyptus	*Eucalyptus citriodora* Hook.	Myrtaceae	Leaf
40	Fennel bitter	*Foeniculum vulgare* Mill.	Apiaceae	Fruit
41	Fennel sweet	*Foeniculum vulgare* Mill.	Apiaceae	Fruit
42	Fir needle	*Abies balsamea* Mill.	Pinaceae	Leaf/twig
43	Gaiac	*Guaiacum officinale* L.	Zygophyllaceae	Resin
44	Galbanum	*Ferula galbaniflua* Boiss.	Apiaceae	Resin
45	Garlic	*Allium sativum* L.	Alliaceae	Bulb
46	Geranium	*Pelargonium* spp.	Geraniaceae	Leaf
47	Ginger	*Zingiber officinale* Roscoe	Zingiberaceae	Rhizome
48	Gingergrass	*Cymbopogon martinii* (Roxb.)	Poaceae	Leaf

Table 3.2 (continued)

S. no.	Common name	Biological source	Family	Parts used
49	Grapefruit	*Citrus × paradisi* Macfad.	Rutaceae	Fruit peel
50	Guaiac wood	*Bulnesia sarmienti* L.	Zygophyllaceae	Wood
51	Gurjum	*Dipterocarpus* spp.	Dipterocarpaceae	Resin
52	Hop	*Humulus lupulus* L.	Cannabaceae	Flower
53	Hyssop	*Hyssopus officinalis* L.	Lamiaceae	Leaf
54	Juniper berry	*Juniperus communis* L.	Cupressaceae	Fruit
55	Laurel leaf	*Laurus nobilis* L.	Lauraceae	Leaf
56	Lavandin	*Lavandula angustifolia* Mill.	Lamiaceae	Leaf
57	Lavender	*Lavandula angustifolia* Miller	Lamiaceae	Leaf
58	Lavender, Spike	*Lavandula latifolia* Medik.	Lamiaceae	Flower
59	Lemon	*Citrus limon* (L.) Burman fl.	Rutaceae	Fruit peel
60	Lemongrass	*Cymbopogon flexuosus*	Poaceae	Leaf
61	Lime distilled	*Citrus aurantiifolia*	Rutaceae	Fruit
62	*Litsea cubeba*	*Litsea cubeba* C.H. Persoon	Lauraceae	Fruit/leaf
63	Lovage root	*Levisticum officinale* Koch	Apiaceae	Root
64	Mandarin	*Citrus reticulata* Blanco	Rutaceae	Fruit peel
65	Marjoram	*Origanum majorana* L.	Lamiaceae	Herb
66	Mugwort common	*Artemisia vulgaris* L.	Asteraceae	Herb
67	Mugwort, Roman	*Artemisia pontica* L.	Asteraceae	Herb
68	Myrtle	*Myrtus communis* L.	Myrtaceae	Leaf
69	Neroli	*Citrus aurantium* L.	Rutaceae	Flower
70	Niaouli	*Melaleuca viridiflora*	Myrtaceae	Leaf
71	Nutmeg	*Myristica fragrans* Houtt.	Myristicaceae	Seed
72	Onion	*Allium cepa* L.	Alliaceae	Bulb
73	Orange	*Citrus sinensis* (L.) Osbeck	Rutaceae	Fruit peel
74	Orange bitter	*Citrus aurantium* L.	Rutaceae	Fruit peel
75	Oregano	*Origanum* spp.	Lamiaceae	Herb

Table 3.2 (continued)

S. no.	Common name	Biological source	Family	Parts used
76	Palmarosa	*Cymbopogon martinii*	Poaceae	Leaf
77	Parsley seed	*Petroselinum crispum* (Mill.)	Apiaceae	Fruit
78	Patchouli	*Pogostemon cablin* (Blanco) Benth.	Lamiaceae	Leaf
79	Pennyroyal	*Mentha pulegium* L.	Lamiaceae	Herb
80	Pepper	*Piper nigrum* L.	Piperaceae	Fruit
81	Peppermint	*Mentha × piperita* L.	Lamiaceae	Leaf
82	Petitgrain	*Citrus aurantium* L.	Rutaceae	Leaf
83	Pimento leaf	*Pimenta dioica* (L.) Merr.	Myrtaceae	Fruit
84	Pine needle	*Pinus silvestris* L. *P. nigra* Arnold	Pinaceae	Leaf/twig
85	Pine needle, Dwarf	*Pinus mugo* Turra	Pinaceae	Leaf/twig
86	Pine silvestris	*Pinus silvestris* L.	Pinaceae	Leaf/twig
87	Pine white	*Pinus palustris* Mill.	Pinaceae	Leaf/twig
88	Rose	*Rosa × damascena* Miller	Rosaceae	Flower
89	Rosemary	*Rosmarinus officinalis* L.	Lamiaceae	leaf
90	Rosewood	*Aniba rosaeodora* Ducke	Lauraceae	Wood
91	Rue	*Ruta graveolens* L.	Rutaceae	Herb
92	Sage, Dalmatian	*Salvia officinalis* L.	Lamiaceae	Herb
93	Sage, Spanish	*Salvia lavandulifolia* L.	Lamiaceae	Leaf
94	Three-lobed sage (Greek and Turkish)	*Salvia fruticosa* Mill.	Lamiaceae	Herb
95	Sandalwood, East India	*Santalum album* L.	Santalaceae	Wood
96	Sassafras, Brazilian (*Ocotea cymbarum* oil)	*Ocotea odorifera* (Vell.)	Lauraceae	Wood
97	Sassafras, Chinese	*Sassafras albidum* (Nutt.) Nees.	Lauraceae	Root bark
98	Savory	*Satureja hortensis* L.	Lamiaceae	Leaf
99	Spearmint, Native	*Mentha spicata* L.	Lamiaceae	Leaf
100	Spearmint, Scotch	*Mentha gracilis* Sole	Lamiaceae	Leaf
101	Star anise	*Illicium verum* Hook fl.	Illiciaceae	Fruit

Table 3.2 (continued)

S. no.	Common name	Biological source	Family	Parts used
102	Styrax	*Styrax officinalis* L.	Styracaceae	Resin
103	Tansy	*Tanacetum vulgare* L.	Asteraceae	Flowering herb
104	Tarragon	*Artemisia dracunculus* L.	Asteraceae	Herb
105	Tea tree	*Melaleuca* spp.	Myrtaceae	Leaf
106	Thyme	*Thymus vulgaris* L.	Lamiaceae	Herb
107	Valerian	*Valeriana officinalis* L.	Valerianaceae	Root
108	Vetiver	*Vetiveria zizanioides* (L.) Nash	Poaceae	Root
109	Wintergreen	*Gaultheria procumbens* L.	Ericaceae	Leaf
110	Wormwood	*Artemisia absinthium* L.	Asteraceae	Herb
111	Ylang-ylang	*Cananga odorata*	Annonaceae	Flower

References

[1] Ansah Herman R. Essential oils and their applications-a mini review. Adv Nutr Food Sci 2019, 4, 2572–5971.

[2] Verma S, Jain A, Gupta VB. Synergistic and sustained anti-inflammatory activity of Guguul with the ibuprofen: A preliminary study. Int J Pharma Bio Sci 2010, 2, 1–8.

[3] Butnariu M, Sarac I. Essential oils from plant. J Biotechnol Biomed Sci 2018, 1(4), 2576–2594.

[4] Kubeczka K-H. History and sources of essential oil research. In: Handbook of Essential Oils Research Gate, 2015, 5–42.

[5] Untapping the Power of Nature: Essential Oil Extraction Methods. 2017 March.

[6] Sharmeen JB, Zengin G. Essential oils as natural sources of fragrance compounds for cosmetics and cosmeceuticals. Pubmed Central 2021, 26(3), 666.

[7] Hamid AA, Usman LA. Essential oils: Its medicinal & pharmacological uses. Int J Curr Res 2011, 3(2), 86–93.

[8] Dagli N, Dagli R. Essential oils, their therapeutic properties, & implication in dentistry: A review. J Int Soc Prev Community Dent 2015, 5(5), 335–340.

[9] Butnariu M, Sarac I. Essential oil from plants. J Biotechnol Biomed Sci 2017, 1(4), 218–226.

[10] Tanu B, Harpreet K. Benefits of essential oil. J Chem Pharm Res 2016, 8(6), 143–149.

[11] Martins VV, Almeida JM. Citral, carvacrol, eugenol & thymol: Antimicrobial activity & its application in food. J Essent Oil Res 2022, 34(3), 181–194.

[12] Fokou JBH, Jazet PM. Essential oil's chemical composition & pharmacological properties. In: Essential Oil – Oils of Nature, January 2020.

[13] Rehman R, Asif Hanif M. Biosynthetic factories of essential oils: The aromatic plants. Nat Prod Chem Res 2016, 4, 227.

R Rushendran, Anuragh Singh, Kumar B Siva and K Ilango*

Chapter 4
Chemical composition of essential oils – fatty acids

Abstract: Essential oils (EOs) are complex, consisting of a variety of volatile and natural bioactive chemicals that it may be difficult to identify all types; they a are frequently used in the food industry as the best alternatives. Essential oils have different chemical compositions based on their source, ambient conditions, and plant species. Since time immemorial, the medicinal potential of these oils have been widely understood in the traditional system. In general, pure essential oils are classified into two chemical ingredient groups: oxygenated compounds (esters, aldehydes, ketones, alcohols, phenols, and oxides) and hydrocarbons (monoterpenes, sesquiterpenes, and diterpenes). Fatty acids make up 94–96% of the weight of different fats and oils. Fatty acids have a significant impact on the physical and chemical characteristics of glycerides, due to their high weight in the molecules and the fact that they make up the reactive fraction of the molecules. Individual oils, on the other hand, can have a large number of straight chain, aromatic, or heterocyclic compounds. Three primary components make up 86–93% of fatty acids: palmitic acid, linoleic acid (omega-6), and oleic acid. In the present scenario, fatty acids of essential oils such as myrcene, linalool, geraniol, bisabolol, palmitic acid, zinziberene, nerolidol, linoleic acid, eicosanol, and erucic acid have the capacity to treat different diseases and are discussed descriptively in this chapter.

Keywords: Essential oils, fatty acids, glycerides, linoleic acid, nervonic acid, polyunsaturated fatty acids

*Corresponding author: Ilango K,** Department of Pharmaceutical Quality Assurance, SRM College of Pharmacy, SRM Institute of Science and Technology, Kattankulathur 603 203, Chengalpattu, Tamil Nadu, India, e-mail: ilangok1@srmist.edu.in
Rushendran R, Anuragh Singh, Department of Pharmacology, SRM College of Pharmacy, SRM Institute of Science and Technology, Kattankulathur 603 203, Chengalpattu, Tamil Nadu, India
Siva Kumar B, Department of Pharmaceutical Chemistry, SRM College of Pharmacy, SRM Institute of Science and Technology, Kattankulathur 603 203, Chengalpattu, Tamil Nadu, India

https://doi.org/10.1515/9783110791600-004

4.1 Introduction

Essential oils are a combination of volatile plant components. Terpenoids and phenolic chemicals constitute the majority of these substances. These flavoring volatile compounds are biosynthesized in specific cell types found in practically every part of the plant, from the leaves to the flowers to the roots, depending on the genus. Organic compounds of essential oils are derived from aromatic plant sources such as roots, flowers, bark, seeds, and leaves. They are extraordinarily complex substances, each oil including hundreds of different chemicals, the great majority of which are present at concentrations of less than 1%. To increase the oil's effectiveness, these compounds act together as a full synergistic unit. The chemical elements of an oil determine its scent, flavor, and therapeutic characteristics, and a single oil can be used for a variety of purposes [1]. Aromatic plants can produce chemicals that are extracted as essential oils and are named by their fragrance. Aromatic plants are found across the plant kingdom and are not restricted to a taxonomic group. It is worth noting that the essential oil makeup varies from one plant taxonomic group to the next. It is also worth noting that chemicals that will eventually create essential oils are classified as secondary metabolites that have a volatile feature in plants [2–4]. Even within the same species, the chemical composition of essential oils might differ from plant to plant. Essential oils are used because of their amazing biological qualities that influence humans, animals, plants, insects, and microorganisms [5–8]. This chapter contains information on the main chemical components, with basic information along with their pharmacological properties and we hope this will be novel to aromatherapy or a useful recap for therapists.

4.2 Compounds of essential oils

The major chemical families are addressed in the following sections. Please note that the percentages stated against the oils are averages based on substantial historical data and the findings of hundreds of gas chromatographic analyses that analyze the chemical make-up of an oil and are obtained from essential oils in color.

4.2.1 Monoterpenes

They are the largest chemical family found in essential oils and are found in almost every essential oil, including limonene and pinene. They have a very small and light molecular structure, and are extremely volatile and free flowing, with strong aroma. They are also more susceptible to oxidation, especially those with high levels of limonene, that is, they will have a shorter shelf life. Examples of oils rich in monoterpenes are grapefruit (96%), black pepper (60%), cypress (75%), lemon (87%), juniper berry (80%), mandarin (90%), lime (72%), nutmeg (75%), orange bitter (90%), pine (70%), orange sweet (85%), and silver fir (90%).

Safety: Skin irritation, especially if oxidized.

4.2.2 Sesquiterpenes

In essential oils, sesquiterpenes are less frequent. As their molecules are larger and heavier than monoterpenes, they are slightly more stable and have a long shelf life. High quantities of sesquiterpenes are obtained from the woods and roots of plants of the Asteraceae family and offer grounding and balancing properties.

Examples of oils rich in sesquiterpenes are cedarwood Virginian (60%), cedarwood Virginian (60%), Cade (60%), German chamomile (35%), Cedarwood Atlas (50%), ginger (55%), *Helichrysum* (40%), patchouli (50%), myrrh (39%), yarrow (45%), ylang-ylang (40%), and vetivert (65%).

4.2.3 Alcohols

Monoterpenes are used to make plant alcohols (monoterpenols). Sesquiterpenes are used to a lesser extent (sesquiterpenols). Most essential oils contain alcohols, which are the most helpful compounds in aromatherapy. They have a strong yet soft presence, are usually rather fluid, and have nice fragrances that are widely tolerated. Antiseptic and pain-relieving qualities are the most common uses. Examples of oils rich in alcohols are catnip (62%), basil (50%), coriander (70%), cedarwood (30%), lavandin (45%), geranium 63%), sweet marjoram (50%), lavender (36%), cornmint (70%), myrrh (40%), peppermint (42%), palmarosa (85%), neroli (40%), Rose Otto (60%), rosewood (90%), tea tree (45%), sandalwood (80%), and vetivert (40%).

4.2.4 Phenols

They are the most potent and irritating chemicals, and they take a long time to dissipate. They have a distinct scent and are chemically active. Phenols have strong bactericidal capabilities as well as a stimulating effect. Examples of oils rich in phenols are cinnamon leaf (86%), basil ct methyl chavicol (90%), clove bud (90%), oregano (70%), fennel (62%), thyme ct thymol (40%), and tarragon (70%).

Safety: Skin and mucous membranes are irritated by phenol-rich oils. Use only at a low dilution and for a short time. Always use alcohol-rich oils when blending. Pregnant women should avoid them.

4.2.5 Aldehydes

The most noticeable component is aldehydes that have a powerful citrus-like scent and are a crucial contributor to the overall aroma of an oil. They are prone to oxidation and are highly unstable. Examples of oils rich in aldehydes are lemon eucalyptus (80%), *Citronella*, *Melissa* (50%), and lemongrass (80%).

Safety: Aldehyde-rich oils should be used in a low dilution (1%) because they can irritate skin and mucous membranes.

4.2.6 Ketones

Ketones (e.g., thujone and pulegone) are among the most poisonous chemicals, which is why oils like Thuga and Pennyroyal are avoided in aromatherapy. However, not all ketones are poisonous, and some have significant therapeutic benefits, notably in the upper respiratory system, where they can help with congestion and mucus flow. Ketones have a characteristic aroma that is commonly described as minty-camphoraceous. They are really penetrating, so use them with caution. Examples of oils rich in ketones are dill (50%), caraway (54%), *Eucalyptus dives* (45%), spearmint (55%), hyssop (46%), sage (35%), rosemary (25%), and *Tagetes* (50%).

Safety: They should be used with caution because they might accumulate in the body. Keep in mind the contraindications of each oil and take extra precautions if you are pregnant.

4.2.7 Esters

They are found in most essential oils; there are not many oils that do not contain them. They have a strong fruity perfume and are generally nontoxic and pleasant to use, making them suitable for use by children, the elderly, and the infirm. Examples of oils rich in esters are bergamot (40%), benzoin (70%), Roman Chamomile (75%), sweet birch (99%), *Helichrysum* (40%), clary sage (70%), lavender (45%), jasmine (54%), bergamot mint (60%), sweet thyme (40%), petitgrain (55%), and wintergreen (99%).

Safety: Sweet birch and wintergreen are strong in methyl salicylate, a compound that is regarded to be more potent than aspirin. Both oils have significant safety concerns and should be used with caution.

4.2.8 Oxides

Oxides contain the most potent scents of all compounds, and they are my favorite components because they are both bracing and breathing! 1,8-Cineole, often known as eucalyptol, is the most prevalent oxide. With its intense expectorant qualities and characteristic aroma, cineole is found in most species of Eucalyptus oil as well as other camphoraceous oils such as Rosemary, Tea Tree, and Cajeput. Oxides are drying molecules that remove moisture from a space, making them ideal for treating respiratory problems. Examples of oils rich in oxides are *Eucalyptus smithii* (78%), German chamomile (35%), wild marjoram (55%), lavender spike (34%), *Ravensara* (60%), Niaouli (60%), and rosemary (30%).

Safety: Cineole can irritate the mucous membranes and skin.

4.2.9 Chemotypes

Essential oils extracted from botanically identical plants can sometimes have vastly diverse chemical compositions and medicinal effects. This could be attributed to a variety of factors like as growing conditions, location, and climate. Red Thyme contains more phenols (thymol), while sweet thyme contains more alcohols (linalool). Chemotypes include herbs like basil, marjoram, and rosemary. Essential oils are used in foods as a preservative or for flavoring, in cosmetics as an odorant component, and in medicine as an active ingredient. The pharmacological effects of essential oils include their effects on both transmissible and nontransmissible diseases. The goal of this chapter is to collect data on the chemical profile of essential oils (fatty acids) as well as their pharmacological potential.

4.3 Arachidonic acid

It is a long-chain polyunsaturated fatty acid, quite abundant in microalgae such as *Porphyridium cruentum*, *Nannochloropsis* sp., *Phaeodactylum tricornutum*, *Monodus subterraneus*, *Crypthecodinium cohnii*, *Chroomonas salina*, *Gracilaria* sp., and *Prunus incisa* comb. It is one of the fatty acids produced in the brain that promotes the growth and repair of skeletal muscle tissue by converting to prostaglandin PGF2 alpha. Fluidity of hippocampal cell membranes and neurodegenerative diseases, such as Alzheimer's disease [9].

4.3.1 Chemistry

Omega-3 (n-3) fractions (with the main synthesis pathway from linolenic acid) and omega-6 (n-6) fractions (with the main synthesis pathway from linoleic acid (LA)) are the two types of polyunsaturated fatty acids (PUFAs). Desaturase converts LA to n-6, which leads to the biosynthesis of gamma-linolenic acid (GLA), dihomo-GLA, and finally, arachidonic acid. It is mostly found in phospholipid membranes, where it competes for metabolism and receptors with n-3 acids and their products [10]. Arachidonic acid was thought to have an all-*cis* double bond arrangement at carbon atoms 5, 8, 11, 14 in an unbranched 20-carbon chain. It is necessary to develop an unambiguous chemical synthesis of arachidonic acid, and the well-known Lindlar catalyst was used to reduce the tetraalkyne, effectively producing cis-double bonds from triple bonds. Arachidonic acid was then carbonylated from the tetra-cis-product [11].

4.4 α-Bisabolol

(−)-α-Bisabolol (6-methyl-2-(4-methylcyclohex-3-en-1-yl) hept-5-en-2-ol), also referred to as levomenol, is an unsaturated, active monocyclic sesquiterpene alcohol produced by direct distillation of an essential oil-containing plants such as *Eremanthus erythropappus*, *Smyrniopsis aucheri*, *Chamomilla recutita*. α-Bisabolol has been found as a key ingredient of the essential oil of *Salvia runcinata*, *Matricaria chamomilla*, and so on. *Salvia runcinata* contains up to 50% and 90%, respectively, of (−)α-bisabolol. Candeia (*Eremanthus erythropappus*) wood, which may contain up to 85% α-bisabolol, is another less explored source. It possesses different biological activities such as anticancer, antioxidant, nephro-, neuro-, cardio-, gastro-protections, anti-inflammatory, and antimicrobial. It also showed a strong protective impact against a variety of illnesses that affect multiple organ systems, including the neurological and cardiovascular systems. (−)-α-Bisabolol has garnered great economic interest due to its exquisite floral odor and antibacterial and gastroprotective properties. Since then, it has been discovered that it may occur in four different stereoisomers. It is a common constituent in cosmetic and dermatological formulations such as hand and body lotions, creams, deodorants, sun-care, lipsticks, after-sun products, sport creams, and infant care products. It is a popular active constituent for skin against repetitive environmental challenges [12].

4.4.1 Chemistry

Bisabolol has the chemical formula $C_{15}H_{26}O$ and is otherwise called alpha-4-dimethyl-alpha-(4-methyl-3-pentenyl)-3-cyclohexene-1-methanol. Bisabolol smells faintly like fragrant vegetation. It is a colorless liquid with a low density (0.93) and a boiling point of 153 °C at 12 Torr. It is a strong lipophilic chemical that oxidizes easily and almost water-insoluble but ethanol-soluble. The primary oxidation products are bisabolol-oxide A and B. The natural enantiomer (+)-bisabolol is uncommon, and the synthetic equivalent is often a combination of (−)-α-bisabolol.

4.5 Eicosonal

It is a waxy substance used in cosmetics as an emollient. Its name is derived from the peanut plant's name and is obtained from many plants such as *Solena amplexicaulis*, *Lonicera japonica*, and *Artemisia baldshuanica*, of which it is a major natural constituent [13]. When compared to primary alcohol, the secondary n-eicosanols were ineffective emulsifiers. The 2- and 4-eicosanols have promising properties that warrant further research, whereas 6-, 8-, and 10-alcohols showed little activity. It is worth noting that the lower the emulsifying efficiency, the greater is the steric hindrance to the hydrophilic group. The possibility that these agents act in the manner described by pink for the mechanism of emulsification by a solid mutually insoluble in both phases is supported by the gel formation of the alcohol–oil solutions at room temperature. The complexities of the emulsification process indicate that more than one factor influences its formation. It has different biological characteristics such as antimicrobial, antioxidant, antitumor, and antifungal [14].

4.5.1 Chemistry

It is a 20-carbon straight-chain fatty alcohol that is usually made by hydrogenating arachidonic acid. 1-Eicosanol is a waxy, water-insoluble, saturated fatty alcohol with a long chain. Fatty alcohols are primary alcohols with a high molecular weight and a straight chain that are derived from natural fats and oils. 1-Eicosanol has thickening, emulsifying, and emollient properties. When it comes to fatty acids, the monolayer of fatty alcohols with different hydrocarbon chain lengths has nearly the same area per molecule [15].

4.6 Farnesol

It is a 15-carbon acyclic sesquiterpene alcohol produced spontaneously by a number of organisms, including *Candida albicans*. This compound was observed in the flowers of *Vachellia farnesiana* L., *Cymbopogon nardus*, *Citrus aurantium*, *Cymbopogon citratus* Stapf., *Polianthes tuberosa*, *Abelmoschus moschatus*, *Myroxylon balsamum*, and so on Wight and Arn. species, also called *"Acacia farnesiana"*; thus, the origin of its name, with the "ol" added to explain the fact that this is a chemical alcohol. It is colorless and oily with a pleasant and mild scent that is hydrophobic and insoluble in water but miscible with oils under ordinary circumstances. Plants and mammals both manufacture farnesol from 5-carbon isoprene molecules. When geranyl pyrophosphate interacts with isopentenyl pyrophosphate, the 15-carbon farnesyl pyrophosphate (FPP) is formed, which is an intermediary in the production of sesquiterpenes like farnesene. Sesquiterpenoids such as farnesol can then be produced, as a result of oxidation. For acyclic sesquiterpenoids, farnesol phosphate-activated derivatives are the precursor compounds. These molecules are duplicated to generate 30-carbon squalene, which is then converted into steroids in plants, animals, and fungus [16]. Farnesol and its derivatives are important beginning points for chemical synthesis, both natural and synthetic. It is also used to draw out the fragrances of pleasant, floral perfumes in perfumery. It enhances the aroma of perfume by acting as a co-solvent that controls the odorants' volatility. It is particularly popular in lilac fragrances and as a deodorant in cosmetic products. Farnesol was mentioned as one of 599 additions to cigarettes as a flavoring element in a 1994 study provided by five major tobacco firms. It is both a natural mite insecticide and a pheromone for a variety of other insects. Due to sensitization, it is restricted from being used in perfumery [17].

It demonstrates this compound's significant pharmacological potential, such as its anti-inflammatory, antioxidant, anxiolytic, chemopreventative, analgesic, antihypertension, depressant, and neuroprotective effects. Farnesol inhibited gram-positive (*Staphylococcus epidermidis*, *Staphylococcus aureus*, *Staphylococcus xylosus*, *Staphylococcus warneri*, *Enterococcus faecium*, and *Enterococcus faecalis*) and gram-negative (*Pseudomonas aeruginosa* and *Escherichia coli*) bacterial growth. It has also been shown to reduce carcinogenesis in animal models, implying that it acts as a chemopreventative and antitumor agent, in vivo.

4.7 Geraniol

It is an isoprenoid monoterpene (acyclic) produced from the essential oils of aromatic plants including *Valeriana officinalis*, *Cinnamomum tenuipilum*, *Phyla scaberrima*, *Perilla frutescens*, *Citrus limon*, *Thymus pulegioides* L., and several other plants. It is also found in minute amounts in *geranium* species and other essential oils such as rose oil, palmarosa oil, and citronella oil. Raspberry, peach, red apple, grapefruit, lime, plum, lemon, orange, pineapple, watermelon, and blueberry are available flavors. Honeybees create geraniol to aid in the identification of nectar-bearing flowers and the location of their hive entrances. It is also a common insect repellent, particularly for mosquitos. It is a common element in these companies' consumer goods, and it is one of the most essential chemicals in the fragrance and flavor industries. It is used as a low-toxicity natural pest management product, as it has insecticidal and repellent properties due to its attractive smell. It has a similar aroma too, but it is chemically unrelated to 2-ethoxy-3,5-hexadiene, popularly called geranium taint, a wine flaw caused by lactic acid bacteria fermenting sorbic acid. It is crucial for the production of other terpenes. Dehydration and isomerization of geraniol, for example, produces myrcene and ocimene. The diverse activities of geraniol suggested that it has pharmacological properties such as antitumor, antifungal, anti-angiogenesis, anti-inflammatory, antiulcerogenic, antioxidative, antimicrobial activities, antidepressant, antinociceptive, hepatoprotective, antidiabetic, anti-apoptosis, antiarrhythmic, cardioprotective, and neuroprotective effects. Geraniol is widely used in the flavor and fragrance industries, and various investigations are being conducted to find an efficient way to biosynthesize it. It decreases K-ras, MAPK, PI3K, and E-catenin transcription, while increasing phosphatase and PTEN, progesterone receptors, and E-cadherin protein expression. Researchers and formulation scientists are interested in the function of geraniol as a penetration enhancer for transdermal drug delivery.

4.7.1 Chemistry

Geraniol is a monoterpenoid that is also an alcoholic compound. It is the major component in citronella oil, as well as rose and palmarosa oils. Despite the fact that commercial samples appear yellow, the oil is colorless. It is transformed to the cyclic terpene-terpineol in acidic solutions. Expected reactions occur in the alcohol category. It has the ability to be transformed to tosylate, which is a precursor to chloride. The Appel reaction, which involves reacting geraniol with triphenyl phosphine and carbon tetrachloride, also produces geranyl chloride. It is possible to hydrogenate it, and it can be oxidized to geranial aldehyde. It is emitted by numerous species of flowers and found in the vegetative tissues of many herbs, where it often coexists with geraniol, geranial, and neral oxidation products. This monoterpene alcohol is a common scent ingredient. It is found in 76% of examined deodorants on the European market, domestic (41%) and household products, and cosmetic (33%) formulations based on natural components, according to a consumer product survey, and its annual output exceeds 1,000 metric tons [18, 19].

4.8 Hentriacontane

It is also called untriacontane. It is found in a variety of plants, including peas (*Pisum sativum*). It is found naturally in a number of common plants and foods including the common pea, gum Arabic, grapes, watermelons, papaya, coconuts, and sunflowers. It has also been found to be a major component of candelilla wax [20]. It has antitumor activity, among other pharmacological effects. It has also been shown to have a cytotoxic effect on lymphoma cells. Hentriacontane, one of the major components discovered, was known to be anti-inflammatory and cytotoxic to lymphoma cells. The importance of macrophages in innate and adaptive immunity is clearly demonstrated in numerous reports on their role in inflammation. Infection, inflammation, atherosclerosis, lupus, cancer, and diabetes are all diseases that macrophages play a role in. Hentriacontane works by inhibiting the NF-kB pathway and lowering pro-inflammatory mediators (NO, PGE2, and LTB4) as well as cytokines (IL-6, TNF-α, IL-1β, and IL-10). As a result, hentriacontane could be considered as a potential treatment for inflammatory diseases [21].

4.8.1 Chemistry

It is a solid-state and long-chain alkane hydrocarbon with the molecular formula $CH_3(CH_2)_{29}CH_3$. A β-diketone, that is, hentriacontane, in which the hydrogens at positions 14 and 16 are replaced by oxo groups. The bark decoction is used to treat a variety of ailments, including skin conditions, diabetes, ulcers, dysentery, piles, gonorrhea, asthma, gleet, urinary diseases, and leucorrhea [22].

4.9 Hexadecanal

Hexadecanal is an organic molecule with 16-carbon free fatty aldehyde and an analogue of palmitic acid, having chemical formula, $C_{16}H_{32}O$ that contains a long-chain fatty aldehyde. *Heliothis maritima*, *Solanum stuckertii*, *Calophyllum inophyllum*, *Dorema ammoniacum* Don, *Dorema aucheri* Boiss, *Trichosanthes huphensis*, human skin, saliva, perspiration, and fecus, all contain it. In 2017, researchers discovered that when neurotypical males are exposed to subliminal amounts of hexadecanal, their electrodermal activity increases; however in men with autism spectrum disorder, it does not [23, 24].

In 2021, it was discovered that inhaling hexadecanal reduced aggressiveness in men but increased aggression in women. It is one of the most common compounds emitted from human newborns' skulls, which may be an evolutionary survival mechanism to persuade moms to safeguard the baby and men to avoid attacking it.

However, it is unknown if the quantity of hexadecanal released by people is sufficient to cause harm to other individuals [25].

4.10 Hexadecanoic acid

It is also called as palmitic acid that has 16-carbon long-chain saturated fatty acid and is present in animals, plants, and microbes. It has the chemical formula CH_3 $(CH_2)_{14}COOH$/ $C_{16}H_32O_2$ and is a major component of palm oil, accounting for up to 44% of total fats and dietary fats, accounting for approximately 13% of the total fatty acid content in peanut oil, 65% in butter, 42% in lard, 53% in tallow, 15% in soybean, 13% in corn oil, and 17% in olive oil. Palmitic acid (PA), β-carotene, and vitamin E are abundant in the oil. PA is generated in low levels by a broad variety of different plants and creatures. It may be found in cheese, butter, beef, milk, olive oil, soybean oil, sunflower oil, and cocoa butter. Karukas have a PA content of 44.90%. Spermaceti contains the cetyl ester of PA (cetyl palmitate) [26].

Excessive carbohydrates in the body are turned to fatty acids; PA is the first fatty acid produced during fatty acid production and the precursor to longer fatty acids. As a result, PA is a substantial component in animal bodies. According to a study, it accounts for 21–30% (molar) of human depot fat and is a substantial, although extremely variable, lipid component of human breast milk. Palmitate has a negative feedback loop with acetyl-CoA carboxylase (ACC), which converts acetyl-CoA to malonyl-CoA, which is then used to add to the expanding acyl chain, limiting further palmitate formation. Palmitoylation is the process by which certain proteins are altered by the addition of a palmitoyl group. PA can change proteins post-translationally in this process, in which PA is covalently attached to the proteins through a thioester bond.

When the thioester bond is severed by thio-esterase, this covalent bond is reversible. As a result, palmitoylation acts as a switch that governs the activity of the protein. Palmitoylation occurs on a wide range of proteins, including intrinsic, mitochondrial, and peripherally associated membrane proteins. Palmitoylation, for example, enhances the hydrophobicity of membrane protein and guides it to certain microdomains inside the cell membrane. It is used to make cosmetics, soaps, and industrial mold release agents. These applications make use of sodium palmitate, which is frequently produced by the saponification of palm oil. To this end, palm oil (*Elaeis guineensis*) is treated with sodium hydroxide (in the form of caustic soda or lye), which triggers hydrolysis of the ester groups, yielding glycerol and sodium palmitate. When PA is hydrogenated, it produces cetyl alcohol, which is used to make cosmetics and detergents. During WWII, aluminum salts of PA and naphthenic acid were used as gelling agents with volatile petrochemicals to generate napalm. The term "napalm" is derived from naphthenic acid and PA [27].

Table 4.1: Physiochemical properties of fatty acids (essential oils).

Compound name	Mol. wt (g)	Color	Odor	Taste	Boiling point	Melting point	Flash point	Density	Log P	TPSA	HBA	HBD	Rotatable bond	Solubility
Myrcene	136.23	Yellow oily liquid	Pleasant	Sweet	167 °C	<−10 °C	103 °F	0.789–0.793	4.17	0	0	0	4	Insoluble in water. Soluble in alcohol, chloroform, ether, benzene, and glacial acetic acid
Linalool	154.25	Colorless to pale yellow liquid	Floral, spicy, wood odor	Spicy, citrus taste	198–200 °C	<25 °C	78 °C	0.858–0.867	2.97	20.23	1	1	4	Soluble in alcohol, ether, fixed oils, propylene glycol; insoluble in glycerine
Geraniol	154.25	Colorless to pale yellow oily liquid	Sweet rose odor	Sweet, citrus taste	230 °C	<−15 °C	108 °C	0.870–0.885	3.56	20.23	1	1	4	Slightly soluble in water; miscible with ether and acetone
Bisabolol	22.37	Colorless liquid	Sweet, mild, floral aroma similar to chamomile tea	Sweet hint of spice or citrus	153–156 °C	25 °C	113.2 ±15.6 °C	0.922–0.931	3.79	20.23	1	1	4	Insoluble in water and slightly soluble in ethanol
Hexadecanal	240.42	White solid	–	–	297–298 °C	33–36 °C	139 °C	0.963	7.05	17.07	1	0	14	Soluble in alcohol
Hexadecanoic acid or palmitic acid	256.42	White crystalline	Virtually odorless	Slight characteristic taste	351–352 °C	61.8 °C	206 °C	0.8527	7.17	37.3	2	1	14	Insoluble in water; soluble in acetone, ethanol, benzene; very soluble in chloroform; miscible with ethyl ether

Hentriacontane	436.8	Leaves from ethyl acetate	–	–	458 °C	67.9 °C	313 °C	0.781	16.4	0	0	0	28	Soluble in water; slightly soluble in ethanol, benzene, chloroform; soluble in petroleum ether
Eicosanol	298.5	White, wax-like solid	–	–	372 °C	66.1 °C	194 °C	0.8405	9.28	20.23	1	1	18	Soluble in acetone, petroleum ether, slightly soluble in ethanol, chloroform
Nerolidol	22.37	Clear pale yellow to yellow	Faint floral odor	–	276–277 °C	122 °C	230 °F	0.872–0.879	4.83	20.23	1	1	7	Soluble in most fixed oils and propylene glycol; slightly soluble in water; insoluble in glycerol
Zingiberene	204.35	Bright yellow	Spicy type	Pungent	134–135 °C	–	107.5 ±13.0 °C	871.3 mg/cm^3	5.77	0	0	0	4	Soluble in alcohol, insoluble in glycerine and propylene glycol
Linoleic acid	280.4	Colorless to straw colored liquid	Oily liquid with low odor	–	365.2 °C	-8.5 °C	112 °C	0.898–0.904	7.05	0	1	2	14	Freely soluble in ether, alcoho1, dimethyl formamide, benzene, insoluble in water
Farnesol	222.37	Slightly yellow to colorless liquid	Slightly yellow to colorless liquid	Weak citrus lime taste	110–113 °C	<25 °C	96 °C	0.884–0.889	5.77	20.23	1	1	7	Soluble in alcohol, insoluble in water
Arachidonic acid	222.37	–	–	–	170 °C	-49.5 °C	113 °C	0.922 g/mL	6.98	20.23	1	2	14	Soluble in ethanol, dimethyl formide, chloroform
Ricinoleic acid	298.5	Viscous yellow liquid	Faint mild odor	Bitter and nauseating taste	245 °C	5.5 °C	224 °C	0.940	6.19	37.3	2	3	15	Soluble in alcohol, acetone, ether, chloroform
β-Ocimene	136.23	Colorless	Floral	–	100 °C	50 °C	116 °F	0.801–0.805	4.3	0	0	0	3	Insoluble in water; soluble in oils, ethanol

4.11 Linalool

Linalool is a floral and spicy unsaturated monoterpene alcohol present in over 200 plants such as *Bursera aloexylon*, *Croton cajucara*, *Cinnamomum osmophloeum*, and spices such as jasmine, lavender, rosewood, basil, and thyme. It is found in essential oils, shampoos, soaps, deodorant, and even food, if you like to cook with a little lavender essence. In fact, it is believed that the average person consumes up to 2 g of linalool each year. Linalool is a scent that can be found in a variety of personal care, household, essential oil, and industrial applications. Linalool autoxidizes when exposed to oxygen, generating hydroperoxides. It is used as a scent in 60–80% of perfumed cleaning agents, detergents, soaps, lotions, hygiene products, and shampoos. It has pharmacological properties such as analgesic, anticonvulsant, antinociceptive, antifungal, antimicrobial, sedative, antidepressant, antiglutamatergic, antileishmanial, anticancer, anxiolytic, and has immune potentiating effects; it additionally ameliorates dyslipidaemia, and is an insecticide against flea, cockroach, and fruit fly. Concentrations of linalool have been shown to give a fruity, hoppy aroma to beer. It is widely acknowledged as a safe compound that has been approved by the FDA as a direct food additive (synthetic flavoring substance) for human and animal consumption (FDA, 1996). In the United States and Europe, the amount of terpene alcohols used in food and beverage flavoring is estimated to be around 75 tons per year by the FAO and WHO Joint Expert Committee on Food Additives. The majority of these terpene alcohols are made up of linalool and its ester. It has been used in natural products like dried herbs, as a grain storage fumigant, and as a bug repellent.

4.11.1 Chemistry

Linalool has high water solubility when compared to other essential oil components. S-(+)- and R-(−)-linalool are the two enantiomeric forms of linalool, and their olfactory and physiological qualities differ. Industrially, it can be made through hemisynthesis from natural pinene or whole chemical synthesis. α-Pinene or β-pinene are used to partially synthesize linalool. *cis*-Pinene is hydrogenated to form *cis*-pinane, which is then oxidized to form a cis/trans-mixture of pinane hydroperoxide, which is then reduced to form the corresponding pinanols that are then pyrolyzed to form d- or l-linalools [28].

4.12 Linoleic acid

LA (18:26; *cis*, *cis*-9,12-octadecadienoic acid) with the chemical formula $C_{18}H_{32}O_2$ is the most abundant PUFA in the human diet. Nuts, vegetable oils, meats, seeds, and eggs are the primary dietary sources of LA, and it may be used as an energy source like fatty acids. Esterification can result in the formation of neutral and polar lipids such as triacylglycerols, phospholipids, and cholesterol esters. Foods with soybean oil as a key component are high in LA. Soybean oil currently contributes 45% of dietary LA. LA, as a component of membrane phospholipids, serves as a structural component to maintain a specific amount of membrane fluidity in the transdermal water barrier of the epidermis. Furthermore, once freed from membrane phospholipids, it can be enzymatically oxidized to a number of cells signaling derivatives, that is, 13-hydroxy or 13-hydroperoxy octadecadienoic acid [29]. It has different biological properties such as anticancer, cardiovascular-protective, anti-osteoporotic, neuroprotective, antioxidative effects, and anti-inflammatory. Although it has been theorized that reducing LA consumption can lower tissue levels of arachidonic acid, this does not appear to be the case in people eating a typical Western diet. LA is converted to dihomo-LA, which is a component of neuronal membrane phospholipids and a substrate for the production of PGE, which appears to be vital for nerve blood flow preservation. The conversion of LA to linolenic acid and other metabolites is hindered in diabetes, which may contribute to diabetic neuropathy aetiology.

4.12.1 Chemistry

LA, being the parent component for the 6-PUFA family, may be lengthened and desaturated to various bioactive 6-PUFAs such as linolenic acid (18:36) and arachidonic acid (20:46). Arachidonic acid can then be transformed into a variety of bioactive chemicals known as eicosanoids, including leukotrienes and prostaglandins. These eicosanoids are necessary for proper metabolic activity of cells and tissues, but when generated in excess, they have been linked to a range of chronic illnesses, including cancer and inflammation. LA has attracted the greatest attention for its potential conversion to arachidonic acid [30].

4.13 Myrcene

It is a natural hydrocarbon (alkene) found in a wide range of plant species; it belongs to the acyclic monoterpene family and is an abundant chemical that occurs naturally as a key constituent in many plants; and it is a component of the hydrocarbon part of many essential oils such as *Cannabis sativa* L., *Humulus lupulus* L. (hops), *Citrus maxima* Merr., *Citrus x aurantium* L., *Pistacia lentiscus* var. *Chia*, *Spondias mombin* L., *Houttuynia cordata* Thunb., *Cymbopogon citratus* L., *Mangifera indica* L., *Pimenta racemosa* L., *Elettaria cardamomum* L. Several terpenes are extracted from hops, with -myrcene constituting the greatest monoterpene portion.

Table 4.2: Phytochemistry of fatty acids (essential oils).

Source	Compound name	Structure	Pharmacological activity
Porphyridium cruentum, Nannochloropsissp., *Phaeodactylum tricornutum, Monodus subterraneus,* *Cryptecodinium cohnii, Chroomonas salina,* *Gracilaria* sp. *Prunus incisa* comb., and microalgae	Arachidonic acid		It plays an essential role in physiological homeostasis, such as repair and growth of cells and neurodegenerative diseases
Eremanthus erythropappus, Smyrniopsis aucheri, *Chamomilla recutita, Salvia runcinata, Matricaria chamomilla*	α-Bisabolol		Antimicrobial, antioxidant, nephro-, neuro-, cardio-, gastro-protections, anticancer, and anti-inflammatory
Solena amplexicaulis, Lonicera japonica, and *Artemisia baldshuanica*	Eicosanol		Antimicrobial, antioxidant, antitumor, and antifungal

Vachellia farnesiana L., *Citronella mucronata*, *Citrus aurantium*, *Cyclamen persicum*, *Cymbopogon citrullus*, *Polianthes tuberosa* L, *Rosa rubiginosa*, musk deer, *Balsamum tolutanum*	Farnesol		Antidepressant, anti-inflammatory, chemopreventative, antioxidant, anxiolytic, neuroprotective, and analgesic effects
Valeriana officinalis, *Cinnamomum tenuipilum*, *Phyla scaberrima*, *Perilla frutescens*, *Thymus pulegioides* L, etc.	Geraniol		Antitumor, antifungal, anti-angiogenesis, anti-inflammatory, antiulcerogenic, antioxidative, antimicrobial activities, antidepressant, antinociceptive, hepatoprotective, antidiabetic, anti-apoptosis, antiarrhythmic, cardioprotective, and neuroprotective effects
Pisum sativum	Hentriacontane		To treat atherosclerosis, lupus, cancer, and diabetes
Heliothis maritima, *Solanum stuckertii*, *Calophyllum inophyllum*, *Dorema ammoniacum* Don, *Dorema aucheri* Boiss, *Trichosanthes huphensis*, human skin, saliva, perspiration, and feces	Hexadecanal		Reduced aggressiveness in men but increased aggression in women; safest chemical for newborn babies.
Annona muricata, *Glycine max*, *Helianthus annuus* L., *Olea europaea* L. butter, cheese, milk, meat, cocoa butter, and karukas	Hexadecanoic acid		Anti-inflammation, bactericidal, personal care products, and cosmetics
Bursera aloexylon, *Croton cajucara*, *Cinnamomum osmophloeum*, and spices such as jasmine, lavender, rosewood, basil, or thyme	Linalool		Analgesic, anticonvulsant, antinociceptive, antifungal, antimicrobial, sedative, antidepressant, antiglutamatergic, antileishmanial, anticancer, anxiolytic, and immune potentiating effects additionally ameliorates dyslipidemia, insecticide

(continued)

Table 4.2 (continued)

Source	Compound name	Structure	Pharmacological activity
Nuts, vegetable oils, seeds, eggs, and meats	Linoleic acid		Anticancer, neuroprotective, anti-osteoporotic, anti-inflammatory, and antioxidative properties.
Cannabis sativa L., *Humulus lupulus* L. (hops), *Citrus maxima* Merr,, *Citrus × aurantium* L., *Pistacia lentiscus* var. Chia, *Spondias mombin* L., *Houttuynia cordata* Thunb., *Cymbopogon citratus* L., *Mangifera indica* L., *Pimenta racemosa* L., *Elettaria cardamomum* L, etc.	Myrcene		Anxiolytic, antioxidant, anti-ageing, suppresses muscle spasms, sleep aid, antibacterial, anti-inflammatory, and analgesic effects
Antirrhinum majus, *Piper aduncum*, *Piper claussenianum*, *Zornia brasiliensis*, *Zanthoxylum hyemale*, *Swinglea glutinosa*, *Baccharis dracunculifolia*, *Ginkgo biloba* L., *Capparis tomentosa* Lam., *Lantana radula* Sw., *Zornia brasiliensis* Vogel, *Achillea millefolium* L., *Magnolia denudata* Desr., *Canarium schweinfurthii* Engl., *Myrocarpus fastigiatus*, *Cinnamomum osmophloeum* Kaneh	Nerolidol		Antioxidant, antimicrobial, antimalarial, insecticidal, anti-parasitic, anti-trypanosomal, antischistosomal, skin penetration enhancer, anti-ulcer, anti-nociceptive, antitumor, anti-NAFLD, antileishmanial, and anti-inflammatory properties

Eupatorium cannabinum, Melanoleuca, Mentha piperita L., *Mentha spicata* L., *Petroselinum crispum, Ocimum basilicum, Artemisia dracunculus, Citrus japonica, Mangifera indica, Lavandula angustifolia, Nepeta nepitella, Xylopia aromatica, Lavandula latifolia,* Orchidaceae, *Humulus lupulus, Citrus bergamia,* etc.	β-Ocimene	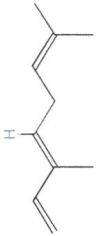	Antioxidant, anti-inflammatory, anticonvulsant, antifungal, antiviral, anticancer, nasal decongestant, expectorant properties
Ricinus communis L., Euphorbiaceae	Ricinoleic acid	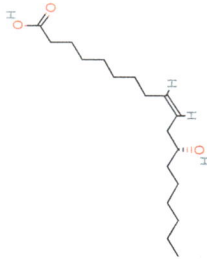	Ricinoleic acid is used in pigment, printing ink, and textile finishing
Zingiber officinale and other plants such as *Casearia sylvestris* Sw., *Solanum lycopersicum, Sorghum bicolor, Ocimum basilicum, Curcuma longa, Elettaria cardamomum,* etc.	Zingiberene	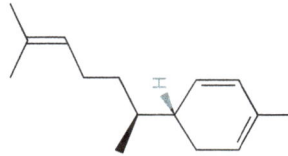	To treat sprains, arthritis, muscular aches, hypertension, sore throats, fever, dementia, helminthiasis, and infectious diseases

Pyrolysis of β-pinene, which is derived from turpentine, produces myrcene. Natural food sources of β-myrcene are believed to be 16,500 times more toxic than its manufactured use as a flavoring ingredient. It is a flavoring and food additive (fragrance agent) that is widely used in the food and beverage industry. Foods that contain myrcene are mango, lemongrass, hops, guava, melon, and thyme; additionally, it contributes a peppery and balsam aroma in beer. β-Myrcene was added to California's Prop 65 list of chemicals that cause cancer or reproductive harm in 2015. The US FDA withdrew authorization for the use of myrcene as a synthetic flavoring substance for use in food in October 2018, despite the agency's continued position that this substance poses no risk to public health under the conditions of its intended use. Many animal studies have demonstrated promising health effects such as anxiolytic, antioxidant, antiaging, suppresses muscle spasms, is a sleep aid, and has antibacterial, anti-inflammatory, and analgesic effects; there are no human studies available on this compound, yet. Although myrcene has a range of potential benefits, some researchers have also linked it with an increased risk of cancer. In 2010, the National Toxicology Program reported that high doses of myrcene had carcinogenic (cancer-causing) effects on rats.

4.13.1 Chemistry

It exists in two isomeric forms, namely, β-myrcene (7-methyl-3-methylene-1,6-octadiene) that plays an important role in cyclization reactions leading to p-menthane and α-myrcene (2-methyl-6-methylene-1,7-octadiene), which is not a natural compound but can be synthesized easily. In β-myrcene, there are three carbon–carbon double bonds (two of them being conjugated) and a gem-dimethyl terminal [31–33].

4.14 Nerolidol

It is also called as peruviol or penetrol, and is a naturally occurring sesquiterpene alcohol with a flowery odor found in a variety of plants. It is made in a step in the process of making (3E)-4,8-dimethy-1,3,7-nonatriene (DMNT), a herbivore-induced volatile that shields plants from harm. It is obtained from plants such as *Antirrhinum majus*, *Piper aduncum*, *Piper claussenianum*, *Zornia brasiliensis*, *Zanthoxylum hyemale*, *Swinglea glutinosa*, *Baccharis dracunculifolia*, *Ginkgo biloba* L., *Capparis tomentosa* Lam., *Lantana radula* Sw., *Zornia brasiliensis* Vogel, *Achillea millefolium* L., *Magnolia denudata* Desr., *Canarium schweinfurthii* Engl., *Myrocarpus fastigiatus*, and *Cinnamomum osmophloeum* Kaneh. It is found in a variety of cosmetics (shampoos and perfumes) as well as non-cosmetic goods (detergents and cleansers). Indeed, the Food and Drug Administration (FDA) has approved nerolidol as a food flavoring agent and a common ingredient in many products, which has prompted researchers to investigate additional biological properties such as antioxidant, anti-microbial, antimalarial, insecticidal, antiparasitic, anti-trypanosomal, antischistoso-mal, skin penetration enhancer, anti-ulcer, anti-nociceptive, antitumor, anti-NAFLD, antileishmanial, and anti-inflammatory. It has a strong, woodsy aroma, but it also has a zesty note to it. Nerolidol is a good choice for assisting sleep because of its sedative and calming properties. In the fabrication of nerolidol-loaded polymeric nanoparticles, sodium alginate and chitosan demonstrated to be a suitable carrier that caused a regulated release of nerolidol, resulting in more efficient distribution into the body.

4.14.1 Chemistry

Nerolidol (3,7,11-trimethyl-1,6,10-dodecatrien-3-ol) exists in two geometric isomers, trans- and cis-forms. It has four different isomeric forms consisting of two enantiom-ers and two geometric isomers. The presence of a double bond at the C-6 position and an asymmetric center at the C-3 position results in the formation of these iso-meric forms. Furthermore, the *cis*- and *trans*-nerolidol are synonymous. To address the increased industrial demand for nerolidol, chemical manufacturing of the sub-stance is necessary. Nerolidol was first produced as a step in the chemical synthesis of geranyl esters from linalool. Nerolidol, like other sesquiterpene compounds, has a high hydrophobicity, allowing it to pass through the plasma membrane and interact with intracellular proteins and/or intra-organelle locations, more easily [34–36].

4.15 β-Ocimene

It is a common plant volatile found in large levels in the leaves and flowers of a wide range of plant species such as *Eupatorium cannabinum*, *Melanoleuca*, *Mentha piperita* L., *Mentha spicata* L., *Petroselinum crispum*, *Ocimum basilicum*, *Artemisia dracunculus*, *Citrus japonica*, *Mangifera indica*, *Lavandula angustifolia*, *Nepeta nepi-tella*, *Xylopia aromatica*, *Lavandula latifolia*, *Orchidaceae*, *Humulus lupulus*, and *Cit-rus bergamia*. This acyclic monoterpene has multiple biological activities in plants, including influencing floral visitation and controlling defensive responses to her-bivory. Ocimene has been examined to work better with myrcene and pinene, among other terpenes. When combined with other cannabinoids, Ocimene produ-ces more potent effects and is also believed to act as a protective agent of plants against harmful elements. Ocimene has a sweet, woodsy scent and a variety of ther-apeutic properties, including antioxidant, anti-inflammatory, anticonvulsant, anti-fungal, antiviral, anticancer, nasal decongestant, and expectorant properties [37].

4.15.1 Chemistry

It consists of octa-1,3,6-triene bearing two methyl substituents at positions 3 and 7 (the 3*E*-isomer). It has two stereoisomers, *cis*- and *trans*-ocimene (or (*Z*)- and (*E*)-ocimene, which are the cis- and trans-versions of the central double bond, respectively. In flower smells, the trans isomer is more prevalent and abundant than the cis isomer. It is produced from compounds like isopentenyl pyrophosphate (IPP) and dimethyl allyl pyrophosphate (DMAPP) through the methyl-erythritol-phosphate (MEP) pathway. β-Ocimene is produced by the enzyme (*E*)-ocimene synthase (also known as geranyl-diphosphate diphosphate-lyase, which converts GPP into *trans*-ocimene and diphosphate [38].

4.16 Ricinoleic acid

It is a fatty acid with the formal name 12-hydroxy-9-*cis*-octadecenoic acid. It is a hydroxy acid and an unsaturated omega-9 fatty acid. It is a big part of the seed oil made from mature castor plants (*Ricinus communis* L., *Euphorbiaceous*). The triglyceride formed from ricinoleic acid accounts for about 90% of the fatty acid content in castor oil. The hydrolysis of castor oil is one of the simplest ways to obtain ricinoleic acid. It can then be separated from hydrolyzed castor oil using urea complexing (i.e., urea addition). Ricinoleic acid is produced by the saponification of castor oil, according to a more recent source. Acid value, moisture level, color, and purity are all factors that influence the use of castor oil. Castor oil, also called ricinus oil, is a fatty acid triglyceride found in the seed of the *Ricinus communis* castor plant (India and Brazil) [39]. Castor oil is unique among all fats and oils in that it is the only source of an 18-carbon hydroxylated fatty acid with one double bond, ricinoleic acid (12-hydroxyoleic acid), which accounts for approximately 90% fatty acid composition. The product uniformity and consistency are relatively high for a naturally occurring material, and it is a nontoxic, renewable resource, and biodegradable [40, 41]. Castor oil's ester linkages, double bonds, and hydroxyl groups serve as reaction sites for a variety of useful derivatives [42].

4.17 Zingiberene

Zingiberene is a monocyclic sesquiterpene found in the root of the Chinese plant *Zingiber officinale*, and in other plants such as *Casearia sylvestris* Sw., *Solanum lycopersicum*, *Sorghum bicolor*, *Ocimum basilicum*, *Curcuma longa*, and *Elettaria cardamomum*. The pungent constituents and active ingredients found in ginger rhizome are numerous. Active compounds in ginger essential oil include oleoresin, terpenes, zingiberone, zingiberol, and zingiberene. Fresh ginger rhizome has a high concentration of gingerol and a low concentration of zingerone. A retro-aldol reaction transforms gingerol into zingerone, when it is cooked or dried [43]. The middle fraction contains 30.5% zingiberene, while the top fraction contains a mixture of sesquiterpene hydrocarbons and sesquiterpene alcohol (50.5%). Sesquiterpenes, pinene, camphor, camphene, and curcumins are all present in small amounts in the oil. β-Phellandrene, sabinene, cineol, borneol, zingiberene, and sesquiterpenes are found in the essential oil of rhizomes. It is the most abundant molecule in ginger oil, accounting for 30% or more of the total volume. It is very likely to protect them from insect pests, possibly by preventing insects from laying eggs and preventing them from feeding on the plant. It has been shown to make tomatoes resistant to insect pests. Along with molecules like nerolidol, geraniol, and curcumene, it is one of the most important volatiles. In addition to its culinary applications, ginger has been shown to be effective in the treatment of a variety of ailments, including sprains, arthritis, sore throats, muscular aches, dementia, hypertension, fever, helminthiasis, and infectious diseases [44].

4.17.1 Chemistry

FPP is converted to zingiberene in the isoprenoid pathway. Nerolidyl diphosphate is formed when FPP undergoes a rearrangement. The ring closes after the pyrophosphate is removed, leaving a carbocation on the ring's tertiary carbon. After that, a 1,3-hydride shift occurs, resulting in a more stable allylic carbocation. The removal of the cyclic allylic proton and subsequent formation of a double bond is the final step in the formation of zingiberene. Zingiberene synthase is an enzyme that catalyzes the reaction that produces zingiberene and other mono- and sesquiterpenes [45].

References

[1]   Rehman R, Asif Hanif M. Nat Prod Chem Res 2016, 4(4), 1–11.
[2]   Tchoumbougnang F, JazetDongmo PM, Sameza ML, NkouayaMbanjo EG, TiakoFotso GB, AmvamZollo PH, Menut C. Biotechnol Agron Soc Environ 2009, 13(1), 77–84.
[3]   Sahari IS, Assim Z, Ahmad FB, Jusoh I. Borneo J Res Sci Tech 2013, 3(1), 43–51.
[4]   Mahmoudi R, Amini K, Hosseinirad H, Valizadeh S, Kabudari A, Aali E. Res J Pharmacog 2017, 4(4), 49–56.
[5]   Pintore G, Usai M, Bradesi P, Juliano C, Boatto G, Tomi F, Chessa M, Cerri R, Casanova J. Flavour Fragrance J 2002, 17(1), 15–19.
[6]   Asressu KH, Tesema TK. J App Pharm 2014, 6, 132–142.
[7]   Hcini K, Sotomayor JA, Jordan MJ, Bouzid S. Asian J Chem 2013, 25(5), 2601–2603.
[8]   Angioni A, Barra A, Cereti E, Barile D, Coïsson JD, Arlorio M, Dessi S, Coroneo V, Cabras P. J Agri Food Chem 2004, 52(11), 3530–3535.
[9]   Bigogno C, Khozin-Goldberg I, Boussiba S, Vonshak A, Cohen Z. Phytochemistry 2002, 60(5), 497–503.

[10] Martin SA, Brash AR, Murphy RC. J Lipid Res 2016, 57(7), 1126–1132.

[11] Szczuko M, Kikut J, Komorniak N, Bilicki J, Celewicz Z, Ziętek M. Int J Mol Sci 2020, 21(24), 1–19.

[12] Eddin LB, Jha NK, Goyal SN, Agrawal YO, Subramanya SB, Bastaki SM, Ojha S. Nutrients 2022, 14(7), 1–30.

[13] Halpern A, Adams JG. J Am Pharmaceu Associat 1949, 38(11), 597–599.

[14] Serafin A, Figaszewski ZA, Petelska AD. J Memb Bio 2015, 248(4), 767–773.

[15] Bergeson LL. Management 2019, 28(4), 1–12.

[16] Khan R, Sultana S. Chem Biol Interact 2011, 192(3), 193–200.

[17] Delmondes GD, Santiago Lemos IC, Dias DD, Cunha GL, Araújo IM, Barbosa R, Coutinho HD, Felipe CF, Barbosa-Filho JM, Lima NT, De Menezes IR. Exp Opin Ther Pat 2020, 30(3), 227–234.

[18] Lei Y, Peng F, Xie J, Cheng P. Planta Med 2019, 85, 48–55.

[19] Chen W, Viljoen AM. S Afr J Bot 2010, 76(4), 643–651.

[20] Samejo MQ, Memon S, Bhanger MI, Khan KM. Pak J Anal Environ Chem 2012, 13(1), 5–10.

[21] Khajuria V, Gupta S, Sharma N, Kumar A, Lone NA, Khullar M, Dutt P, Sharma PR, Bhagat A, Ahmed Z. Biomed Pharmacother 2017, 92, 175–186.

[22] Malakar C, Choudhury PP. Asian J Pharm Clin Res 2015, 8(2), 60–63.

[23] Endevelt-Shapira Y, Perl O, Ravia A, Amir D, Eisen A, Bezalel V, Rozenkrantz L, Mishor E, Pinchover L, Soroka T, Honigstein D, Sobel N. Nat Neurosci 2018, 21(1), 111–119.

[24] Whalley K. Nat Rev Neurosci 2018, 19(1), 3.

[25] H. Underwood E. Science 2021.

[26] Mba OI, Dumont MJ, Ngadi M. Food Biosci 2015, 10, 26–41.

[27] Fatima S, Hu X, Gong RH, Huang C, Chen M, Wong HL, Bian Z, Kwan HY. Cellular Mol Life Sci 2019, 76(13), 2547–2557.

[28] An Q, Ren JN, Li X, Fan G, Qu SS, Song Y, Li Y, Pan S. Food Fun 2021, 12, 10370–10389.

[29] Whelan J, Fritsche K. Adv Nutr 2013, 4(3), 311–312.

[30] Kim KB, Nam YA, Kim HS, Hayes AW, Lee BM. Food Chem Toxicol 2014, 70, 163–178.

[31] Surendran S, Fatimah Q, Surendran G, Lilley D, Heinrich M. Front Nutr 2021, 8, 1–14.

[32] Weems AC, DelleChiaie KR, Yee R, Dove AP. Biomacromol 2019, 21(1), 163–170.

[33] Behr A, Johnen L. Chem Sustain Energy Mat 2009, 2(12), 1072–1095.

[34] Keong Chan W, Teng-Hern Tan L, Gan Chan K, Han Lee L, Hing Goh B. Molecules 2016, 21, 1–40.

[35] Ahmad RM, Greish YE, Maghraby E, Lubbad L, Makableh Y, Hammad FT. Nanomaterials 2022, 12(7), 1–11.

[36] Nigmatov AG, Serebryakov P, Yanovskaya LA. Pharm Chem J 1987, 21, 529–533.

[37] Farre Armengol G, Filella I, Llusia J, Penuelas J. Molecules 2017, 22(7), 1–9.

[38] Erskine Hawkins J, William Burris A. J Org Chem 1959, 24(10), 1507–1511.

[39] Johnson W. Int J Toxicol 2007, 26(3), 31–77.

[40] Kula J. Curr Med Chem 2016, 23(35), 4037–4056.

[41] Khan Marwat S, Khan EA, Baloch MS, Sadiq M, Ullah I, Javaria S, Shaheen S. Pak J Pharm Sci 2017, 30, 5.

[42] Sudheer PD, Seo D, Kim EJ, Chauhan S, Chunawala JR, Choi KY. Enzyme Microb Tech 2018, 119, 45–51.

[43] Breeden DC, Coates RM. Tetrahedron 1994, 50, 11123–11132.

[44] Bou DD, Lago JH, Figueiredo CR, Matsuo AL, Guadagnin RC, Soares MG, Sartorelli P. Molecules 2013, 18(8), 9477–9487.

[45] Zhang S, Zhao X, Hu D, Qi Y, Zhou M, Li D, Hua Q, Wu Y, Liu Z. BiochemEng J 2021, 176, 108188.

Saeed Mollaei, Poopak Farnia and Saeid Hazrati

Chapter 5
Essential oils and their constituents

Abstract: Essential oils are aromatic and volatile liquids which are produced from plant parts such as roots, fruits, herbs, wood, seeds, bark, buds, twigs, flowers, and leaves. They are a mixture of fragrant substances or a mixture of aromatic and odorless substances. Under normal conditions, the essential oil compounds are volatile, but due to geographical origin, climate, rainfall, and so on, the constituent of essential oils is different. The compounds of essential oils are mainly highly volatile and lipophilic, and are categorized into two groups: terpenoids (mono-, sesqui-, and di-terpenes) and phenylpropanoids. But other types of compounds such as oxygen-, sulfur-, or nitrogen-containing compounds may also be present. These compounds include fats, alkaloids, anthraquinones, and coumarins. These compounds can be extracted from essential oils by methods other than distillation. The purpose of this chapter is to go over the introduction of essential oils, and their biosynthesis pathways and chemical constituents, as well as their biological activities.

Keywords: Biosynthesis, essential oil, medicinal plants, terpenoids, volatile oil

5.1 Introduction

Because of unique and valuable properties of herbal plants, they are extremely important. Plant-derived essential oils have long been applied as a flavoring agent in beverages and foods, as well as in herbal medicine [1].

Historical discoveries in Iran, Egypt, China, and India indicate that essential oils have been used by people in various forms. For example, in ancient Egypt, essential oils were obtained by brewing. Later, Romans and Greeks applied hydrodistillation for extraction, and a special place was given to aromatic plants. With the rise of Islamic civilization, essential oil extraction methods evolved rapidly. During the Renaissance, Europeans improved the extraction methods with the advancement of knowledge and conducted studies on essential oils. Today, the essential

Saeed Mollaei, Phytochemical Laboratory, Department of Chemistry, Faculty of Sciences, Azarbaijan Shahid Madani University, Tabriz, Iran, e-mail: s.mollaei@azaruniv.ac.ir
Poopak Farnia, Mycobacteriology Research Centre (MRC), National Research Institute of Tuberculosis and Lung Disease (NRITLD), Shahid Beheshti University of Medical Sciences, Tehran, Iran
Saeid Hazrati, Department of Agronomy, Faculty of Agriculture, Azarbaijan Shahid Madani University, Tabriz, Iran, e-mail: saeid.hazrati@azaruniv.ac.ir

https://doi.org/10.1515/9783110791600-005

oils of mint, *Lavendula*, *Pelargonium*, chamomile, bergamot orange, rose, eucalyptus, and *Santalum* are among the most popular and widely used essential oils [2].

Since essential oils evaporate faster than other oils, they are also called volatile oils or ethereal oils. They are aromatic and volatile substances that are present in only 10% of plants and are stored in special structures, such as secretory trichomes, resin ducts, secretory structures, glands, secretory tubes, and secretory spaces [3]. The essential oils obtained from plants are very low and rarely more than 1%; however, in several plants, such as cloves and nutmeg, they reach more than 5% [4].

Essential oils are highly hydrophobic and are soluble in alcohols, nonpolar and slightly polar solvents, oils, waxes, and very little water. Most of them are colorless or pale yellow (except for the blue essential oils of chamomile) and are liquids with a lower density than water (except for the extracts of sassafras, *Chrysopogon zizanioides*, cinnamon, and cloves) [5]. The molecular structure of essential oils (double bonds, functional groups of hydroxyl, aldehydes, esters, etc.) makes them easily oxidized by light, heat, and air [6].

The production and consumption of essential oils have quickly increased across the world. Despite their high price (since a large amount of plants is required to obtain a small amount of essential oils), the production of essential oils has been increased. Hence, their functional properties dominated their prices. The annual production of essential oils is about 40,000–60,000 tons, which is worth more than 700 million dollars [7].

Many of the essential oils in the industry are obtained from eucalyptus, oranges, *Nardostachys jatamansi*, peppers, and lemons; however, chamomile, lavender, peppers, tea, eucalyptus, *Pelargonium*, sandalwood, rosemary, lemon, rose, jasmine, and *Boswellia* are mostly used at home. Brazil, China, the United States, Indonesia, India, and Mexico have the largest share of the world's essential oil production. Major consumers include the United Kingdom, the United States, Japan, and the European Union, particularly France and Germany [8].

Essential oils are a complex mixture of volatile compounds containing 60 or more separate compounds. The main volatile compounds include esters (granyl acetate), oxides (1,8-cineole), phenolic ethers (anethole), phenols (eugenol), lactones (bergapten), ketones (umbellunone), cyclic aldehydes (cuminal), aldehydes (neral), acids (benzoic acid), alcohols (linoleic acid and santalol), and hydrocarbons (menthane, bisabolene, pinene, and limonene). All of these compounds are generally divided into two groups: phenylpropanoids and terpenoids. In addition, in another classification, they are divided into two categories: hydrocarbons and oxygenated compounds [7, 9, 10]. Although this classification may seem simple, it is used in this chapter.

The biological activities of essential oils can be dependent on one component or a mixture of all components. The following presents different categories of essential oil components and their properties. The main constituents of some essential oils are given in Table 5.1.

Table 5.1: The major constituents of various essential oils.

Rosmarinus officinalis	a- Pinene	1,8-Cineole	Camphor	Verbenone	Myrcene	β-Pinene
Dracocephalum polychaetum	Limonene	Methyl cyclogeranate	–	–	–	–
Origanum vulgare	Thymol	γ-Terpinene	Terpinene-4-ol	–	–	–
Salvia macrosiphon	Linalool	Manool	–	–	–	–
Salvia officinalis	1,8-Cineole	α-Thujone	Camphor	–	–	–
Ocimum basilicum	Linalool	1,8-Cineole	Eugenol	Camphor	–	–
Cinnamomum zeylanicum	1,8-Cineole	o-Cymene	α-Phellandrene	α-Pinene	–	–
Foeniculum vulgare	trans-Anethole	α-Pinene	–	–	–	–
Laurus nobilis	1,8-Cineole	Linalool	α-Terpinyl acetate	Sabinene	–	–
Cymbopogon flexuosus	Geranial	Neral	β-Myrcene	–	–	–
Mentha piperita	Menthol	Menthone	Menthofuran	Menthyl acetate	–	–
Hypericum helianthemoides	α-Pinene	β-Pinene	(E)-β-Ocimene	β-Caryophyllene	Germacrene D	–
Artemisia dracunculus	trans-Anethole	Limonene	Methyl chavicol	trans-Ocimene	–	–
Mentha arvensis	Menthol	p-Menthan-3-one	–	–	–	–
Mentha pulegium	Pulegone	Menthone	1,8-Cineole	Isomenthone	Piperitenone	Isomenthone
Citrus sinensis	Sabinene	Linalool	(E)-β-Ocimene	Citronellal	–	–

(continued)

Table 5.1 (continued)

Rosmarinus officinalis	a- Pinene	1,8-Cineole	Camphor	Verbenone	Myrcene	β-Pinene
Origanum vulgare	γ-Terpinene	(Z)-β-Ocimene	(E)-β-Ocimene	o-Cymene	–	–
Nepeta asterotricha	4aβ,7α,7aβ-Nepetalactone	1,8-Cineole	Linalool	4-Terpineol	–	–
Mentha longifolia	Carvone	Piperitone	Pulegone	1,8-Cineole	Menthol	Menthone
Nepeta bornmuelleri	4aβ,7α,7aβ-Nepetalactone	–	–	–	–	–
Nepeta depauperata	α-Pinene	1,8-Cineole	β-Caryophyllene	Spathulenol	Caryophyllene oxide	–
Artemisia biennis	cis-β-Ocimene	trans-β-Farnesene	Acetylenes (Z)	(E)-En-yn-Dicycloethers	–	–
Stachys byzantina	Germacrene D	Menthone	1,8-Cineole	α-Terpineol	α-Cadinol	Linalool
Nepeta gloeocephala	β-Pinene	1,8-Cineole	(Z)-β-Ocimene	–	–	–
Zingiber officinale	Zingiberene	β-Bisabolene	β-Sesquiphellandrene	Ar-Curcumene	–	–
Hypericum perforatum	α-Pinene	β-Pinene	(E)-β-Ocimene	β-Caryophyllene	Germacrene D	–
Wedelia chinensis	Carvocrol	t-Caryophyllene	–	–	–	–
Citrus aurantium	Limonene	β-Myrcene	α-Pinene	–	–	–
Nepeta binaloudensis	4aα,7α,7aα-Nepetalactone	–	–	–	–	–
Hypericum scabrum	α-Pinene	β-Pinene	(E)-β-Ocimene	β-Caryophyllene	Germacrene D	–
Artemisia annua	Artemisia ketone	Camphor	Linalool	–	–	–

	Terpin-4-ol	Caryophyllene oxide	Sabinene	β-Pinene	trans-Caryophyllene	
Mentha officinalis	Terpin-4-ol	Caryophyllene oxide	Sabinene	β-Pinene	trans-Caryophyllene	–
Artemisia absinthium	Myrcene	β-Thujone	trans-Sabinyl acetate	–	–	–
Zanthoxylum armatum	Linalool	Limonene	Undecan-2-one	–	–	–
Thymus vulgaris	Thymol	Terpinene	p-Cymene	Carvacrol	–	–
Lavandula angustifolia	1,8-Cineole	Linalool	Camphor	Linalyl acetate	–	–
Coriandrum sativum	Linalool	Geranyl acetate	Geraniol	–	–	–
Ziziphora tenuior	Pulegone	Limonene	–	–	–	–
Cuminum cyminum	3-Carn-10-al	p-Cement	p-Comic aldehyde	γ-Terpinene	β-Pinene	2-Carn-10-al
Oliveria decumbens	Thymol	Carvacrol	–	–	–	–
Satureja sahendica	p-Cymene	γ-Terpinene	Thymol	β-Pinene	β-Myrcene	Carvacrol
Stachys schtschegleevi	Spathulenol	Germacrene D	α-Pinene	α-Cadinol	β-Eudesmol	–
Ziziphora clinopodioides	Pulegone	Menthone	–	–	–	–
Heracleum persicum	Limonene	α-Pinene	γ-Terpinene	Hexyl butyrate	Anethole	Myristicin
Rosa damascene	Limonene	2-Phenylethyl alcohol	Citronellol	Geraniol	Methyleugenol	–
Zosima absinthifolia	Camphor	β-Caryophyllene	Caryophyllene oxide	Octyl acetate	α-Pinene	–
Nigella sativa	Thymoquinone	–	–	–	–	–
Achillea millefolium	d-Cadinol	1,8-Cineole	Transnerolidole	Germacrene D	α-Pinene	Borneole

(continued)

Table 5.1 (continued)

Rosmarinus officinalis	a- Pinene	1,8-Cineole	Camphor	Verbenone	Myrcene	β-Pinene
Satureja hortensis	γ-Terpinene	α-Terpinolene	Carvacrol	p-Cymene	–	–
Zataria multiflora	Carvacrol	Thymol	–	–	–	–
Ferula asafoetida	(E)-1-Propenyl sec-butyl disulfide	10-epi-γ-Eudesmol	β-Pinene	α-Pinene	–	–

5.2 The chemistry of essential oils

Plants produce two groups of metabolites, including primary and secondary metabolites. Primary metabolites are compounds found in all plants and include nucleic acids, lipids, carbohydrates, and proteins. Secondary metabolites are biosynthesized only in several plants and are categorized as alkaloids, polypeptides, shikimates, and terpenoids. Essential oils consist of different chemical constituents. The compounds of plant's essential oils are mostly divided into two different groups: phenylpropanoids and terpenes. Although terpenes and terpenoids (the oxygenated derivatives) are abundant in essential oils, some living organisms have high amounts of chemical compounds called phenylpropanoids. The presence of phenylpropanoids in the plant gives it a special smell and taste [11, 12].

5.2.1 Terpenes

Terpenes and terpenoids are formed from the condensation of five-carbon unit with two unsaturated bonds (2-methyl-1,3-butadiene, i.e., isoprene). So, they are often named isoprenoids. These compounds have several isomeric linear or cyclic structures and have varying degrees of unsaturations, substitutions, and oxygenated derivatives commonly named terpenoids. The units of isoprene are connected from the head and tail of the molecule. The branched end of the chain is called the molecule head and the other end is called the tail. Hence, this structural order is named head-to-tail binding. This binding pattern can be described by the biosynthesis of terpenoids. Also, these compounds are the main and most diverse class of volatile compounds. These compounds are structurally and functionally categorized as different groups. Based on the number of isoprene units and the structure, terpenes are classified into four groups including hemiterpene (one isoprene unit), monoterpenes (two isoprene units), sesquiterpenes (three isoprene units), and diterpenes (four isoprene units). Many of the essential oils are complex mixtures of phenolic compounds (cinnamates and phenylpropanes), monoterpenes ($C_{10}H_{16}$), and sesquiterpenes ($C_{15}H_{24}$). Also, heavier terpenes such as diterpenes (four units of isoprene) may be present in essential oils at small amounts; however, they do not usually play a main role in the smell of essential oils like diterpenes in ginger oil. Moreover, according to functional groups, these compounds are categorized into ketones, aldehydes, ethers, alcohols, and carbohydrates [13, 14].

5.3 Biosynthetic pathways of essential oil constituents

The biosynthesis of essential oils' main constituents is done by three pathways: mevalonate, methylerythritol, and shikimic acid. Sesquiterpenes are produced by the mevalonate pathway, monoterpenes and diterpenes are biosynthesized by the methylerythritol pathway, and phenylpropenes are produced by the shikimic acid pathway.

Isopentenyl diphosphate, isopentenyl pyrophosphate, and dimethylallyl diphosphate could produce plant terpenoids (Figure 5.1). "Active isoprene units" are biosynthesized from the biosynthetic pathways of both methylerythritol phosphate and mevalonic acid. Then the geranyl diphosphate, ten-carbon compound and precursor of monoterpenoids, was produced by the reaction between isopentenyl diphosphate and dimethylallyl diphosphate. The reason for this naming is the existence of a single pair of carbon dioxide [15].

5.3.1 Mevalonate pathway

The mevalonic acid pathway produces terpenoid molecules. In this pathway, in the first step, mevalonic acid with six carbon atoms is formed. Then, isopentenyl pyrophosphate, which consists of a branched five-carbon molecule (one isoprene unit), is biosynthesized by a series of enzymatic transformations, and finally the isoprene is bonded to two phosphate groups (Figure 5.1). Isoprene is the major compound of essential oils that begins to form terpenoid compounds [15–17].

5.3.2 Methyl erythritol pathway

In the non-mevalonate pathway, 1-deoxy-D-xylulose 5-phosphate and 2-C-methyl-D-erythritol-4-phosphate (MEP) are involved, which these compounds are obtained from the condensation of pyruvate and glyceraldehyde phosphate (Figure 5.1) [15–17].

5.3.3 Shikimic acid pathway

Since shikimic acid is the main precursor to flavonoids and lignin, it is the main synthetic mediator for plants. Moreover, the aromatization of shikimic acid leads to production of benzoic acid derivatives in several essential oils. Flavonoids are protective agents against ultraviolet rays and act as antioxidants. Lignin is also the main constituent in the structural materials of the plant (Figure 5.1) [15–17].

5.4 The structure of organic compounds in essential oils

Essential oils are composed of organic compounds whose atoms are mainly carbon, hydrogen (hydrocarbons), oxygen, nitrogen, and sulfur, and have low molecular weight. The compounds of essential oils are mostly composed of neutral atoms such as carbon and hydrogen, which can attach to one or more functional groups and produce polar compounds [18].

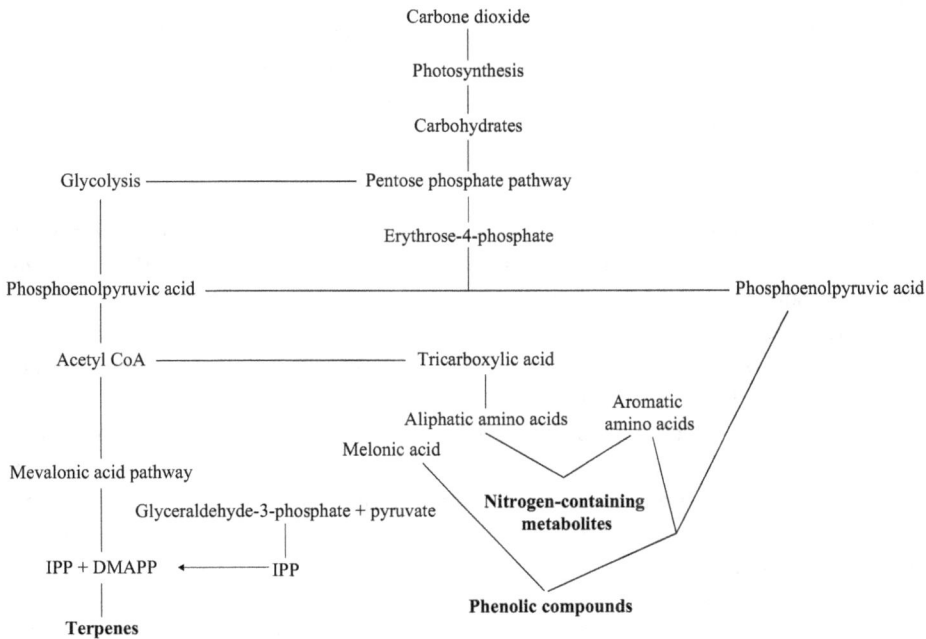

Figure 5.1: Biosynthetic pathways of secondary metabolites in plants.

Obtaining information about the chemistry of essential oil compounds is highly valuable to understand their properties. To evaluate their effects, it is necessary to be familiar with the chemical structure of essential oils.

5.5 Classification of essential oil compositions

5.5.1 By biosynthesis pathway

The compounds of plant essential oils are mostly divided into three separate chemical categories: terpenes, phenylpropanoids, and others. Compounds of terpenes can be separated into two major groups: (1) terpenes with a hydrocarbon structure, mostly monoterpenes, sesquiterpenes, and diterpenes; and (2) their oxygenated derivatives, for example, lactones, esters, acids, phenols, ketones, aldehydes, oxides, and alcohols. Some familiar compounds in essential oils are categories based on chemical functional groups [19]. Table 5.2 shows the classification of essential oil compounds based on biosynthesis pathway and functional groups.

5.5.1.1 Hydrocarbons

The first major group of compounds is hydrocarbons. They are formed completely from carbon and hydrogen atoms that differ considerably in complexity and size. These compounds are highly soluble in lipids (lipophils) and insoluble in water [18].

Benzenoids, alkenes, and alkanes as simple hydrocarbons are named nonterpenoid hydrocarbons, since they are not biosynthesized through mevalonate or non-mevalonate (MEP) pathways.

Hydrocarbons which have no ring are categorized as alicyclic and encompass alkynes, alkenes, and alkanes. The word "alicyclic" expresses molecules composed of carbon chains which are in a straight line and do not have aromatic or closed rings. Examples of these groups include the eight-, nine-, and ten-carbon aldehydes with a sharp smell (found in low amounts in citrus essential oils), and the six-carbon compounds with a pleasant smell in some flower oils, such as rose and jasmine. The compound octanal aldehyde ($C_8H_{10}O$) is detected in sweet orange oil. Alicyclic molecules are detected only in essential oils, but if oxygenated functional groups are bound to them, their smell is usually noticeable despite their small amount [20, 21].

In alkanes, the atoms are bonded by single bonds. Methane and ethane are simple examples of alkanes. Alkenes are compounds with at least one carbon–carbon double bond, while alkynes have at least one carbon–carbon triple bonds. However, alkynes are not usually detected in essential oils. Several essential oil compounds had one or more rings. For this reason, they are called monocyclic, bicyclic, tricyclic, tetracyclic, and so forth.

Another group of hydrocarbons is named aromatics. They typically have a benzene ring, and include phenylpropyl, phenylethyl, benzyl, phenyl compounds, and polycyclic structures (benzo[α]pyrene and naphthalene). The name "aromatic" is derived from the first benzene derivatives extracted from plants that had a pleasant smell. However, later, compounds with an unpleasant smell were also discovered [18].

Table 5.2: Category of essential oil compounds.

Category	Subcategory		Functional group	Compounds	Plant source
Terpenoids	Hemiterpenes	Acyclic	Hydrocarbon groups	Isoprene	Conifers, poplars, oaks, willows, Schoenocaulon officinale, Archangelica officinalis, Levisticum officinale, Valeriana officinalis, Senecio sp., Ligularia sp., Tuber melanosporum, Peucedanum ostruthium
			Hydroxyl groups	Prenol, isoamyl alcohol	
			Carbonyl-containing groups	trans-Tiglic acid, cis-tiglic acid, isovaleric acid, senecioic acid, angelic acid	
		Monocyclic	Carbonyl-containing groups	Utililactone, epiutililactone	
	Monoterpenes	Acyclic	Hydrocarbon groups	2,6-Dimethyloctane, myrcene, ocimene, β-citronella	Lippia citriodora, Cymbopogon citratus, Laurus nobilis, Cymbopogon flexuosus, Coriandrum sativum, Cinnamomum camphora, lemon (Citrus limon), orange (Citrus sinensis), grapefruit (Citrus paradisi), mandarin (Citrus reticulata), lime (Citrus aurantifolia), Mentha spicata, Carum carvi, Eucalyptus globulus, Rosmarinus officinalis, Artemisia absinthium, Thymus vulgaris, Pinus palustris, Pinus caribaea, Pinus pinaster, Mentha piperita, Blumea balsamifera, Kaempferia galanga, Cinnamomum camphora
			Hydroxyl groups	Linalool, citronellol, geraniol	
			Carbonyl-containing groups	Citronellal, neral, geranial	
		Monocyclic	Hydrocarbon groups	Limonene, menthane, α-phellandrene, β- phellandrene, terpinolene, p-cymene, sabinene	
			Hydroxyl groups	α-Terpineol, terpinen-4-ol, l-menthol, isopulegol, δ-terpineol	
			Carbonyl-containing groups	Thymol, carvone, menthone, (+)-pulegone, (-)-piperitone, campholenic aldehyde	
			Oxygen-bridged groups	cis-Rose oxide	

(continued)

Table 5.2 (continued)

Category	Subcategory		Functional group	Compounds	Plant source
		Bicyclic	Hydrocarbon groups	Thujane, carane, sabinene, Δ-3-carane, α-pinene, camphene, isocamphane, ferchane	
			Hydroxyl groups	Sabinol	
			Carbonyl-containing groups	Thujone, fenchone, piperitone oxide, diosphenol, camphor, umbellunone	
			Oxygen-bridged groups	Ascaridole, cineole, 1,8-cineole	
Terpenoids	Sesquiterpenes	Acyclic	Hydrocarbon groups	Farnesene and farnesane	*Cymbopogon* sp., *Cyclamen* sp., *Polianthes tuberoza*, *Citrus aurantium*, *Jasminum officinale*, *Zingiber officinale*, *Piper cubeba*, *Zingiber officinale*, *Curcuma longa*, *Humulus lupulus*, *Juniperus oxycedrus*, *Juniperus* and *Cupressus* sp., *Pogostemoncablin*, *Artemisia cina*, *Artemisia chamaemelifolia*, *Artemisia maritima*, *Thapsiagarganica*
			Hydroxyl groups	Farnesol and β-nerediol	
		Monocyclic	Hydrocarbon groups	β-Bisobolene, α-zingiberene, α-humulene, germacrene D, elemane, xanthane	
			Hydroxyl groups	Bisobolol	
			Carbonyl-containing groups	Abscisic acid	
		Bicyclic	Hydrocarbon groups	Chamazulene, seslinene, valencene, eremophitane, himalachane, chamigrane, β-caryophyllene, δ-selinene, β-elemene, bicyclogermacrene	
			Hydroxyl groups	β-Santalol, Cyperol, selin-11-en-4-ol, α-murool	

			Group	Example	Sources
			Carbonyl-containing groups	Noothatone, khusimone, zizanal	
			Oxygen-bridged groups	Caryophyllene oxide	
		Tricyclic	Hydrocarbon groups	Thujopsene, aromadendrene, bourbonane	
			Hydroxyl groups	Khushimol and patchoulol	
			Carbonyl-containing groups	Matricin and santonin	
			Oxygen-bridged groups	Aromadendrene oxide	
	Diterpenes	Acyclic	Hydrocarbon groups	Phytane	*Helichrysum heterolasium*, Clary sage (*Salvia sclarea*), white horehound (*Marrubium vulgare*), Diviner's sage (*Salvia divinorum*), resin from coniferous trees, e.g., pine tree (*Pinus sylvestris*), common sage (*Salvia officinalis*), rosemary (*Rosmarinus officinalis*), Danshen (*Salvia miltiorrhiza*), *Stevia rebaudiana*, castor bean (*Ricinus communis*)
			Hydroxyl groups	Phytol	
		Monocyclic	Hydrocarbon groups	Camphorene and cambrene	
			Hydroxyl groups	Vitamin A, 9-geranyl-α-terpineol	
		Bicyclic	Hydrocarbon groups	Labdane	
			Hydroxyl groups	Manool, selareol, larixol	
Terpenoids	Diterpenes	Tricyclic	Hydrocarbon groups	Miltiorins	
			Carbonyl-containing groups	Nimbiol and pimara-7,15-diene-3-one	
			Oxygen-bridged groups	Epimanoyl oxide	

(continued)

Table 5.2 (continued)

Category	Subcategory	Functional group	Compounds	Plant source
	Tetracyclic	Hydrocarbon groups	Kaurane and kaur-16-ene	
		Hydroxyl groups	Ent-16,17-β-dihydroxykaurane	
	Macrocyclic	Hydrocarbon groups	Casbene	
Phenylpropene		Hydrocarbon groups	Isoprene	*Cinnamomum osmophloeum, Ocotea quixos, Angelica sinensis, Lonicera japonica, Bupleurum chinense,* carrot, basil, cinnamon, sparsely nutmeg (seed of *Myristica fragrans), Asiasarum sieboldii, Illicium verum*
		Hydroxyl groups	Eugeol and isoeugenol	
		Carbonyl-containing groups	Cinnamaldehyde, coniferyl aldehyde, cinnamyl acetate, phenethyl alcohol	
		Oxygen-bridged groups	Safrole, anethole, eugenol epoxide, estragole, dillapiole, dihydrodillapiole, methyleugenol, myristicin, elimicin, asarone	
Others		Sulfur compounds	Methyl isothiocyanate, allyl isothiocyanate, iberin, 2-butyl isothiocyanate, isobutyl isothiocyanate, 3-methylbutylisothiocyanate, cyclopentyl isothiocyanate, sulforaphane, benzyl isothiocyanate, phenethyl isothiocyanate, (*E*)-propenyl *sec*-butyl disulfide, (*Z*)-propenyl *sec*-butyl disulfide	*Capparis spinosa, Ferula asafetida, Armoracia rusticana, Citrus aurantium, Jasminum* sp., *Chrysopogon zizanioides,* bitter almond oil
		Nitrogen-containing compounds	Coniine, nicotine, methyl anthranilate, skatole	
		Inorganic compounds	Hydrocyanic acid	

5.5.1.2 Terpenoids

5.5.1.2.1 Hemiterpenes

Hemiterpenoids are the simplest groups of compounds among the terpenoids. Isoprene (boiling point 34 °C) is the main and prominent hemiterpene, which is produced from the leaves of several trees (e.g., oaks, conifer willows, and poplars) and medicinal plants (e.g., *Hamamelis japonica*). Isoamyl alcohol, senecioic acids, tiglic, angelic, and isovaleric are the other known hemiterpenoids, which are extracted from natural sources [22]. Figure 5.2 depicts the structure of some important hemiterpenes.

5.5.1.2.2 Monoterpenes

Monoterpenes are the simplest terpene compounds in essential oils. They are divided into three: noncyclic (linear), monocyclic, and bicyclic. Each category includes two groups of hydrocarbons and oxygenated derivatives.

The structure of most noncyclic monoterpenes contains 10 carbons. Bicyclic monoterpenes vary widely in carbon structures, and the most important of which are camphene, pinene, thujane, careen, and fenchane [23].

Addition of the double bond (oxidation), its removal (reduction), and combination with oxygen in the form of alcohol, aldehyde, ketone, and ester can cause further changes in the structure. Figure 5.2 depicts the structure of some important monoterpenes.

5.5.1.2.2.1 Iridoids

Iridoids are monoterpene compounds characterized by a cyclopenta[c]pyranoid skeleton, which is well known as the iridane skeleton. These compounds are often present in plants in combination with sugar units [24].

5.5.1.2.3 Sesquiterpenes

Terpenoid compounds with 3 units of isoprene (15 carbon atoms) are prominent as sesquiterpenes. Sesquiterpenes are divided into hydrocarbon and oxygenated groups. Hydrocarbon sesquiterpenes include linear, monocyclic, bicyclic, tricyclic, and tetracyclic [25]. Figure 5.2 depicts the structure of some important sesquiterpenes.

5.5.1.2.4 Diterpenes

Diterpenes are produced from geranylgeraniol pyrophosphate. These compounds are of plant origin and are detected in resins, resinous secretions, and high-boiling gum components that are the remaining of the hydrodistillation of essential oils [26]. Figure 5.2 shows the chemical structure of some diterpenes found in some plant's essential oils.

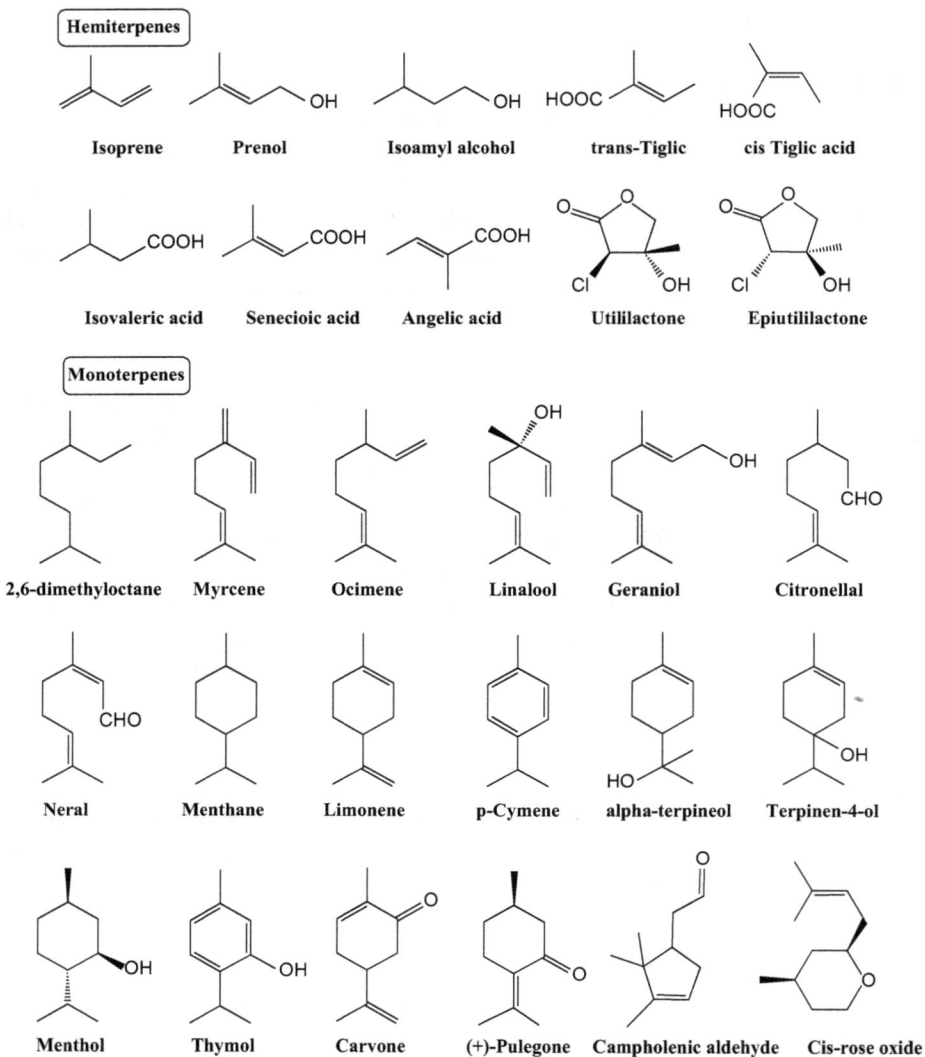

Figure 5.2: Chemical structures of some terpenoids.

Monoterpenes

Thujane

Sabinene

alpha-Pinene

Camphene

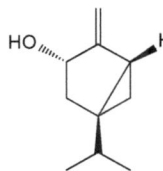
Sabinol

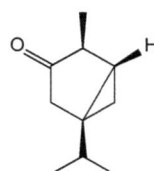
Thujone

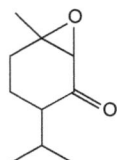
Piperitone oxide

Camphor

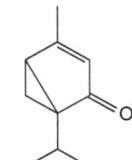
Umbellunone

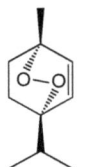
Ascaridole

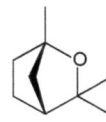
Cineole

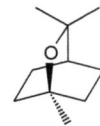
1.8-Cineole

Sesquiterpenes

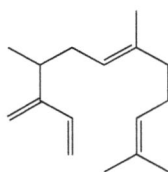
Farnesene

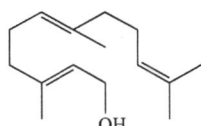
Farnesol

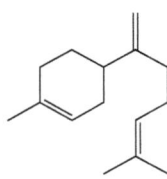
beta-Bisobolene

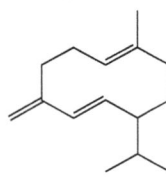
Germacrene D

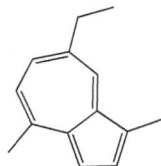
Chamazulene

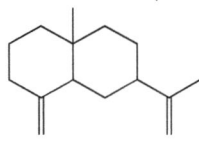
Seslinene

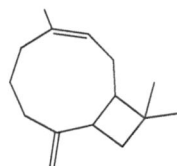
beta-Caryophyllene

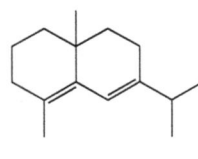
Selinene

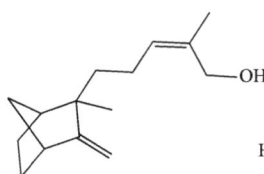
beta-Santalol

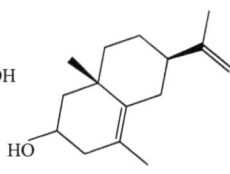
Cyperol

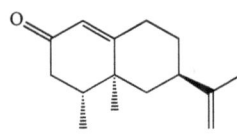
Noothatone

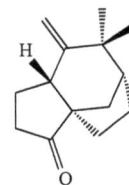
Khusimone

Figure 5.2 (continued)

Sesquiterpenes

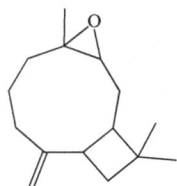

Caryophyllene oxide

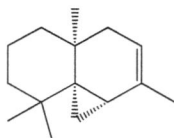

Thujopsene

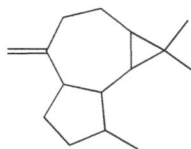

Aromadederene

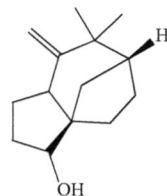

Khushimol

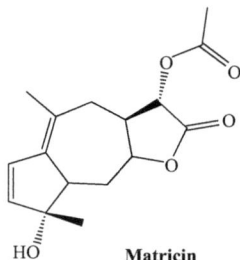

Matricin

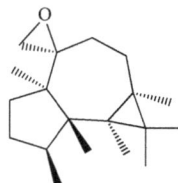

Aromadederene oxide

Diterpenes

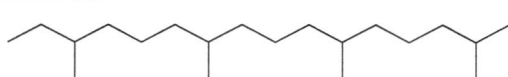

Phytane

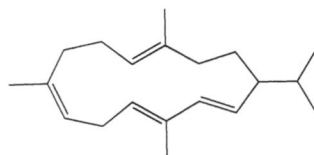

Cambrene

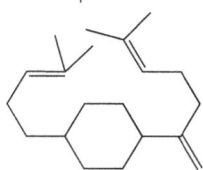

Camphorene

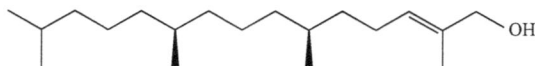

Phytol

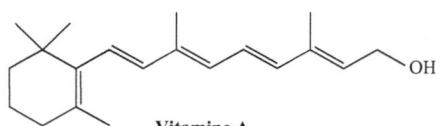

Vitamine A

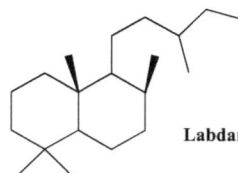

Labdane

Figure 5.2 (continued)

Diterpenes

| Manool | Selareol | Miltiorins | Nimbiol |

| Epimanoyl oxide | Kaurane | Ent-16,17-beta-dihydroxykaurane | Casbene |

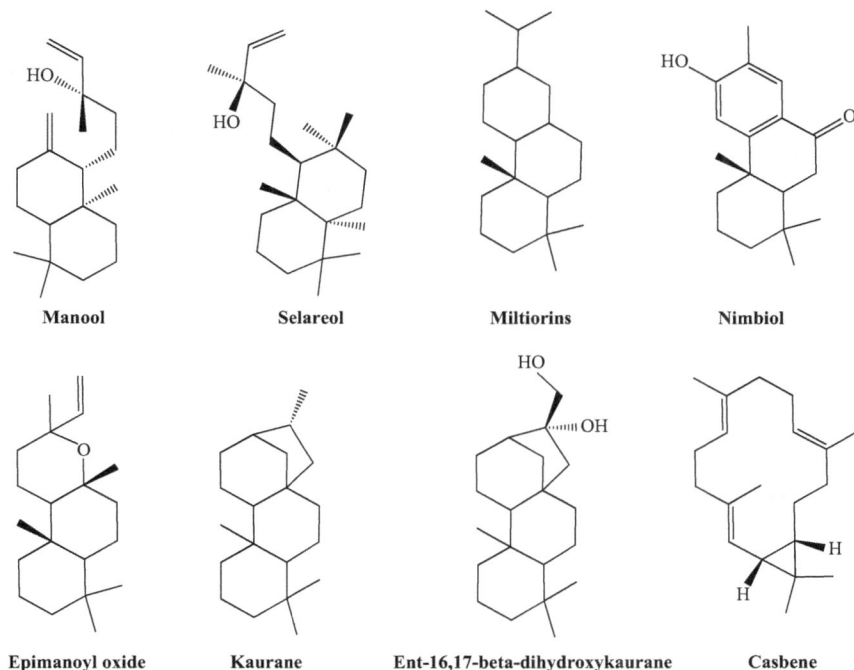

Figure 5.2 (continued)

5.5.1.3 Phenylpropenes

Phenylpropenes had a six-carbon aromatic ring with a three-carbon side chain. The side chain has usually a double bond, but oxygen is rarely found in it. By replacing one to four atoms of oxygen on the aromatic ring, other compounds are formed [27]. Figure 5.3 displays the structure of some compounds of phenylpropenes.

5.5.2 By functional groups

A functional group is an atom or group of atoms which mainly determines the chemical properties of the molecule having it. In essential oils, most functional groups are atoms other than carbon (mostly oxygen) and are known as heteroatoms. In aromatherapy, six categories of functional groups are very important. The functional groups replace the hydrogen atoms in a hydrocarbon. It does not mean that the hydrocarbon part of molecules has no role in the physical and chemical properties of compounds,

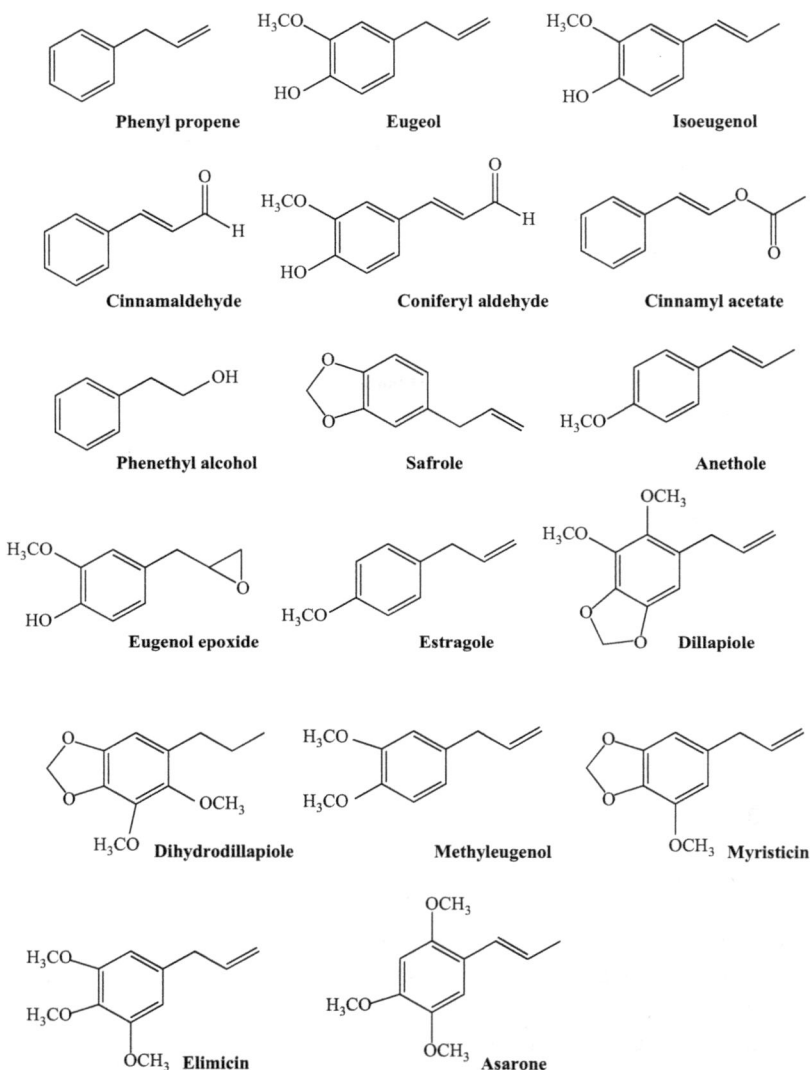

Figure 5.3: Chemical structure of some phenylpropenes.

but rather has an important effect on the solubility and volatility of a compound, which are main factors in enhancing access to taste and smell receptors. It may be better to know that functional groups play a main role in intermolecular interactions, while the structural framework plays a moderately nonspecific role [18].

According to the classification of essential oil constituents, the second main group of constituents is oxygenated compounds containing oxygen, carbon, and hydrogen, which are categorized as different groups based on the type of functional groups. These groups are phenols, oxides, lactones, ketones, ethers, esters, alde-hydes, alcohols, acids, and peroxides. The molecular weights of terpenes are almost the same as their oxygenated products [28, 29].

In the following, the main functional groups in the components of essential oils are examined.

5.5.2.1 Hydrocarbon groups

5.5.2.1.1 Alkenes
Due to the electron density of carbon–carbon double bond, it has special chemical properties and can be considered a functional group containing heteroatom. Al-kenes are the common compounds of essential oils recognized as monoterpene hy-drocarbons [18]. Limonene and terpinolene are the examples of this group.

5.5.2.2 Hydroxyl groups

5.5.2.2.1 Alcohols
The compounds that have hydroxyl group are recognized as alcohols. These com-pounds are the greatest group of terpene derivatives detected in essential oils. The name of all alcohols ends in "ol." Alcohols have a hydroxyl group bounded to a carbon atom [18]. The numbers of monoterpene alcohols are not abundant, but are present in a large amount of essential oils. These compounds are several sesquiter-pene alcohols; however, most of them are detected in a small number of essential oils. They are somewhat nonmutagenic and nontoxic, and have low sensitivity and irritability. Monoterpene alcohols (monoterpenols) are good disinfectants and have antifungal activities [21]. Terpinen-4-ol and α-terpineol are examples of alcohols and are one of monoterpene alcohols.

5.5.2.2.2 Phenols
Phenols have a hydroxyl group and their name generally ends in "ol." In these com-pounds, however, hydroxyl group is bonded to a benzene ring and, generally, an isopropyl tail (the three-carbon chain bounded to the ring in the middle carbon atom), makes the hydroxyl group very weak and relatively reactive acid [18]. Com-mon phenols in essential oils are carvacrol, thymol, and eugenol in *Origanum vul-gare*, *Syzygium aromaticum*, and *Thymus vulgaris*, respectively.

5.5.2.3 Groups containing carbonyl

5.5.2.3.1 Aldehydes

Aldehydes have the -CHO functional group, which is one of some examples of a carbonyl-containing group. In these compounds, one oxygen atom is bounded to the carbon at the end of a carbon chain, and its fourth bond is often hydrogen. They may be considered semioxidized primary alcohols, and are generally spread as natural compounds of essential oils. These compounds have a considerably fruity smell. The name "aldehydes" ends in "al" or "aldehyde" [18]. The aldehydes in *Pelargonium* and cumin are examples of important aldehydes.

5.5.2.3.2 Esters

Ketones have a carbonyl group and are structurally similar to aldehydes. These compounds can be synthesized by the oxidation of alcohol. A ketone is produced by oxygenation of alcohol, and has an oxygen atom which is bonded to carbonyl group. In ketones, the carbonyl group is bounded to two other carbons. Ketones are moderately stable compounds and do not oxidize easily. The names of ketones end in "one," with one exception: camphor [18]. Menthone, (+)-pulegone, and (-)-piperitone are examples of important ketones.

5.5.2.3.3 Carboxylic acids

Carboxylic acid is another group which is rarely in essential oils. They are produced from the oxidation of aldehydes. Owing to their low volatility, they are rarely detected in essential oils. Carboxylic acids are weak acids and frequently have a pungent smell. They are named after their hydrocarbon fraction and have the suffix "ic" followed by the word "acid." These compounds have a low volatility rate and are highly reactive. They can react with alcohols and readily produce esters (or lactones, when the reaction between alcohol and carboxylic acid groups is intramolecular). Also, these compounds can produce amides by reaction with amines [18, 21]. Benzoic and caffeic acids are examples of this group.

5.5.2.3.4 Carboxylic esters

Carboxylic esters commonly have a very sweet and fruity smell and can be synthesized from the reaction between carboxylic acid and terpene alcohol. The highest amount is obtained at fruit/plant maturity or full bloom of the flower. In bergamot and mint, monoterpene alcohols such as (-)-menthol and linalool are converted to (-)–menthyl acetate and linalyl acetate, respectively [18].

The names of the esters usually end as "yl" and "ate" which are derived from the parent alcohol and carboxylic acid, respectively. For example, linalool and acetic acid formed linalyl acetate. Some esters have antifungal properties [18, 21].

5.5.2.3.5 Lactones

Lactones are cyclic esters derived from lactic acid. These compounds have an oxygen atom bonded to carbonyl group at a closed ring. They are cyclic esters. Several simple and more complex molecules which have low volatility are found in essential oils in the form of five-membered cyclic rings which are known as γ-lactones. γ-Decalactone belongs to this group and has a peach-like taste. δ-Decalactone is a compound that has a six-membered cyclic rings and is known as δ-lactone. This compound has a creamy-coconut smell. Lactones such as sedanolide and butylphthalide which are in the form of benzofuran derivatives are detected in plants such as *Apium graveolens* and *Angelica* sp. Allantolactone, dehydrocostus lactone, and massoia lactone are all theoretically allergenic. The names of lactones are relatively variable. Although the suffixes "oolid" and "lactone" are relatively common, the name of lactones usually ends in "ine" or "lactone" [18, 30].

Other lactones, including bergaptene, scopoletin, and coumarin, are detected in some essential oils, while neptalactones are the characteristic of nepta oils. Coumarin and its derivatives are benzenoid lactones detected in some essential oils, and are responsible for the smell of freshly picked alfalfa. Citropten belongs to the derivative of coumarin and is phototoxic.

5.5.2.4 Oxygen-bridged groups

5.5.2.4.1 Ethers

These compounds are molecules in which one oxygen atom in a compound is attached to two carbon atoms. Ethers are formed when an alkyl group such as methyl is bonded to a benzene ring via an oxygen atom. These compounds are also present in cyclic forms, where an oxygen atom is the part of a ring. Furthermore, they are well known as oxides. Cineole is the most important oxide in essential oil, which exists in two forms. 1,4-Cineole is seldom found in essential oils, while 1,8-cineole is the most common form. Epoxide is another type of cyclic ether. In this case, three-membered ring has one oxygen and two carbon atoms. They often form during the oxidative metabolism of alkenes and are highly reactive, since the ring is highly strained. Ethers may too be formed by the degradation of products like 1,2-epoxide and limonene [18]. *cis*-Rose oxide and piperitone oxide are other examples of ethers.

5.5.2.4.2 Peroxides

The peroxide molecules are formed by attaching two oxygen atoms to two carbon atoms. Due to high reactivity of peroxides, on prolonged exposure to air or water and at high temperatures (sometimes explosively), these decompose easily. A typical example is the toxic ascaridol that is detected in the wormseed oil [18]. There are some other peroxides in essential oils.

5.5.2.4.3 Furans

Furans are made up of a five-membered ring which contains one oxygen atom. This ring has chemical properties like benzene, and is considered as aromatic ring. Also, they can be considered as cyclic ether [18]. A small number of essential oils contain furans; however, mentofurans are present in most mint oils.

5.5.2.4.4 Furanocoumarins

There are some compounds in essential oil constituents which are known as furano-coumarins. These compounds are naturally phototoxic and are formed from furan. Bergapten and methoxsalen are examples of this group, which is produced in several essential oils such as citrus. The moiety of furanocoumarin may help as a functional group or, due to its size and rigidity, merely as a structural framework. It may participate in strong or weak intermolecular interactions [18].

5.5.2.5 Other heterocyclic compounds

These compounds have atoms other than carbon in closed rings. They are produced only in a small number of essential oils and contain compounds, such as oxygen-containing furanoids, coumarins, lactones, methyl anthranilate, and nitrogen-containing indole compounds. These compounds consist of carbon atoms arranged in a ring with an oxygen or nitrogen atom as part of the ring. They are unusual in essential oils and are found mostly in flowering oils, such as narcissus, neroli, and jasmine. Alkaloids are heterocyclic compounds having a nitrogen atom as part of a closed ring. Since these compounds are soluble in water, they are rarely extracted in hydrodistillation of essential oils.

5.5.2.6 Sulfur compounds

These relatively pungent and reactive compounds are produced only in a small number of essential oils. These compounds have a relatively simple structure and have no phenylpropanoid hydrocarbon or terpene components. The chemical names of sulfur-containing molecules generally include "sulf" or "thio" sequences and isothiocyanates, sulfoxides, disulfides, sulfides, and trisulfides. Allyl isothiocyanate is one of the recognized compounds in this group (Figure 5.4). In addition, (Z)-propenyl *sec*-butyl disulfide and (E)-propenyl *sec*-butyl disulfide are major compounds produced in the essential oil of *Ferula asafoetida* [18, 31].

They seem to be necessary in plant protection and in plant nitrogen detoxification. Although the compounds containing sulfur atom have an unpleasant pungent smell, the organosulfur compounds in essential oils can be aromatic. Sulfur compounds have also been found to be used in beverages, processed foods, fruits, and flavoring vegetables [18].

5.5.2.7 Nitrogen-containing compounds

Nitrogen-containing compounds are produced only in a small number of essential oils. They have a very interesting aroma; however, in high concentrations, they smell like feces. These compounds are utilized as a flavoring agent in cigarettes and ice cream and a stabilizer in floral fragrances. Indole is a white crystalline powder and is an alkaline-like compound. It turns red when exposed to air. This compound is found in essential oils of some citrus and floral absolutes, jasmine, and neroli (Figure 5.4). The compound consists of a benzene ring attached to a heterocyclic pyrrole ring. It smells like skatole and is used in various perfumes. Examples include pyrazine, pyridine, indole, skatole, and methyl anthranilate. Methyl anthranilate is produced in essential oils of some citrus including bergamot, lemon, and orange. Skatole is a compound in the form of powder or large crystals and is found in the blooms of *Jasminum* sp. and *Citrus aurantium*. Furthermore, pyrazines and pyridines are present in the essential oils of *Chrysopogon zizanioides*, *Citrus sinensis*, and *Piper nigrum* [18, 32].

5.5.2.8 Mineral compounds

Hydrogen cyanide or hydrocyanic acid (Figure 5.4) is a highly toxic mineral acid produced in bitter almond oil. This compound is produced during the distillation process, but must be removed before using the essential oil [18].

5.6 Application of essential oils

Essential oils are used as flavorings and aromatizers in medicine and foods. They are widely used in cosmetics industry to make perfumes and fragrance sprays, and in the health industry as fragrances in soaps and toothpaste. One of the uses of essential oils in addition to the above is their therapeutic effects. Today, the use of essential oils in the treatment of diseases is recognized as a science. This science is known as "aromatherapy." Essential oil is used to treat diseases of the skin and hair; respiratory system; gastrointestinal tract; circulatory disorders; common diseases and disorders in women; diseases related to muscles, joints, and neurological disorders; and to modify mental and psychological changes. For example, camphor's essential oil has anti-irritant, pain-relieving, and anti-itching effects. *Peucedanum officinale*'s essential oil affects the nervous system [33].

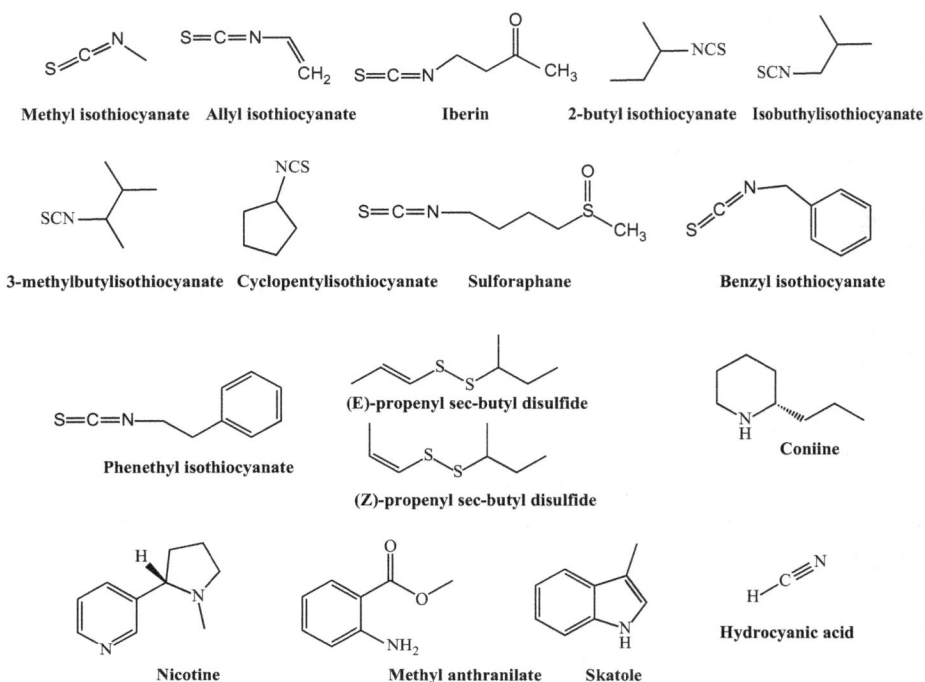

Figure 5.4: Other heterocyclic compounds or molecules.

5.7 Biological properties of essential oils

Until now, more than 10,000 natural compounds have been recognized in grains, fruits, vegetables, and other plants. These natural compounds are mainly categorized as essential oils, nitrogen-containing compounds, vitamins, alkaloids, phenolics, carotenoids, and organosulfur compounds. Among these, essential oil compounds have attracted a lot of attention due to their wide variety and biological activities such as antioxidant, antifungal, antibacterial, anticancer, and antimalarial [33–37]. In the following, some biological properties of essential oils are described.

5.7.1 Antimicrobial properties

Since essential oils contain many antimicrobial compounds, they are considered strong antimicrobials. Even small amounts of compounds can play a crucial role in antimicrobial properties. Essential oils and their compounds are widely known as antimicrobial agents and are used for various purposes, such as food preservation, pharmaceutical supplements, and therapy. Aromatic plants have inhibitory properties against bacteria, fungi, and yeast, and many of their properties are owing to the essential oil produced by their secondary metabolites [33, 34].

Essential oils and plant extracts of various species can control microorganisms related to the skin, tooth decay, and food waste, including gram-positive and gram-negative bacteria. One of the main problems in developing countries is infectious diseases. Microorganisms are resistant to many antibiotics; consequently, there are many problems in treating infectious diseases. Repeated use of commercial antimicrobial drugs to treat infectious diseases increases the resistance of microorganisms. Hence, researchers began looking for new antimicrobials from various plant sources. Out of 2,600 plant species, approximately 700 species have been reported for medicinal uses. In traditional medicine, herbs and plant products are used to treat various infections and other diseases [35–37].

The antimicrobial activities of essential oils have been demonstrated by the use of various microorganisms, including food poisoning microorganisms, fungi, yeasts, and plant and animal viruses. Various research works have been conducted to determine which of the functional groups or spatial structure of the constituents of essential oils is responsible for their biocidal property. In homologous series of aliphatic compounds, chain length plays a role in determining the amount of antimicrobial activity. Although there is much information about the activity of functional groups, limited information is available on the mechanism of action of their combination [34, 35].

Various research works have been conducted on the antimicrobial activities of essential oils at the SAC Auchincruive Institute. One of the important results of the study on natural essential oils in this institute was that the biocidal property of an essential oil indicated its effect on gram-negative and gram-positive bacteria. Studies have shown that *Limnophila gratissima*'s essential oils, such as streptomycin and chloramphenicol, are active against pathogenic bacteria, such as *Bacillus cereus*, *Escherichia coli*, *Pseudomonas aeruginosa*, and *Staphylococcus aureus* [33–37].

5.7.2 Antifungal properties

Studies have demonstrated that a large number of essential oils have antifungal properties. Plants are a considerable source of bioactive molecules with therapeutic potential. Although there are effective drugs against yeasts and skin fungi, plant extracts, as well as essential oils, are of interest, owing to their antifungal properties. Some essential oils in very low concentrations can prevent the growth of fungi in the culture medium. The fungicidal power of essential oils is not affected by factors, such as temperature and concentration of the injectable substance [34, 38].

The antifungal properties of essential oils of 15 medicinal plants against *Aspergillus fumigatus* and *Aspergillus niger* were studied. *Ferulago angulata* is one of the herbal plants that several studies have investigated the antifungal activities of its essential oil against *Candida albicans*. Laboratory studies of five plant species in West Africa have demonstrated strong fungicidal properties against *Rhizopus*, *Ustilago maydis*, *Curvularia lunata*, and *Ustilaginoidea virens* [38].

5.7.3 Antioxidant properties

There are three reasons to seek and use natural antioxidants: (1) Numerous clinical studies indicate that eating fruits and vegetables reduces the risk of diseases, such as cancer, cardiovascular disorders, and diabetes; (2) synthetic antioxidants have harmful effects on foods and beverages; (3) the general understanding is that natural, diet-related antioxidants are healthier than synthetic analogues. Accordingly, there is a considerable interest in using aromatic spices and herbs as a source of alternative antioxidants to synthetic antioxidants. Essential oil is a substance in herbs and spices that may act as an antioxidant. It has been mentioned that the presence of hydroxyl groups in phenolic compounds causes the antioxidant properties of aromatic plants [34].

From old times, antioxidants have been added to foods to prevent food spoilage, owing to the oxidation of unsaturated fats and other factors. Today, instead of synthetic antioxidants, natural antioxidants like essential oils are used, and the main component of which are phenolic substances, such as essential oils of *Salvia officinalis*, *Thymus*, and *Satureja*. The essential oil of *Thymus spathulifolius* also has antioxidant properties because of the high amount of carvacrol and thymol. Due to the high amount of carvacrol and thymol, Egyptian corn silk's essential oil indicates high antioxidant property [39].

Many studies have been conducted on essential oils in terms of antioxidant properties. The antioxidant properties of *Tanacetum macrophyllum* were measured using the Ferric reducing antioxidant power (FRAP) method. *Tanacetum argenteum* exhibited poor antioxidant properties in studies [33–36].

5.7.4 Anticancer properties

Essential oils are found in leaves, barks, flowers, and fruits, and play a main role as chemopreventive agents. For instance, the essential oil constituents of *Pistacia lentiscus* L. have been related to inhibition of lung, colon, and prostate cancer in vitro. Numerous essential oil compounds have been stated to have strong antioxidant properties and to have antimutagenic, anticarcinogenic, antiproliferative, and anticancer properties. Many studies have shown that essential oils have potential therapeutic applications in cancer prevention [33–35].

Antimutagenic effect of essential oil is attributed to certain mechanisms, including inactivation of mutagens by direct scavenging, activation of cell antioxidant enzymes, activation of enzymatic detoxification of mutagens, inhibition of penetration of mutagens into cells, and antioxidant capture of radicals produced by mutagens. Phytochemical compounds isolated from medicinal plants, such as α-terpineol, camphor, citral, α-terpinene, d-limonene, citronellal, and 1,8-cineole, modulated hepatic mono-oxygenase action by interacting with procarcinogen or promutagen xenobiotic biotransformation [40].

Previous research showed that essential oils could reduce apoptosis, necrosis, and mitochondrial damage in the yeast *Saccharomyces cerevisiae*. Some reports indicated that the constituents of essential oils showed antimutagenicity toward mutation caused by UV lights [36, 37, 40].

5.7.5 Antimalarial properties

Plant metabolites are of interest for use as repellents and insecticides. Among these metabolites, essential oils had great attention because of the diversity of their biological activities including antimalarial and pesticide. Most of the insecticides and repellents are produced from plant's essential oils. Lower molecular weight components of essential oils such as phenylpropanoids, monoterpenes, and sesquiterpenes are very significant class of antimalarial natural products. In vitro studies, some of active monoterpene constituents of essential oils indicate antiplasmodial property. Importantly, mechanistic researches have showed that the biosynthesis of the terpenoid side chains of ubiquinones in the trophozoite and/or schizont stages, and some compounds such as dolichols were stopped by linalool and limonene (low-molecular-weight monoterpenes), and nerolidol and farnesol (sesquiterpenes). Meanwhile, isochavicol and isochavicol, which are low-molecular-weight volatile phenylpropanoids, exhibit antiplasmodial property. Therefore, essential oils, which were produced by medicinal plants, are of specific interest due to their potential antimalarial and antiplasmodial properties [41].

References

[1] Shibamoto K, Mochizuki M, Kusuhara M. Aroma therapy in antiaging medicine. Anti-Aging Med 2010, 7(6), 55–59.

[2] Singh R, Shushni MA, Belkheir A. Antibacterial and antioxidant activities of *Mentha piperita* L. Arab J Chem 2015, 8(3), 322–328.

[3] Geng S, Cui Z, Huang X, Chen Y, Xu D, Xiong P. Variations inessential oil yield and composition during Cinnamomum cassia bark growth. Ind Crops Prod 2011, 33(1), 248–252.

[4]   Bnouham M. Medicinal plants with potential galactagogue activity used in the Moroccan pharmacopoeia. J Complement Integrative Med 2010, 7(1), 1–13.

[5]   Gupta V, Mittal P, Bansal P, Khokra SL, Kaushik D. Pharmacological potential of Matricaria recutita: A review. Int J Pharm Sci Drug Res 2010, 2(1), 12–16.

[6]   Sabzghabaee AM, Nili F, Ghannadi A, Eizadi-Mood N, Anvari M. Role of menthol in treatment of candidial napkin dermatitis. World J Pediatr 2011, 7(2), 167–170.

[7]   de Sousa DP. Analgesic-like activity of essential oils constituents. Molecules 2011, 16(3), 2233–2252.

[8]   Hunter M. Essential Oils: Art, Agriculture, Science, Industry, and Entrepreneurship (A focus on the Asia-Pacific), Nova Science Publishers, New York, 2009.

[9]   Andrade EHA, Alves CN, Guimarães EF, Carreira LMM, Maia JGS. Variability in essential oil composition of Piper dilatatum LC Rich. Biochem Syst Ecol 2011, 39(4), 669–675.

[10] Ben-Arye E, Dudai N, Eini A, Torem M, Schiff E, Rakover Y. Treatment of upper respiratory tract infections in primary care: a randomized study using aromatic herbs. Evid Based Complement Alternat Med 2011, 2011, 690346.

[11] Baser KHC, Buchbauer G. Handbook of Essential Oils: Science, Technology, and Applications. second CRC Press, Boca Raton, FL, 2015.

[12] Zuzarte M, Salgueiro L. Essential Oils Chemistry, Springer, Switzerland, 2015.

[13] Mann J, Davidson RS, Hobbs JB, Banthorpe D, Harborne JB. Natural Products: Their Chemistry and Biological Significance, Longman Scientific and Technical, Harlow, 1994.

[14] Zuzarte M, Gonçalves M, Cruz M, Cavaleiro C, Canhoto J, Vaz S, Pinto E, Salgueiro L. Lavandula luisieri essential oil as a source of antifungal drugs. Food Chem 2012, 135(3), 1505–1510.

[15] Davis EM, Croteau R. Cyclization enzymes in the biosynthesis of monoterpenes, sesquiterpenes, and diterpenes. Biosynthesis 2000, 209, 53–95.

[16] Rehman R, Hanif MA, Mushtaq Z, Al-Sadi AM. Biosynthesis of essential oils in aromatic plants: A review. Food Rev Inter 2016, 32(2), 117–160.

[17] Said-Al Ahl HA, Hikal WM, Mahmoud AA. Essential oils: Biosynthesis, chemistry and biological functions. J Chem Pharm Res 2017, 9, 190–200.

[18] Tisserand R, Young R. Essential Oil Safety: A Guide for Health Care Professionals, Elsevier Health Sciences, United Kingdom, 2013.

[19] Formisano C, Delfine S, Oliviero F, Tenore GC, Rigano D, Senatore F. Correlation among environmental factors, chemical composition and antioxidative properties of essential oil and extracts of chamomile (Matricaria chamomilla L.) collected in Molise (South-central Italy). Ind Crop Prod 2015, 63, 256–263.

[20] Bowles EJ. Chemistry of Aromatherapeutic Oils, Allen and Unwin, Australia, 2003.

[21] McGuinness H. Aromatherapy Therapy Basics, Hodder and Stoughton, London, 2003.

[22] Wijesekera K, Dissanayake AS. 4 Terpenes. Chemistry of Natural Products: Phytochemistry and Pharmacognosy of Medicinal Plants 2022, 19, 65.

[23] Tavera-Loza H. Monoterpenes in Essential Oils; Biosynthesis and Properties. Chemicals via Higher Plant Bioengineering. Shahidi E (Ed), Kluwer Academic Publishers and Plenum Press, New York, 1999.

[24] Wang C, Gong X, Bo A, Zhang L, Zhang M, Zang E, Zhang C, Li M. Iridoids: Research advances in their phytochemistry, biological activities, and pharmacokinetics. Molecules 2020, 25(2), 287.

[25] de Cássiada Silveira E Sá R, Andrade LN, De Sousa DP. Sesquiterpenes from essential oils and anti-inflammatory activity. Nat Prod Com 2015, 10(10), 1934578X1501001033.

[26] Stanetic D, Buchbauer G. Biological activity of some volatile diterpenoids. Current Bioactive Comp 2015, 11(1), 38–48.

[27] Sadgrove NJ, Padilla-González GF, Phumthum M. Fundamental chemistry of essential oils and volatile organic compounds, methods of analysis and authentication. Plants 2022, 11(6), 789.

[28] Blande JD, Holopainen JK, Niinemets Ü. Plant volatiles in polluted atmospheres: Stress responses and signal degradation. Plant Cell Environ 2014, 37(8), 1892–1904.

[29] Niinemets Ü, Kännaste A, Copolovici L. Quantitative patterns between plant volatile emissions induced by biotic stresses and the degree of damage. Front Plant Sci 2013, 4, 262.

[30] Marongiu B, Piras A, Porcedda S, Falconieri D, Maxia A, Frau M, Gonçalves M, Cavaleiro C, Salgueiro L. Isolation of the volatile fraction from Apium graveolens L. (Apiaceae) by supercritical carbon dioxide extraction and hydrodistillation: Chemical composition and antifungal activity. Natural Prod Res 2013, 27(17), 1521–1527.

[31] Moghaddam M, Farhadi N. Influence of environmental and genetic factors on resin yield, essential oil content and chemical composition of Ferula asafoetida L. populations. J Appl Res Med Arom Plant 2015, 2(3), 69–76.

[32] Berger RG. Flavours and Fragrances: Chemistry, Bioprocessing and Sustainability, Springer Science and Business Media, Heidelberg, 2007.

[33] Bakkali F, Averbeck S, Averbeck D, Idaomar M. Biological effects of essential oils–a review. Food Chem Toxicol 2008, 46(2), 446–475.

[34] Adorjan B, Buchbauer G. Biological properties of essential oils: An updated review. Flav Frag J 2010, 25(6), 407–426.

[35] Figueiredo AC. Biological properties of essential oils and volatiles: Sources of variability. Nat Volatil Essent Oil 2017, 4(4), 1–3.

[36] Kar S, Gupta P, Gupta J. Essential oils: Biological activity beyond aromatherapy. Nat Prod Sci 2018, 1, 24(3), 139–147.

[37] Herman RA, Ayepa E, Shittu S, Fometu SS, Wang J. Essential oils and their applications-a mini review. Adv Nutr Food Sci 2019, 4(4), 1–13.

[38] Nazzaro F, Fratianni F, Coppola R, Feo VD. Essential oils and antifungal activity. Pharmaceuticals 2017, 10(4), 86.

[39] Amorati R, Foti MC, Valgimigli L. Antioxidant activity of essential oils. J Agric Food Chem 2013, 61(46), 10835–10847.

[40] Andrade MA, Braga MA, Cesar PH, Trento MV, Espósito MA, Silva LF, Marcussi S. Anticancer properties of essential oils: An overview. Current Cancer Drug Target 2018, 18(10), 957–966.

[41] Dell'Agli M, Sanna C, Rubiolo P, Basilico N, Colombo E, Scaltrito MM, Ndiath MO, Maccarone L, Taramelli D, Bicchi C, Ballero M. Anti-plasmodial and insecticidal activities of the essential oils of aromatic plants growing in the Mediterranean area. Malaria J 2012, 11(1), 1–10.

Shivendra Singh Dewhare*, Nagendra Kumar Chandrawanshi,
Shekhar Verma, Sonam Soni and Pramod Kumar Mahish

Chapter 6
Extraction, production, and encapsulation of essential oils

Abstract: This chapter describes various traditional and modern technologies that are available for the extraction of essential oils from medicinal and aromatic plants. This chapter also describes the principles and mechanisms of various essential oil extraction techniques, feasibility, factors affecting the extraction process, benefits, merits, and demerits. This chapter further describes various encapsulation techniques required for imparting stability to essential oils.

Keywords: Medicinal plants, Essential oils, Extraction methods, Encapsulation of oils

6.1 Introduction

6.1.1 Biodiversity hotspots

Biodiversity hotspots are those geographic areas that encompass rich collection of diversified species, but are endangered. Biodiversity hotspots are usually poorly maintained, and need conservation and preservation. A biodiversity hotspot is defined as an area that has 30% or less of its original natural vegetation found nowhere else on the planet. There are currently 36 such biodiversity hotspots in the world. Although these hotspots represent just 2.3% of the Earth's land surface, they support more than half of the world's plant species and nearly 43% of the bird, mammal, and reptile and amphibian species [1, 2].

Interestingly, India, which has just 2.4% of the world's land area, accounts for nearly 7–8% of all the recorded species, including over 91,000 species of animals and 45,000 species of plants. Due to the diverse climatic conditions, different varieties of the ecosystem, such as wetlands, forests, grasslands, marine and desert

---
*Corresponding author: Shivendra Singh Dewhare, School of studies in Life Science, Pt. Ravishankar Shukla University, Raipur, Chhattisgarh, India, e-mail: shivendraprsu@gmail.com
Nagendra Kumar Chandrawanshi, School of studies in Biotechnology, Pt. Ravishankar Shukla University, Raipur, Chhattisgarh, India
Shekhar Verma, Sonam Soni, University College of Pharmacy, Pt. Deendayal Upadhyay Memorial Health Science, India & Ayush University of Chhattisgarh, India
Pramod Kumar Mahish, Government Digvijay Autonomous Post Graduate College, Rajnandgaon, Chhattisgarh, India

https://doi.org/10.1515/9783110791600-006

ecosystem, maintain rich biodiversity. Four of the 36 globally identified biodiversity hotspots: The Himalayas, the Western Ghats, the Northeast, and the Nicobar Islands, are found in India [2].

6.2 Medicinal plant extract

Extraction from raw plant extract involves the separation of medicinally active portions of plant tissues from the mixture of inactive components by using appropriate solvents known as *menstruum*, utilizing standard extraction procedures [3].

The general methods of extraction of medicinal plants include maceration, infusion, digestion, decoction, percolation, hot continuous extraction, counter current extraction, ultrasound extraction, supercritical fluid extraction, etc.

6.3 Essential oils

Essential oils are obtained commercially from different parts of aromatic plants and are widely used in consumer goods such as perfumes, cosmetics, food products, soaps, beverages, and pharmaceuticals [4–6]. Essential oils are volatile in nature, and appropriate extraction procedures must be employed to ensure the quality of the essential oil.

The techniques used to extract essential oils are highly dependent on the nature, characteristics, and components that are required from the extracts. Several extraction methods had been employed to obtain essentials oils form various parts of the raw material. Broadly, these extraction methods are classified into two categories, namely, conventional or classical methods and advances or innovative methods [7–11].

6.3.1 Conventional or classical extraction methods

The traditional oil extraction techniques include hydrodistillation, hydrodiffusion, steam distillation, solvent extraction, cold-pressing, and hydrolytic maceration distillation,

6.3.1.1 Hydrodistillation

Hydrodistillation method, first discovered by Avicenna, is the oldest and the simplest oil extraction method that is still being used today. Hydrodistillation involves soaking the raw material in water/solvent and bringing it to a boil. The vapors produced are liquefied in a condenser to separate the essential oils from the water/solvent system. Hydrodistillation method is frequently used for extracting essential oils from raw materials that have a high boiling point. This method is able to protect the oils from being overheated (below 100 °C) as it is surrounded by water, and this process usually keeps the temperature below 100 °C [11–13].

6.3.1.2 Hydrodiffusion

Hydrodiffusion extraction method is a technique used to isolate essential oils form a dried plant sample. In this method, steam is introduced from the top of the generator into the container that holds the plant sample [11, 14].

6.3.1.3 Steam distillation

Steam distillation is one of the traditional methods to isolate essential oils from aromatic plant materials. The injected steam passes in the bottom-to-top direction through the plant material placed on an alembic, without maceration. Hot steam results in the opening of the pores in the plant material, which results in the extraction of essential oils. The vapor mixture containing essential oils is condensed with the help of condenser to obtain the essential oils [11, 12, 15, 16].

Some of the advantages of steam distillation include steam flow controllability and the absence of a need for vacuum to recover the essential oils, thus making it economical, to some extent [10–12, 15, 17, 18].

6.3.1.4 Solvent extraction

Essential oils from fragile and delicate plant materials cannot be extracted using direct heat and steam method. Organic solvents, such as methanol, ethanol, acetone, and petroleum ether, are used to extract essential oils. Mild heating is applied to the solvents containing the plant material, followed by filtration and evaporation of the solvents. The filtrate (resinoid), also called as "concrete," thus obtained contains a mixture of essential oils, fragrance, and wax. Alcohol is added to the filtrate mixture, followed by distillation at low temperature [10, 12, 19]. The alcohol absorbs fragrance and is evaporated, while the aromatic absolute essential oil remains in the pot residue. Solvent extraction includes multiple steps, which makes it complex, time-consuming, and relatively more expensive [3].

6.3.1.5 Cold-pressing or scariform method

Cold-pressing method is one of the oldest techniques used for extracting essential oils. It is a mechanical extraction process, also known as expression or mechanical separation. Cold pressing utilizes a special kind of sponge that is used to absorb the plant oil. It is then transferred to a vice-like device over a container. After a few rounds, the collected oil is separated from the total plant extract. Cold-pressing method is predominantly employed for extracting essential oils form citrus plants such as lemon, oranges, grapefruit, and tangerines. Essential oils are extracted at very low temperature and pressure, which helps in maintaining all the natural properties of the plant extract [10, 20].

6.3.1.6 Hydrolytic maceration distillation

Some of the plant material necessitates maceration in warm water for the extraction of essential oils as their volatile compounds are in bounded form (glyosidic bonds). Some of the examples include brown mustard, wintergreen plant, garlic, and bitter almonds [3, 8, 15].

6.3.1.7 Essential oil extraction with fat

In some parts of the world, essential oil extraction using fat is still employed (also known as enfleurange). The procedure is relatively simple. Freshly picked flowers from orchards are layered onto a semihard fat base (corps), which readily absorbs the emitted perfume. The flowers are kept for 24 h or more, depending upon the variety and then carefully removed. This cycle is repeated several times until the fat base is saturated with flower oil. The fat base is removed from the chassis, carefully melted (pomade), and then the essential oils are extracted from the fat base with alcohol and further processed.

Another variation of enfleurange includes the immersion of flowers into melted fat (45–60 °C) for 1–2 h, which significantly reduces the time. The flowers are separated from the molten fat by filtration. This cycle is repeated multiple times and the essential oils are extracted and concentrated under reduced pressure from the absolute of maceration. This method is generally used for highly delicate flowers whose fragrance is rapidly lost after harvest [21, 22].

6.3.2 Advance or innovative extraction methods

Essential oil extraction using classical methods sometimes results in chemical alteration, such as oxidation and isomerization due to the high temperature and long duration. These drawbacks necessitated the development of new extraction methods that could maintain the natural composition in its native state. Some of the main methods are discussed below:

6.3.2.1 Supercritical fluid extraction (SFE)

Supercritical fluid extraction employs supercritical fluids as extraction solvents. Supercritical fluids have unique properties of solubility similar to liquid and diffusion properties similar to gas. One of the important aspects of SF is that their solvating properties radically change near their critical points due to the small temperature and pressure variations.

Carbon dioxide (CO_2) is one of the most widely used solvent for essential oil extraction due to its many properties such as inertness, nontoxicity, low critical temperature, low cost, and most importantly, its ability to extract thermolabile compounds. It can also be used for extracting lipophilic compounds [11, 17, 23].

Supercritical fluid extraction involves the repeated steps of compression and decompression cycle. CO_2 is compressed to achieve its supercritical state and it is then passed over the plant raw material. Next, the decompression mixture of CO_2 and plants extract is fed into two separators where the fluid is decomposed to separate the plant extracts from CO_2.

The second separator channels the CO_2 into a storage tank. The final extract doesn't contain any CO_2 residue as it can spontaneously revert to its gaseous state under normal temperature and atmospheric pressure [11, 17, 23].

Several studies have indicated that essential oils isolated using supercritical fluid extraction exhibited superior performance and pharmacological activities, such as enhanced antibacterial and antifungal properties, as compared to essential oils isolated from other techniques such as the hydrodistillation method [15, 17].

6.3.2.2 Ultrasonic-assisted extraction (UAE)

Ultrasonic-assisted extraction (UAE) utilizes energy from ultrasonic waves for the extraction of volatile compounds. This technique utilizes ultrasound waves, ranging from 20 to 2,000 kHz. The mechanical energy generated by the ultrasound waves results in the formation of cavitation, which are small vacuum bubbles. These bubbles implode near samples, which results in the formation of localized high pressure and temperature. Collective factors cause the destruction of cell membranes, resulting in the release of intracellular material in the solution (water, methanol/ethanol), followed by subsequent extraction.

UAE is employed for the extraction of unstable and thermolabile compounds. It offers many advantages such as low energy consumption, reduced time, high yield, and enhanced pharmacological activities [3, 13, 24].

6.3.2.3 Microwave-assisted extraction (MAE)

Microwave-assisted extraction (MAE) utilizes microwave energy (300 MHz to 300 GHz) for the extraction of essential oils from raw materials. This technique involves the amalgamation of two techniques wherein the raw material is heated using microwaves, causing the polar compounds to align themselves along the direction of the electric field. This phenomenon leads to compounds rotating at a high speed, resulting in heat generation and ultimately leading to the disruption of membrane/cells. Once released from the cells, the essential oils can be extracted using distillation.

MAE can be adapted based on the type of compounds to be extracted. Usually solvent-free extraction is used for extracting volatile compounds and solvent extraction is carried out for nonvolatile compounds.

MAE has many advantages such as less extraction time, low energy consumption, reduced cost of operation, and less CO_2 emission, which makes this technique to be considered as a green technology [3, 10, 25–27].

Further advances in the MAE method has led to the development of different variations such as the solvent-free microwave extraction, microwave hydrodistillation, compressed air microwave distillation, vacuum microwave hydrodistillation, and microwave hydrodiffusion and gravity method [28, 29].

6.3.2.4 Pulsed electric field (PEF) extraction

Pulsed electric field (PEF) extraction is a relatively new technique for the isolation of essential oils. It involves the treatment of plant material with short bursts of high voltage electric field for 1 s, resulting in the accumulation of specific energy (usually 10–20 kJ/kg). This causes the cell membrane pores to expand, resulting in increased permeability, which assists in the efficient extraction of essential oils. The main advantage is that this method does not utilize heat, thus minimizing the degradation of thermolabile compounds present in the plant extract [30, 31].

Nowadays, PEF extraction methods are used as a pretreatment step, followed by a conventional method, for the efficient extraction of essential oils. Studies have also indicated that a significantly higher yield can be obtained using PEF extraction method as compared to MAE or UAE [30].

6.3.2.5 Enzyme-assisted extraction

Enzyme-assisted extraction involves the use of hydrolytic enzymes such as pectinase, amylase, hemicellulases, and cellulases to degrade the polymers such as cellulose, xylan, pectin, and lignin present in the cell wall. Pretreatment of the plant extract with a hydrolytic enzymes mixture, followed by the conventional distillation method, results in a significant increase in essential oil yield. Extraction of essential oil cumin (*Cuminum cyminum* L.) by pretreatment with hydrolytic enzymes has been shown to increase total yield up to 20% [32–34].

6.4 Encapsulation of essential oils

Essential oil exhibits various important pharmacological effects, making it highly invaluable. But due to its volatility, it is highly unstable as it quickly evaporates and emits a strong odor. It also risks degradation upon direct exposure to heat, oxygen, humidity, and various other environmental factors [35–37].

To overcome these limitations, encapsulation of essential oils in a support matrix by chemical or physical interaction can help retain essential oils for a longer period of time.

Generally, encapsulation of essential oils can be by broadly classified into different techniques which are discussed below.

6.4.1 Emulsification

Emulsification is a process of making a stable colloidal system, usually prepared by mixing two immiscible liquids, which are stabilized with the help of surfactants. This technique basically encapsulates hydrophobic compounds so that they retain their bioactivity and stability. Based on the size, emulsion can be categorized into nanoemulsion and microemulsion, which can range from 75 nm to 400 μm. Some examples include the encapsulation of finger citron (*Citrus medica* L. var. sarcodactylis), carvacrol oil, D-limonene, *Zanthoxylum limonella* oil, peppermint oil, thyme oil, cinnamon oil, and clove oil. Encapsulation of these essential oils has helped in their stability and bioactivity [36].

6.4.2 Coacervation

Coacervation technique is based on the colloidal phenomenon, in which two phases of a biopolymer having opposite polarity separate from the solution and deposit around the essential oil to form coacervates. The capsule thus formed can be further stabilized by the addition of glutaraldehyde, which results in the formation of emulsion. Coacervation technique can be used for encapsulating essential oils such as thyme (*Thymus vulgaris*) essential oil and pimento (*Pimenta dioica* (L) Merr) essential oil [37, 38].

6.4.3 Spray drying

Spray drying is one of the earliest encapsulation methods. It involves the preparation of an aqueous emulsion and the use of a homogenization process. Emulsion is then atomization by a hot air stream in a drying chamber, resulting in the formation of microcapsules, whose sizes range from 0.2 to 400 μm. One of the advantages of this method is that it is a short and continuous process, which makes it economical to use. Essential oils that are encapsulated using spray drying include *Nigella sativa* oil, lavender oil, and almond oil [39–41].

6.4.4 Inclusion complexation

Inclusion complexation method involves the encapsulation or entrapment of essential oils within a polymer cavity through hydrogen bonding and van der Waals forces. Basil (BEO) and *Pimenta dioica* (PDEO) essential oils are some of the examples of the inclusion complexation method [35, 42].

6.4.5 Ionic gelation

Ionic gelation encapsulation is based on the cross-linking between the polyelectrolyte ions and the counter ions present in essential oils, to form a hydrogel. This method is preferable for encapsulating volatile essential oils as it does not require a high temperature. Encapsulation of clove and *Schinus molle* essential oil is by the ionic gelation method [35, 43].

6.4.6 Liposomes

Liposomes comprise one or more phospholipid bilayer and an aqueous core in the form of vesicles. Liposome-mediated encapsulation is a suitable carrier for hydrophobic, hydrophilic, and amphipathic compounds, thus augmenting its usability. *Melaleuca alternifolia* essential oil, thymol, carvacrol, and geraniol essential oil are some of the examples derived from the liposome-mediated encapsulation method [37, 40, 44].

6.4.7 Lyophilization

Lyophilization or freeze drying is the simplest and most widely used technique for encapsulating essential oils. Lyophilization works on the principle of sublimation, wherein the solid phase directly gets converted to the vapor phase. Lyophilization method preserves the natural flavor and color, and prevents degradation of the essential oils [42, 45, 46].

6.5 Conclusion

Essential oils play an important role and have become an irreplaceable commodity. Various extraction methods of essential oils, that is, traditional and modern methods have been discussed, while simultaneously highlighting their important characteristics. Traditional methods focus on the age-old proven techniques, which have been very well established on a large scale and are still most widely used. They offer high reproducibility and easy maintenance, but suffer from low yield and reduced bioactivity. Modern extraction techniques offer significant advancement in efficient extraction of essential oils, enhanced bioactivity, high yield, but currently suffer from high initial cost and issues in maintenance.

Encapsulation of essential oils by various methods has become all important to preserve their bioactivity and prevent their degradation over time, thus making them available for a longer period of time.

References

[1]   Marchese C. Biodiversity hotspots: A shortcut for a more complicated concept. Glob Ecol Conserv 2015, 3, 297–309.
[2]   Hrdina A, Romportl D. Evaluating global biodiversity hotspots-very rich and even more endangered. J Landsc Ecol Republic 2017, 10(1), 108–115.
[3]   Abubakar AR, Haque M. Preparation of medicinal plants: Basic extraction and fractionation procedures for experimental purposes. J Pharm Bioallied Sci 2020, 12(1), 1–10.
[4]   Adebayo SA, Amoo SO, Mokgehle SN, Aremu AO. Ethnomedicinal uses, biological activities, phytochemistry and conservation of African ginger (Siphonochilus aethiopicus): A commercially important and endangered medicinal plant. J Ethnopharmacol 2021, 266, 113459.
[5]   Govindaraj S. Extraction and therapeutic potential of essential oils: A review. Ther Prop Med Plants 2020, 10(2), 73–102.
[6]   Butnariu M, Sarac I. Essential oils from plants. J Biotechnol Biomed Sci 2018, 1(4), 35–43.
[7]   Khan MF, Dwivedi AK. A review on techniques available for the extraction of essential oils from various plants. Int Res J Engg Tech 2018, 5(5), 1100–1103.

[8]   Balakrishna T, Vidyadhara S, Sashidhar R, Ruchitha B, Venkata Prathyusha E. A review on extraction techniques. Indo Am J Pharm Sci 2016, 3(8), 880–891.

[9]   Yadav R, Yadav N. Review on extraction methods of natural essential oils. Int J Pharm Drug Anal 2016, 4(6), 266–273.

[10]   Rassem HHA, Nour AH, Yunus RM. Techniques for extraction of essential oils from plants: A review. Aust J Basic Appl Sci 2016, 10(16), 117–127.

[11]   Aziz ZAA, Ahmad A, Setapar SHM. et al., Essential oils: Extraction techniques, pharmaceutical and therapeutic potential – A review. Curr Drug Metab 2018, 19(13), 1100–1110.

[12]   Charles DJ, Simon JE. Comparison of extraction methods for the rapid determination of essential oil content and composition of Basil. J Am Soc Hortic Sci 2019, 115(3), 458–462.

[13]   Dalla Nora FM, Borges CD. Pré-tratamento por ultrassom como alternativa para melhoria da extração de óleos essenciais. Cienc Rural 2017, 47(9), 1–9.

[14]   Elyemni M, Louaste B, Nechad I. et al., Extraction of essential oils of rosmarinus officinalis l. by two different methods: Hydrodistillation and microwave assisted hydrodistillation. Sci World J 2019, p.2019 Article ID 3659432.

[15]   Ranjitha J, Vijiyalakshmi S. Facile Methods for the Extraction of Essential Oil From the Plant Species-a Review. Int J Pharm Sci Res 2014, 5(4), 1107.

[16]   Lopes P, Carneiro F, De Sousa A, Santos S, Oliveira E, Soares L. Technological evaluation of emulsions containing the volatile oil from leaves of Plectranthus Amboinicus Lour. Pharmacogn Mag 2017, 13(49), 159–167.

[17]   Tamboli P, Undire N, RaihanA Shaikh M, Mandake MB. Extraction and formulation of perfume from plants: A review. J Emerg Technol Innov Res 2021, 8(5), 663–666.

[18]   Golmohammadi M, Borghei A, Zenouzi A, Ashrafi N, Taherzadeh MJ. Optimization of essential oil extraction from orange peels using steam explosion. Heliyon 2018, 4(11) e00893.

[19]   McNerney R, Clark TG, Campino S, et al. Removing the bottleneck in whole genome sequencing of Mycobacterium tuberculosis for rapid drug resistance analysis: A call to action. *Int J Infect Dis*. Published online 2017.

[20]   Çakaloğlu B, Özyurt VH, Ötleş S. Cold press in oil extraction. A review. Ukr Food J 2018, 7(4), 640–654.

[21]   Ghasemy-Piranloo F, Kavousi F, Dadashian S. Comparison for the production of essential oil by conventional, novel and biotechnology methods. Authorea 2020, 1(1), 1–34.

[22]   New Directions Aromatic. Untapping the power of nature: Essential oil extraction methods. *A Compr Guid to Essent Oil Extr Methods*. Published online 2017:1–7.

[23]   Ibáñez E, Mendiola JA, Castro-Puyana M. Supercritical fluid extraction. Encycl Food Heal 2015, 8(2), 227–233.

[24]   Chen G, Sun F, Wang S, Wang W, Dong J, Gao F. Enhanced extraction of essential oil from Cinnamomum cassia bark by ultrasound assisted hydrodistillation. Chinese J Chem Eng 2021, 36, 38–46.

[25]   Movaliya SK. Extraction of essential oil by microwave-assisted extraction: A review. Int J Adv Res Innov Ideas Educ 2017, 3(2), 5312–5321.

[26]   Akhtar I, Javad S, Yousaf Z, Iqbal S, Jabeen K. Microwave assisted extraction of phytochemicals an efficient and modern approach for botanicals and pharmaceuticals. Pak J Pharm Sci 2019, 32(1), 223–230.

[27]   Cardoso-Ugarte GA, Juárez-Becerra GP, Sosa-Morales ME, López-Malo A. Microwave-assisted extraction of essential oils from herbs. J Microw Power Electromagn Energy 2013, 47(1), 63–72.

[28]   Zill-e-huma H *Microwave Hydro-Diffusion and Gravity: A Novel Technique for Antioxidants Extraction Zill-e-Huma.*; 2010.

[29] Asofiei I, Calinescu I, Gavrila AI, Ighigeanu D, Martin D, Matei C. Microwave hydrodiffusion and gravity, a green method for the essential oil extraction from ginger – Energy considerations. UPB Sci Bull Ser B Chem Mater Sci 2017, 79(4), 81–92.

[30] Carpentieri S, Režek Jambrak A, Ferrari G, Pataro G. Pulsed electric field-assisted extraction of aroma and bioactive compounds from aromatic plants and food by-products. Front Nutr 2022, 8(January), 1–12.

[31] Ranjha MMAN, Kanwal R, Shafique B. et al., A critical review on pulsed electric field: A novel technology for the extraction of phytoconstituents. Molecules 2021, 26(16), 1–23.

[32] Shimotori Y, Watanabe T, Kohari Y. et al., Enzyme-assisted extraction of bioactive phytochemicals from Japanese peppermint (Mentha arvensis l. cv. 'hokuto'). J Oleo Sci 2020, 69(6), 635–642.

[33] Miljanović A, Bielen A, Grbin D. et al., Effect of enzymatic, ultrasound, and reflux extraction pretreatments on the chemical composition of essential oils. Molecules 2020, 25(20), 10.3390/molecules25204818.

[34] Amudan R, Kamat DV, Kamat SD. Enzyme-assisted extraction of essential oil from. Syzygium Aromaticum South Asian J Exp Biol 2011, 1(6), 248–254.

[35] Tiwari S, Singh BK, Dubey NK. Encapsulation of essential oils – a booster to enhance their bio-efficacy as botanical preservatives. J Sci Res 2020, 64(01), 175–178, doi: 10.37398/jsr.2020.640125.

[36] Reis DR, Ambrosi A, Luccio MD. Encapsulated essential oils: A perspective in food preservation. Futur Foods 2022, 5(February), 100126.

[37] Murtadza SAA, Jai J, Md Zaki NA, Hamzah F. Essential oils encapsulation performance evaluation: A review on encapsulation parameters. Malaysian J Chem Eng Technol 2021, 4(2), 114.

[38] López A, Lis MJ, Bezerra FM. et al., Recent study on production and evaluation 2539 of antimicrobial microcapsules with essential oils using complex coacervation. Adv Asp Eng Res Vol 6 2021, 12(8), 50–66.

[39] Tomazelli Júnior O, Kuhn F, Padilha PJM. et al., Microencapsulation of essential thyme oil by spray drying and its antimicrobial evaluation against vibrio alginolyticus and vibrio parahaemolyticus. Brazilian J Biol 2018, 78(2), 311–317.

[40] Nguyen TTT, Le TVA, Dang NN. et al., Microencapsulation of essential oils by spray-drying and influencing factors. J Food Qual 2021, p.2021 Article ID 5525879.

[41] Mohammed NK, Tan CP, Manap YA, Muhialdin BJ, Hussin ASM. Spray drying for the encapsulation of oils – a review. Molecules 2020, 25(17), 1–16.

[42] Sousa VI, Parente JF, Marques JF, Forte MA, Tavares CJ. Microencapsulation of essential oils: A review. Polymers 2022, 14(9) 1730.

[43] Jamil B, Abbasi R, Abbasi S. et al., Encapsulation of cardamom essential oil in chitosan nano-composites: In-vitro efficacy on antibiotic-resistant bacterial pathogens and cytotoxicity studies. Front Microbiol 2016, 7.1580.

[44] Van Vuuren SF, Du Toit LC, Parry A, Pillay V, Choonara YE. Encapsulation of essential oils within a polymeric liposomal formulation for enhancement of antimicrobial efficacy. Nat Prod Commun 2010, 5(9), 1401–1408.

[45] Maes C, Bouquillon S, Fauconnier ML. Encapsulation of essential oils for the development of biosourced pesticides with controlled release: A review. Molecules 2019, 24(14) 2539.

[46] Ozkan G, Franco P, De Marco I, Xiao J, Capanoglu E. A review of microencapsulation methods for food antioxidants: Principles, advantages, drawbacks and applications. Food Chem 2019, 272, 494–506.

Nagendra Kumar Chandrawanshi*, Deepali, Anjali Kosre,
Shivendra Singh Dewhare, Shekhar Verma,
Pramod Kumar Mahish and Ashish Kumar

Chapter 7
Bioactivity of essential oils – anticancer, anti-HIV, antiparasitic, anti-inflammatory, and other activities

Abstract: Mushrooms are macrofungus and they have been widely used since ancient times as food. It has wealthy resources for various biologically active components; it is immensely valued for its functional and medicinal purposes because of its distinctive flavors and textures. Mushrooms have well-known nutritional value that can be compared to many vegetables and act as a substitute for meat. Some edible mushrooms are well-known sources of volatile oils. With various sophisticated equipment and standard procedures, a library of volatile and nonvolatile mushrooms has been generated. They play a significant role in antimicrobial, anticancer, anti protozoan, and other dimensions, benefitting the human health. The present chapter focuses on mushroom oils, and their extraction and quantification process – and their biological importance, in detail. The chapter will be the pathway for current research development and for future research.

Keywords: Biologically active components, nutritional value, volatile oils, anticancer

7.1 Introduction

Mushrooms are macrofungi, with spore-bearing structures, and are commonly found on the soil surface as saprophytes or parasites over their host. Mushrooms exist in

*corresponding author: **Nagendra Kumar Chandrawanshi,** School of Studies in Biotechnology, Pt.
Ravishankar Shukla University, Raipur, Chhattisgarh, India, e-mail: chandrawanshi11@gmail.com
Deepali, Anjali Kosre, School of Studies in Biotechnology, Pt. Ravishankar Shukla University,
Raipur, Chhattisgarh, India
Shivendra Singh Dewhare, School of Studies in Life Science, Pt. Ravishankar Shukla University,
Raipur, Chhattisgarh
Shekhar Verma, University College of Pharmacy, Pandit Deendayal Upadhyay Memorial Health
Science, India and Ayush University of Chattisgarh, India
Pramod Kumar Mahish, Government Digvijay Autonomous Post Graduate College,
Rajnandgaon, India
Ashish Kumar, Department of Biotechnology, Sant Gahira Guru University, Sarguja, Ambikapur,
Chhattisgarh

https://doi.org/10.1515/9783110791600-007

two fungal phyla, Basidiomycota and Ascomycota, and over 14,000 species are identified [1]. Mushrooms have been widely used since ancient times as food or flavoring materials for medicinal or functional purposes because of their distinctive flavors and textures [2]. Mushrooms have well-known nutritional value that can be compared to many vegetables, and they act as a substitute for meat [1]. Also, the saprophytic nature of mushrooms has contributed to their recycling capabilities, allowing them to convert agro-wastes, food remains, and dead bodies into highly nutritious food [1]. Among the mushroom species–genus *Pleurotus* comprises a diverse group of cultivated mushroom species, with high nutritional values and significant pharmacological properties. The commercial cultivation of *Pleurotus* has increased because the growth substrates are relatively inexpensive. They can be obtained abundantly from agricultural, forestry, and related industrial activities. The *Pleurotus* species is now the third most cultivated mushroom in the world, with annual production approaching one megaton [2]. Similarly, another mushroom, *Cordyceps*, comes from the Latin words cord, which means "club," and ceps, referring to "head." The fruiting bodies of these fungi often erupt from the head of the larva and the adult stages of many insect species. *Cordyceps* are entomophagous fungi from the phylum Ascomycota, family Ophiocordycipitaceae, and order Hypocreales. They are known to parasitize many orders of insects at different life stages, from larva to adult stages. Numerous species within the genus have a golden reputation due to their long history of safe use in traditional medicines [1]. *Scleroderma verrucosum* (Bull.) Pers. (Sclerodermataceae), *Cortinarius infractus* Berk. (Cortinariaceae), *Hypholoma capnoides* (Fr.) P. Kumm. (Strophariaceae), and *Hypholoma fasciculare* (Huds.) P. Kumm. (Strophariaceae) grow in Turkey. One of the mushrooms, *H. capnoides*, is edible; two of them, *S. verrucosum* and *H. fasciculare*, are not edible, and the last one, *C. infractus*, is not known as a food source mushroom [3]. Generally, *Cordyceps* species feed on insect larvae, and sometimes they also parasite on mature insects. *Cordyceps* grow on all groups of insects – crickets, cockroaches, bees, centipedes, black beetles, and ants, to name a few. Although several species are known to have medical value, only a few are cultivated, and the most popular and well-known are *Cordyceps sinensis* and *Cordyceps militaris*. However, *Cordyceps* is not limited to insects. It may grow on other arthropods and on the fungi, *Elaphomyces* Nees. This group belongs to the order Hypocreales, which includes 912 known species, assigned to the families of Cordycipitaceae, Ophiocordycipitaceae, and partial Clavicipitaceae [4]. *Cordyceps* refers only to the macrofungi, and these macrofungi were previously placed in the old genus, *Cordyceps* Fr. (Clavicipitaceae and Clavicipitales). Due to their particular edible and medicinal values, *Cordyceps* is very popular in China, where a substantial domestic market exists. By most Chinese people's standards, *Cordyceps* refers only to 'Dongchong Xiacao (worm in winter, a herb in summer); *O. sinensis*, is the most expensive type and is produced only in the Tibetan Plateau; other *Cordyceps* species in the marketplace are termed "fake Dongchong Xiacao," some of which may not be consumed in this way. Previously, most studies focused on specific species, especially those that are traditionally used in the Asian culture. However, there is recent attention to exploring mushrooms' chemical composition to investigate its generous rich macrofungi's potential for medical and pharmaceutical uses [1].

7.1.1 Bioactive compounds

Mushrooms are rich sources of compounds such as terpenes, phenolic compounds, fatty acids, glucans, polysaccharides, and proteins. Such compounds are responsible for various biological activities. The potential biological functions of many mushroom species include anticancer, antiviral, antitumor, antidiabetic, antidepressive, antioxidant, immunomodulatory, anti-inflammatory, neuroprotective, hepatoprotective, nephroprotective, osteoprotective, hypotensive, cholesterologenic, antiallergic, antimicrobial, antihyperlipidemic, and hypocholesterolemic activity [1, 2]. Moreover, *Cordyceps* contains an abundance of polysaccharides, representing 3–8% of the overall weight, and commonly originate from fruiting bodies. *Cordyceps* polysaccharide is one of the main bioactive components and nucleotides (including adenosine, uridine, and guanosine). Investigations show that guanosine is the most abundant nucleotide in both natural and cultivated *Cordyceps*. *C. sinensis* is the most expensive and the most extensively studied *Cordyceps* species. *C. sinensis* contains crude fats, proteins, fiber, carbohydrates, cordycepin (30-deoxyadenosine), cordycepic acid (D-mannitol), polysaccharides, and a series of vitamins. Of them all, *Cordyceps* species, *C. militaris* have been most successfully cultivated and most intensively studied. Most *Cordyceps* products in the market place are developed from the fruiting bodies of cultivated *C. militaris*. According to chemical analysis, *C. militaris* contains cordycepin, adenosine, polysaccharide, mannitol, trehalose, polyunsaturated fatty acids, δ-tocopherol, *p*-hydroxybenzoic acid, and β-(1→3)-D-glucan [1].

    P. sajor-caju (PSC), also known as gray oyster mushroom, has high nutritional values. Eleven components in the volatile oil from *P. sajor-caju* have been reported [2]. The glucans isolated from mushrooms (Basidiomycetes and Ascomycetes) till date include a-glucans, which is the most common branched (1R4),(1R6)-a-D-glucans; and b-glucans, which can be linear (1R3)- or (1R6)-linked, or branched (1R3),(1R6)-b-D-glucans. Many fungal glucans exhibit biological activity and belong to a group of physiologically active compounds, named biological response modifiers (BRMs). The glucans that exhibit b-configuration have shown to be the most effective BRMs, and their activity may vary according to their molecular weight, degree of branching, and conformation in solution. However, these data require further investigation [5]. Recent interest in utilizing mushroom-based bioactive compounds has increased due to their potential bioactivities and as alternatives to reducing high concentrations of chemical utilization [6]. The species complex around the medicinal fungus, *Ganoderma lucidum* Karst. (Ganodermataceae), is widely known in traditional medicine and used in modern applications such as in functional food or in nutraceuticals. Many publications reflect its abundance and variety in biological actions, triggered by primary metabolites, such as polysaccharides, or by secondary metabolites, such as lanostane-type triterpenes [7]. *G. lucidum* is one of the most famous medicinal fungi and is a traditional Chinese medicine with various biological activities used in Asian countries. To clarify its pharmacodynamic material basis, 15 lanostane triterpenoids were obtained from the fruiting bodies of *G. lucidum*, including 8 previously undescribed lanostanoids. Their structures, including absolute configuration, were based on ultraviolet, infrared, high-resolution electrospray ionization mass spectrometry, 1D and 2D nuclear magnetic resonance, and X-ray crystallographic analysis [8]. Many potential compounds have been successfully extracted from *G. lucidum*. Significant components include triterpenoids, meroterpenoids, sesquiterpenoids, steroids, alkaloids, and polysaccharides. They are rich in triterpenes, polysaccharides, and meroterpenoids. Thus, *Ganoderma* always has shown a wide range of biological activities [9]. *Calocybe indica* mushroom is well known for being a good source of proteins (12.48%) and fiber (6.87%). Its structural characterization revealed the mushroom powder's porous and small crystalline structure. The ethanolic compounds extracted are identified as total phenolics and flavonoids, caffeic acid, syringic acid, *p*-coumaric acid, and rutin, respectively. Furthermore, *C. indica* has ethyl tridecanoate, hexadecanoic acid ethyl ester, pentadecanoic acid ethyl ester, undecanoic acid ethyl ester, *N*, α, α′-trimethyl diphenethylamine, nicotinonitriles, phosphonic acid decyl-, 1-hexyl-2-nitrocyclohexane, diallyl divinylsilane, and 3-phenyl-pyrrole(2,3-β) pyrazine, which was confirmed during gas chromatography-mass spectrometry (GC-MS) analysis [6].

7.2 Bioactive components' (including oils) extraction and analysis procedure

Edible mushrooms are a well-known source of volatile oils. Various techniques for isolating food flavor; hydrodistillation (HD) is a common technique for separating a volatile component from nonvolatile materials because of its versatility. The solvent-assisted flavor exparotation (SAFE) method is another versatile method to isolate volatile components in food by minimizing the formation of artifacts. The high vacuum distillation technique allows the isolation of volatiles from complex food matrices under mild conditions due to low temperatures. In flavor analysis, gas chromatography-olfactometry (GC-O) is the most widely used method for evaluating odorants. In particular, GC-O, including aroma extract dilution analysis, is a valuable method to estimate the contributions of the most odor-active components by sniffing analysis. In the sniffing analysis of serial dilutions of volatile oils, the volatile components can be ranked according to their odor potency. Odor potency is expressed as the flavor dilution (FD) factor, which is the ratio of the initial concentration of a component to its most diluted concentration at which GC-O can still detect the odor [10].

7.2.1 Hydrodistillation extraction

Crude essential oils of *S. verrucosum, C. infractus, H. capnoides,* and *H. fasciculare* were obtained from the fruiting bodies of mushrooms (*ca.* 40 g, each) by hydrodistillation in a Clevenger-type apparatus with a cooling bath (–15 °C) system (3 h) (yields: 0.15%, 0.10%, 0.10%, and 0.12% (v/w), respectively). The oils' chemical constituents were determined by GC-MS analysis (Agilent-6890N/5973 Network System) [11]. The volatile oil obtained from the fresh fruiting bodies of *G. pfeifferi* Bres. by the hydrodistillation extraction method and analyzed by GC-MS. There were four volatile compounds representing 90.5% of the total oil identified. The majority of essential oils was dominated by 1-octen-3-ol (amyl vinyl carbinol) 1 (73.6%), followed by 1-octen-3-ol acetate 2 (12.4%), phenylacetaldehyde 3 (3.0%), and 6-camphenol 4 (1.5%) [12].

7.2.2 Solid-phase microextraction

Solid-phase microextraction (SPME) techniques are techniques for obtaining volatile oil compounds from mushrooms. *Boletus edulis, Lactarius camphoratus, Cantharellus cibarius,* and *Craterellus tubaeformis* are SPME that successfully analyzed the bioactive compounds. They are also odor-contributing compounds. They indicated that the saturated and unsaturated aldehydes and ketones contributed to the odor of the wild mushrooms [13].

7.2.3 Supercritical fluid extraction

Supercritical fluid was applied to the volatile compounds extracted from *Craterellus tubaeformis*. The effects of extraction pressure, temperature, and supercritical carbon dioxide volume on the extraction yield and the content of mushroom alcohols in the extracts were investigated from 80 to 95 bar, and 35 to 55 °C. Hence, the present study provides a valuable reference for the extraction of volatile compounds from mushrooms using supercritical fluid for further industrial applications [14].

7.2.4 Gas chromatography

GC-MS analyses were performed using an Agilent-5973 Network System. A mass spectrometer with an ion trap detector in full scan mode under electron impact ionization (70 eV) was used. The chromatographic HP-5 capillary column was used for the analysis (30 m × 0.32 mm i.d., film thickness 0.25 µm). Helium was the carrier gas used at a flow rate of 1 mL/min. The injections were performed in splitless mode at 230 °C. One µL essential oil solution in hexane (HPLC grade) was injected and analyzed, with the column held initially at 60 °C for 2 min and then increased to 260 °C with a 5 °C/min heating ramp, and subsequently kept at 260 °C for 13 min. A computerized integrator calculated the relative percentage amounts of the separated compounds from the total ion chromatograms [11]. Similarly, *H. Capnoides,* and *H. fasciculare* have shown the presence of different natural compounds, including triterpene compounds, aryl alcohols, fatty acids, chlorinated aromatics, etc. The chemical composition and antimicrobial activity of the essential oils of *S. verrucosum, C. infractions, H. capnoides,* and *H. fasciculare* were studied. As part of this systematic research, the essential oil constituents of the mushrooms were obtained by the widely used hydrodistillation method in a Clevenger-type apparatus. The GC-MS technique then investigated the obtained crude essential oils. Identification of the compounds was made by a typical library search (NIST, WILLEY) and the available literature was used for comparison [9].

7.2.5 Gas chromatography-olfactometry

Tricholoma matsutake Sing. mushrooms are characterized by aroma-active compounds from raw and cooked mushrooms. This was verified by GC-O using aroma the extract dilution analysis. The analyzed result of the mushroom-extracted compounds showed 1-octen-3-one as the major aroma-active compound in raw pine mushrooms. This compound had the highest FD factor, followed by ethyl 2-methyl butyrate (floral and sweet), linalool (citrus-like), methional (boiled potato-like), 3-octanol (mushroom-like and buttery), 1-octen-3-ol (mushroom-like), (*E*)-2-octen-1-ol (mushroom-like), and 3-octanone (mushroom-like and buttery). In contrast, methional, 2-acetylthiazole (roasted) an unknown compound (chocolate-like), 3-hydroxy-2-butanone (buttery), and phenylacetaldehyde less flavor dilution (FD) values due to unidentified compounds [15]. Two different techniques were employed for the evaluative study of *T. matsutake* Sing. mushroom by GC-O and sensory analysis. The 23 potential aroma-active compounds that are responsible for the sensory attributes of *T. matsutake* Sing. are such as (*E*)-2-decenal, alpha-terpineol, phenyl ethyl alcohol, and 2-methylbutanoic acid ethyl ester. They were responsible for most of these characteristics. These aroma characteristics were strongly associated with 1-octen-3-one, 1-octen-3-ol, 3-octanol, 3-octanone, (*E*)-2-octen-1-ol, and methional [16]. The chemical analysis of a mycelial extract of *P. lipsiensis* by GC-MS identified 73 molecules, such as steroids, terpenes, and lipids. According to literature data, among these molecules, 11 possess APA. The study revealed the diversity of compounds with anti-protozoal potential produced by the mycelia of *Agaricomycetes* mushrooms, particularly *P. lipsiensis* against *G. duodenalis* [17]. Various mushrooms had a variety of volatile and semi volatile components. Among them are 11 wild edible mushrooms, such as *Suillus bellini, S. luteus, S. granulatus, Tricholomopsis rutilans, Hygrophorus agathosmus, Amanita rubescens, Russula cyanoxantha, Boletus edulis, T. equestre, Fistulina hepatica,* and *Cantharellus cibarius* that were extracted by headspace SPME and by liquid extraction, combined with GC-MS. Fifty mushrooms having potential volatile and nonvolatile components were formally identified, and 13 others were tentatively identified. Using sensorial analysis, the descriptors "mushroom-like," "farm-feed," "floral," "honey-like," "hay-herb," and "nutty" were obtained. One was rich in C8 derivatives, such as 3-octanol, 1-octen-3-ol, trans-2-octen-1-ol, 3-octanone, and 1-octen-3-one; another one was rich in terpenic volatile compounds, and the last one was rich in methional. The presence and the content of these compounds contribute considerably to the sensory characteristics of the analyzed species [18]. Nordic wild edible mushrooms have different critical odors, thus characterizing the aroma compounds of four *B. edulis*, *Lactarius camphoratus*, *C. cibarius*, and *Craterellus tubaeformis* mushrooms species. This was verified by GC-O as well as by GC-MS [13]. The aroma compounds of *B. edulis* extracted from cooked and uncooled and in cooked oyster mushrooms *P. ostreatus* have been successfully verified by capillary GC and GC-MS. It is found that unsaturated alcohols and ketones containing eight atoms of carbon determine the aroma of raw mushrooms and they take part in the aroma formation of cooked mushrooms as well. The content of these compounds was the highest in canned cepes. The concentration of these alcohols and ketones in oyster mushrooms was lower than in cepes. The content of aliphatic and aromatic aldehydes was much higher in oyster mushrooms. The volatile aliphatic and heterocyclic Maillard reaction products, isomeric octenols, and octenones formed the aroma of cooked and canned mushrooms [19].

7.3 Chemical constituents

The hydrodistillation of the *Pleurotus* and *P.sajor caju* afforded pale yellowish oils in 0.044% and 0.040% (w/w) yields, respectively. The SAFE oils furnished yellowish oils in yields of 0.029% *P. sajor caju* and 0.042% (w/w). In the HD and SAFE oils, an impressive total of 84 components were identified, which composed 80.3% (PS) and 88.9% (PSC) of the total HD oil, 92.2% (PS), and 83.0% (PSC) of the total SAFE oils. In the PS HD oil, 1-octen-3-ol (12.0%) was the main component, followed by hexadecanoic acid (11.5%) and linoleic acid (9.9%). The classification of the oil is based on the functional group as follows: four hydrocarbons, six alcohols, two aldehydes, four ketones, one ether, six esters, two acids, two lactones, two nitriles, and one miscellaneous. In contrast, the main component in the PS SAFE oil was 11-(1-ethylpropy)-heneicosane (12.9%), followed by 1-octen-3-ol (11.3%) and butoxyethanol (11.0%). A total of 31 components (6 hydrocarbons, 9 alcohols, 1 aldehyde, 2 ketones, 1 ether, 5 esters, 1 acid, 2 lactones, 2 nitriles, and 2 sulfides) were identified. 11-(1-Ethylpropyl)–heneicosane has been reported as a volatile constituent in isodon amethystoiddes, *Pericarpium citrireticulate virisde* and *Urtica angustifolia*. This component was identified in mushrooms for the first time. Comparing the two extraction methods in PS,

HD efficiently furnished acids (21.4%), esters (18.7%), and alcohols (17.9%), which are the three classes with relatively high content in the oil. The SAFE method resulted in a relatively high percentage of alcohols (42.9%), hydrocarbons (31.6%), and ketones (4.7%). The HD and SAFE methods results indicated that the significant volatiles in PS were alcohols. Moreover, since two sulfides were identified in the SAFE oil, which were detected as a small amount in the HD oil, the SAFE method displayed an advantage over the HD method for the extraction of sulfides in the oil of PS as displayed in Table 7.1. In the *P. sajor caju* hydroxy distillation oil, ethyl linoleate (36.9%) was the main component, followed by hexadecanoic acid (23.4%) and diphenyl propane (9.0%). The classification of the oil, based on the functional group, is summarized as follows: ten hydrocarbons, nine alcohols, six aldehydes, three ketones, one ether, five esters, six acids, one lactone, and four sulfides. In contrast, in the PSC SAFE oil, the main component was methional (27.6%), followed by 6-undecanone (23.2%) and dimethyl sulfone (10.2%). A total of 15 components (4 hydrocarbons, 2 alcohols, 2 aldehydes, 1 ester, 2 acids, and 3 sulfides) were identified as shown in Table 7.1. PSC contains characteristic components such as 2,3,-di-methyl-1-butanol and 1,4-cyclooctadiene. It had been reported that 2,3-dimethyl-1-butanol was present in *Litsea fulva*, camellia oil, and *G. sinense*. The main constituents of the investigated essential oils of mushrooms are the following: 3-octanone (49.1%) and 3-octanol (26.8%) in *S. verrucosum*; musk ambrette (62.3%) and DL-limonene (5.5%) in *C. infractus*; 1-octen-3-ol (21.7%) and 3-octanol (14.3%) in *H. capnoides*; and 1-octen-3-ol (18.2%) and DL-limonene (14.2) in *H. fasciculare* [11].

7.4 Bioactivity of essential oils

7.4.1 Anticancer

Various biologically active compounds, exerting an anticancer activity, were extracted from varieties of mushrooms and analyzed. The mushroom of *Cordyceps* species has potential sources of anticancer bioactive compounds, and Cordycepin has been successfully extracted and analyzed for antitumor activity against B16 melanoma cells. Also, Cordycepin induced apoptosis in mouse Leydig tumor cells in vitro. Also, it inhibits cell proliferation and further apoptosis of human colorectal carcinoma using SW480 and SW620 in vitro. In gall bladder cancer cells, cordycepin causes loss of cancer cell viability and apoptosis via inhibiting the mammalian target of rapamycin complex 1. *C. militaris* was found to inhibit U937 cells grown in a dose-dependent manner. It is also used in the treatment of human leukemia. *Cordyceps* has shown promising activities in inhibiting the growth of cancer cells and also in reducing cancer cell sizes [1, 22]. Similarly, ethanolic extract of *C. militaris* showed a potent antitumor effect in RMA cell-derived tumors in a xenograft mouse model.

Table 7.1: Mushroom bioactive components' (essential oils) extraction and their characterization.

S. no.	Particular compound	Pleurotus salmoneostramineus		Pleurotus sajor-caju		Scleroderma verrucostum		Cortinarius infractus		Hypolama capnoides		Hypolama fasciculare		Exp. RI	References
		HD	SAFE	HD	SAFE	Area (%)	Q (%)	Area (%)	Q (%)	Area (%)	Q (%)	Area (%)	Q (%)		
1	Butyl acetate	–	–	0.5	–	–	–	–	–	–	–	–	–	812	10
2	Lactonitrile	–	1.8	–	–	–	–	–	–	–	–	–	–	816	10
3	2,3-Dimethyl-1-butanol	–	–	–	5.9	–	–	–	–	–	–	–	–	821	10
4	Methyl ethyl disulfide	–	–	–	0.5	–	–	–	–	–	–	–	–	828	10
5	2-(Methylthio)-ethanol	–	–	–	–	–	–	–	–	–	–	–	–	835	10
6	Hexanol	2.2	5.8	0.8	–	–	–	–	–	–	–	–	–	868	10
7	Pentanoic acid	–	–	–	–	–	–	–	–	–	–	–	–	887	10
8	2-Butoxyethanol	–	–	11	–	–	–	–	–	–	–	–	–	907	10
9	γ-Butyrolactone	–	1.3	–	–	–	–	–	–	–	–	–	–	910	10
10	Methinol	–	–	0.5	–	–	–	–	–	–	–	–	–	913	10
11	2,3-Dihydro-4methyl-furan	–	3.7	–	–	–	–	–	–	–	–	–	–	914	10
12	Diethyl sulfide	–	–	–	–	–	–	–	–	–	–	–	–	919	10
13	Dimethyl sulfone	–	3.7	0.3	–	–	–	–	–	–	–	–	–	945	10
14	Benzaldehyde	–	0.2	0.5	–	–	–	–	–	–	–	–	–	956	[10]
15	2-Pentylfuran	–	7.6	–	–	–	–	–	–	–	–	–	–	990	[10]
16	Ethyl hexanoate	–	–	–	–	–	–	–	–	–	–	–	–	998	[10]
17	Octanal	–	–	–	–	–	–	–	–	–	–	–	–	1003	[10]
18	Phenol	–	4.7	2	–	–	–	–	–	–	–	–	–	1004	[10]
19	2-(2-Ethoxyethoxy)-ethanol	–	–	–	–	–	–	–	–	–	–	–	–	1017	[10]
20	2-Acetylthiazole	–	–	–	–	–	–	–	–	–	–	–	–	1020	[10]
21	3-Ethyl-2-1,3 hexadiene	–	–	–	–	–	–	–	–	–	–	–	–	1031	[10]
22	2-Ethylhexanol	–	0.6	0.3	–	–	–	–	–	–	–	–	–	1034	[10]

(continued)

Table 7.1 (continued)

S. no.	Particular compound	Mushroom species												Exp. RI	References
		Pleurotus salmoneostramineus		Pleurotus sajor-caju		Scleroderma verrucostum		Cortinarius infractus		Hypolama capnoides		Hypolama fasciculare			
		HD	SAFE	HD	SAFE	Area (%)	Q (%)	Area (%)	Q (%)	Area (%)	Q (%)	Area (%)	Q (%)		
23	Phenylacetaldehyde	–	–	–	–	–	–	–	–	–	–	–	–	1047	[10]
24	1,4-Cyclooctadiene	–	–	1.4	–	–	–	–	–	–	–	–	–	1051	[10]
25	1-Octanol	–	1.7	–	–	–	–	–	–	–	–	–	–	1072	[10]
26	Succinimide	–	–	–	–	–	–	–	–	–	–	–	–	1089	[10]
27	Nonanal	–	–	0.3	–	–	–	–	–	–	–	–	–	1103	[10]
28	1,1-Dimethylpropyl ester hexanoic acid	–	–	–	–	–	–	–	–	–	–	–	–	1146	[10]
29	2-Ethylhexanoic acid	–	–	–	–	–	–	–	–	–	–	–	–	1159	[10]
30	6-Undecanone	–	–	–	–	–	–	–	–	–	–	–	–	1175	[10]
31	3-Ethyl-3-methyl-pentane	–	–	0.1	–	–	–	–	–	–	–	–	–	1185	[10]
32	Butyldiglycol	–	–	0.5	–	–	–	–	–	–	–	–	–	1193	[10]
33	Thiobenzoic acid	–	–	–	–	–	–	–	–	–	–	–	–	1239	[10]
34	2-Undecanone	1.7	–	–	–	–	–	–	–	–	–	–	–	1293	[10]
35	Tridecane	–	–	–	–	–	–	–	–	–	–	–	–	1300	[10]
36	(2E,4E)-Decadienal	0.9	–	0.3	–	–	–	–	–	–	–	–	–	1315	[10]
37	Dihydroisojasmone	–	2.1	0.2	–	–	–	–	–	–	–	–	–	1342	[10]
38	γ-Nonalactone	1	–	–	–	–	–	–	–	–	–	–	–	1362	[10]
39	Skatole	–	–	–	–	–	–	–	–	–	–	–	–	1385	[10]
40	Tetradecane	–	–	–	–	–	–	–	–	–	–	–	–	1400	[10]
41	β-Chamigrene	–	–	0.3	–	–	–	–	–	–	–	–	–	1478	[10]
42	β-Bisabolene	–	–	–	–	–	–	–	–	–	–	–	–	1506	[10]
43	Dicyclohexyl ketone	–	–	0.5	–	–	–	–	–	–	–	–	–	1519	[10]

No.	Compound							RI	Ref
44	Calamenone	–	–	0.1	–	–	–	1524	[10]
45	1-Phenyl-1,2,3-triazole	–	–	–	–	–	–	1531	[10]
46	Aromadendrene	0.8	–	–	–	–	–	1554	[10]
47	Nerolidol	–	–	0.2	–	–	–	1563	[10]
48	Dodecanoic acid	–	–	–	–	–	–	1568	[10]
49	Ethyl dodecanoate	–	–	–	–	–	–	1594	[10]
50	1,6-Dimethyl 1-4-(1-methylethyl)-1,2,3,4,4a,7-hexahydronaphthalene	–	–	–	–	–	–	1627	[10]
51	(Z)-6-Dodecen-4-olide	–	–	–	–	–	–	1629	[10]
52	Espatulenol	–	–	–	0.6	–	–	1630	[10]
53	α-Copaene	0.9	0.1	–	–	–	–	1648	[10]
54	Tetradecanal	–	–	0.3	–	–	–	1712	[10]
55	Tetradecanoic acid	–	–	0.5	–	–	–	1777	[10]
56	Nimbiol	–	–	0.2	–	–	–	1787	[10]
57	Pentadecanoic acid	–	–	2.4	–	–	–	1878	[10]
58	Dibutyl phthalate	1.9	–	–	–	–	–	1960	[10]
59	Hexadecanoic acid	11.5	23.4	–	–	–	–	1977	[10]
60	Ethyl hexadecanoate	0.9	1.5	–	–	–	–	1992	[10]
61	Geranyllinalool	–	–	–	0.8	–	–	2028	[10]
62	Octadecanoic acid	–	–	–	0.8	–	–	2052	[10]
63	Linoleic acid	9.9	–	–	–	–	–	2144	[10]
64	Ethyl linoleate	8.5	36.9	–	–	–	–	2156	[10]
65	Diphenylolpropane	–	9	–	–	–	–	2173	[10]
66	Docosane	–	2.3	–	–	–	–	2200	[10]
67	Linoleic acid	–	0.2	–	–	–	–	2274	[10]
68	8-Hexylpentadecane	10.59	–	–	–	–	–	2320	[10]
69	Octyl adipate	6	–	–	–	–	–	2396	[10]
70	Squalene	0.7	–	–	–	–	–	2819	[10]
71	Nonacosane	7.2	–	–	–	–	–	2900	[10]
72	Santene	–	–	–	–	0.2	81	886	[11]

(continued)

Table 7.1 (continued)

S. no.	Particular compound	Pleurotus salmoneostramineus		Pleurotus sajor-caju		Scleroderma verrucostum		Cortinarius infractus		Hypolama capnoides		Hypolama fasciculare		Exp. RI	References
		HD	SAFE	HD	SAFE	Area (%)	Q (%)	Area (%)	Q (%)	Area (%)	Q (%)	Area (%)	Q (%)		
73	Anisole	–	–	–	–	0.3	93							922	[11]
74	Tricyclene	–	–	–	–			0.5	95					930	[11]
75	α-Pinene	–	–	–	–	1.6	94	3.2	94	8.7	94	7.2	94	940	[11]
76	Camphene	–	–	–	–	2	97	3.7	95	8.5	95	8.9	95	956	[11]
77	Benzaldehyde	–	–	–	–	0.3	80	–	–	–	–	–	–	961	[11]
78	β-Pinene	–	–	–	–	–	–	2.4	95	–	–	–	–	980	[11]
79	1-Octen-3-ol	12.5	11.3	0.5	–	–	–	1.9	90	21.7	90	18.2	90	982, 983	[10, 11]
80	3-Octanone	4.9	4.7			49.1	87	0.7	91	8	91	5.3	91	986, 988	[10, 11]
81	Myrcane					0.7	86	–	–	–	–	–	–	993 989,	[11] [10, 11]
82	3-Octanol	1.5	7.3	–	–	49.1	87	0.7	91	8	91	5.3	91	993	[11]
83	α-Terpinene	–	–	–	–	–	–	0.4	93	–	–	–	–	1017	[11]
84	o-Cymene	–	–	–	–	–	–	0.1	94	–	–	–	–	1029	[11]
85	DL-Limonene	–	–	–	–	2.6	96	5.5	96	10.8	96	14.2	96	1032	[11]
86	Benzene acetaldehyde	–	–	–	–	0.5	91	0.3	91	–	–	–	–	1043	[11]
87	Cumene	–	–	–	–	–	–	–	–	1.6	80	–	–	1064	[11]
88	n-Octanol	–	–	–	–	5.1	90	–	–	–	–	–	–	1067	[11]
89	Terpinolene	–	–	–	–	0.2	96	–	–	–	–	–	–	1086	[11]
90	E-Pinocarveol	–	–	–	–	–	–	0.5	87	–	–	–	–	1136	[11]
91	Camphor	–	–	–	–	–	–	0.6	95	–	–	–	–	1143	[11]

No.	Compound				A (%)	A	B (%)	B	C (%)	C	D (%)	D	RI	Ref
92	Pinocarvone	–	–	–	–	–	0.4	86	–	–	–	–	1161	[11]
93	Borneol	–	–	–	0.2	90	0.5	90	–	–	–	–	1165	[11]
94	Terpinen-4-ol	–	–	–	–	–	0.2	94	–	–	–	–	1174	[11]
95	α-Terpineol	–	–	–	–	–	0.2	80	–	–	–	–	1187	[11]
96	Myrtenol	–	–	–	–	–	0.2	80	–	–	–	–	1193	[11]
97	(E, 4E)-Nonadienal	–	–	–	–	–	0.3	80	–	–	–	–	1213	[11]
98	Bornyl acetate	–	–	–	3.2	96	6.9	96	8.4	98	13.8	96	1286	[11]
99	E-Caryophyllene	–	–	–	–	–	0.3	98	–	–	1.3	80	1420	[11]
100	(2E, 4E)-Decadienal	–	–	–	–	–	0.2	87	–	–	–	–	1316	[11]
101	E-Sabinyl acetate	–	–	–	–	–	0.2	80	–	–	–	–	1294	[11]
102	α-Copaene	–	–	–	–	–	0.2	98	–	–	–	–	1377	[11]
103	α-Humulene	–	–	–	–	–	0.2	98	–	–	–	–	1455	[11]
104	Germacrene D	–	–	–	–	–	0.2	93	–	–	–	–	1481	[11]
105	Delta-Cadinene	–	–	–	–	–	0.1	97	–	–	–	–	1525	[11]
106	E-Nerolidol	–	–	–	–	–	0.3	87	–	–	–	–	1565	[11]
107	Caryophyllene oxide	–	–	–	–	–	0.4	80	–	–	1.5	87	1585	[11]
108	Hexadecane	–	–	–	–	–	0.4	94	–	–	–	–	1600	[11]
109	α-Cadinol	–	–	–	–	–	0.3	87	–	–	1.5	91	1657	[11]
110	Musk ambrette	–	–	–	–	–	62.3	80	–	–	–	–	1926	[11]
111	Manool	–	–	–	–	–	0.4	80	–	–	–	–	2058	[11]
112	Dehydroabietal	–	–	–	–	–	0.2	99	–	–	–	–	2271	[11]
113	4-epi-Abietal	–	–	–	–	–	0.3	86	–	–	–	–	2303	[11]

Moreover, some *Cordyceps* species have anti-leukemia activities and ameliorate suppressive effects of chemotherapy on bone marrow function, as a model for cancer treatment [1]. The scientific community and research have provided plenty of data showing that mushroom extracts demonstrate interesting biological properties such as antitumor. These extracts contain different molecules such as steroids, polyphenols, hydroquinones, triterpenes, proteins, glycoproteins, and polysaccharides, and they are involved in biological effects [5, 23]. *G. lucidum* is the primary source of the hydroxyl groups on the lanostane triterpenes compounds. They have the C-3, C-24, and C-25 positions. They found potential for developing the anti-HIV-1 virus. A previous study has shown that lanostane triterpenes could initiate response against human immunodeficiency virus-1 protease. It has latent anti-HIV-1 activity. This is possible due to the biological activity of polysaccharides, leading to the production of tumor necrosois factor-α and interferon-γ by macrophages and spleen cells in mice. They intensely destroy cancer cells [11, 24].The fruiting body of *G. calidophilum* has the potential to develop active anticancer constituents – isolated four novel spiro-lactone lanostane triterpenoids (spiroganocalitones A-D). The structures of the compounds are derived and elucidated, based on 1D, 2D-NMR, and HRMS analyses. Among them, Ganoderone A showed moderate cytotoxic activity against K562, BEL7402, and SGC790 cell lines, with IC50 values of 7.62, 6.28, and 3.55 μM, respectively [20]. *G. lucidum* has the potential for bioactive compounds such as triterpenoids; it is known as Ganoderic Acids and is, referred to as ganoderic acid. In particular, it is known as Ganoderic Acid-DM (GA-DM). It is extracted from the *G. lucidum* mushroom. It has a promising therapeutic agent, in that, GA-DM can induce cell death in cancer cells, while exhibiting minimal toxicity to normal bystander cells. Furthermore, GA-DM's capability to stimulate an immune response in the tumor microenvironment potentially provides long-term protection from malignant tumors [21]. *Ganoderma* has several medicinal effects, which is confirmed by in vitro and in vivo studies, but there are some limitations. Clinical trials mainly face a lack of pure constituents. Accurate identification of the compounds obtained is also problematic. These studies have shown that *Ganoderma* has valuable potential for the prevention and treatment of cancer. In any case, *G. lucidum* cannot be used as the first-line therapy for cancer [25].

7.4.2 Anti-HIV

The modern computational *in sillico* approach subjects a 3D multi-conformational molecular dataset to screening against an in-house collection of validated structures, including the ligand-based 3D pharmacophore models. In silico optimization and evaluation processes generated predictive validated data, interpreted results have been more relevant for reducing the procedure of clinical testing procedure, and it will be significant for static drug development. *Ganoderma* compounds were predicted as ligands of at least one of the selected pharmacological targets in antiviral and the metabolic syndrome screening. Only a minority of the individual compounds (around 10%) have been investigated on these targets for the associated biological activity. Accordingly, this study displays putative ligand-target interactions for a plethora of *Ganoderma* constituents in the empirically manifested field of viral diseases and metabolic syndrome, which serve as a basis for future applications to assess the yet undiscovered biological actions of *Ganoderma* secondary metabolites at a molecular level [7]. Similarly, Triterpenes are naturally occurring, and other derivatives have biosynthesized the Ruzicka's isoprene rule. Triterpenes have been reported to possess many therapeutic applications, including antiviral properties [26]. Antiviral activities were observed with an IC50 of 39.19, 26.17, and 10.3.3 µg/mL. The selectivity index (SI) for *P. ostreatus*, *L. edodes*, and *A. bisporus* was 4.34, 3.44, and 1.5, respectively. The docking analysis revealed that catechin from three mushroom isolates – chlorogenic acid from *A. bisporus*, kamperferol of *P. ostreatus*, and quercetin from *L. edodes* – showed C-DOCKER interaction energy in the range of 22.8–37.61 (kcal/mol) with protease, compared to boceprevir ligand of 41.6 (kcal/mol). Docking of superoxide dismutase, catalase from the three mushrooms, and tyrosinase from *A. bisporus* showed ligand contact surface area with the protein as 252.74 Å2, while the receptor contact surface area was 267.23 Å2. Novel farnesyl hydroquinone, ganomycin A and ganomycin B were isolated with chloroform solvent from *Ganoderma colossum*. Their compounds potentiated HIV-1 protease with IC50 values of 7.5 and 1.0 µg/mL, respectively [27].

7.4.3 Antiparasitic

The antiparasitic activity (APA) of mycelial extracts from seven medicinal agaricomycetous mushrooms was assessed. The prominent mushroom species including *Polyporus lipsiensis*, *G. applanatum*, *P. ostreatus*, *P. flabellatus*, *Oudemansiella canarii*, *L. edodes*, and *Pycnoporus sanguineus* extracted compounds shown potential anti-protozoal. The extracts of mycelia and fermented culture broths of the tested mushroom species were evaluated against *G. duodenalis* by biological assays. *P. lipsiensis* showed the highest APA [17]. Similarly, the parasite lactate, dehydrogenase (pLDH), assay evaluated the mushroom and plant organic extracts for antiplasmodial activity. The chemical compounds found, such as saponins, terpenoids, anthraquinones, coumarins, alkaloids, and flavonoids, were determined by their potent characteristics of antiplasmodial compounds using GC-MS. Four mushrooms and thirteen plant species used in this study were used from the north-central Namibian regions, namely Kavango East, Kavango West, Ohangwena, and Oshikoto. Of these, only two plants were active against 3D7 strains of the potent malaria parasite, *Plasmodium falciparum*. Two antiplasmodial compounds, Npk1 F70 and NPk1 F78, were isolated successfully from the Npk1 (*Pechuel-loeschea leubnitiziae*) dichloromethane (DCM) extracts [28].

7.4.4 Anti-inflammatory

Inflammation is an immunological response; it is beneficial for the host's response to infection and tissue injury, but ultimately it has to restore the typical tissue structure and function. A normal inflammatory response is self-limiting, although prolonged inflammation contributes to many inflammatory diseases' pathogenesis and cancer [5]. The possible mechanism of the inflammatory action of molecules is based on the recognition of b-glucans by cell surface receptors, called pattern recognition receptors (PRRs). The PRRs recognize the b-glucans as pathogen-associated molecular patterns (PAMPs) and initiate immune responses. Various human receptors have been identified, like dectin-1. They complement receptor 3 (CR3), scavenger receptors, lactosyl-ceramide, and toll-like receptors (TLRs). The immune responses initiated by these polymers are by the activation of leucocytes. They stimulate phagocytic and cytotoxic activities and produce proinflammatory mediators by the immune system cells. However, some mushroom extracts have anti-inflammatory properties, the most studied effects of antitumor and immune stimulating activities. Ganoluciduone A was an unusual octonorlanostane isolated from *Ganoderma* for the first time. In addition, the anti-inflammatory activities of all isolates were evaluated by observing their inhibitory effects on nitric oxide production in RAW264.7 cells activated by lipopolysaccharide. Ganoluciduone B exhibited moderate inhibitory activity on nitric oxide production, with an inhibition rate of 45.5% at a concentration of 12.5 μM [8]. *Ganoderma* meroterpenoids have shown significant anti-inflammatory activity [9]. There were 17 triterpenoids, including some novel lanostane triterpenoids (1–3 and 5) isolated from *G. lucidum*. Their structure was elucidated from 1D-NMR, 2D-NMR, and HRESIMS; they had shown the potential application for anti-neuroinflammation and inhibited the hydrolysis activity of fatty acid amide hydrolase (FAAH) [29]. *G. applanatum* has potential bioactive compounds such as lanostane-type triterpenoids, including two triterpenoids, such as applanoic acids E and F, and another one, C21 nortriterpenoid. It also has two highly oxygenated lanostane triterpenoids. These components, including applanoic acid E, 16,17-dehydroapplanone E, and methyl applaniate B, have shown inhibitory effects on the release of NO by LPS-induced BV-2 cells [30].*G. lucidum* is another primary source of 12 lanostane triterpenes. They have the richness of seven triterpenes, such as butyl lucidenate E2, butyl lucidenate D2 (GT-2), butyl lucidenate P, butyl lucidenate Q, Ganoderiol F, and methyl ganodenate J. They inhibited oxygenase (HO)-1 expression and suppressed lipopolysaccharide (LPS)-induced nitric oxide (NO) production. Thus, they are prominent for HO-1 in the anti-inflammatory effects of these triterpenes [31]. Similarly, *Calocybe indica* has shown potent anti-inflammatory activity, including good membrane stabilization activity [6]. Cordycepin, extracted from the *Cordyceps* species, is partly responsible for the anti-inflammatory effects. Although shown by other compounds, b-(1R3)-D-glucan can also exhibit potent anti-inflammatory activity [1]. In recent years, some natural bioactive products isolated from the genus *Ganoderma*'s fungi have been found and they have successfully demonstrated to have antitumor, liver protection, anti-inflammatory, immune regulation, antioxidation, antiviral, anti-hyperglycemic, and anti-hyperlipidemic effects [11].

7.4.5 Antimicrobial activities

The volatile oil of the mushroom of basidiomycete *G. pfeifferi* Bres. was analyzed for antimicrobial activities of the oil. The oil's antimicrobial activity was evaluated against five bacteria strains, including *Staphylococcus aureus* and *Candida albicans* fungi strains, using disk diffusion and broth microdilution methods. They showed that the gram-positive bacteria species are more sensitive to essential oil than the gram-negative bacteria. The oil showed strong antimicrobial potent against test microorganisms [12]. Two new unique farnesyl hydroquinones, such as ganomycin A and ganomycin B, were extracted from *G. pfeifferi*, and spectroscopic approaches analyzed their structures. They had potential antimicrobial potent components and have shown activity against many gram-positive and gram-negative bacteria [32]. *C. indica* mushroom extract showed effective antimicrobial activity against gram-positive (23.67 mm) and harmful bacteria (20.33 mm) in terms of the zone of inhibition [6]. *G. boninense* is extracted from various numerous and complex profiles of natural compounds. The extraction made with normal-phase SPE and its metabolites were analyzed using liquid chromatography-mass spectrometry (LCMS). Furthermore, the antibacterial assay method, such as broth microdilution and disk diffusion, were analyzed. Strong susceptibility was observed in methicillin-resistant *S. aureus* (MRSA) eluted fraction, with high zone inhibition [33]. *Trametes* species, laccase (EC 1.10.3.2.), mediates the oxidative coupling of antibiotics with sulfonamide or sulfone structures with 2,5-dihydroxybenzene derivatives to form new heterodimers and heterotrimers. These heteromolecular hybrid products are formed by the nuclear amination of *p*-hydroquinones with the primary amino group of the sulfonamide or sulfone antibiotics, and they inhibit in vitro the growth of *Staphylococcus* species, including multidrug-resistant strains [34]. Antibacterial activity of the Lingzhi mushroom, *G. lucidum*, was evaluated against four bacterial pathogens. Ethyl alcohol and water extracts of the fruit body powder were tested using the disk diffusion method against *S. aureus*, *Klebsiella pneumoniae*, *Bacillus cereus*, and *Pseudomonas aeruginosa*. It was noted that the aqueous extract inhibited the growth of pathogenic bacteria [35]. *G. lucidum* beverage was most effective against the tested bacteria, with the lowest minimum inhibitory concentration (0.04 mg/mL) against *S. epidermidis* and *Rhodococcus equi* and a minimum bactericidal concentration (0.16 mg/mL) against *B. spizizenii*, *B. cereus*, and *R. equi*. The vacuum-dried sample was less effective, with the lowest minimum bactericidal concentration against the gram-positive bacteria, *R. equi* (1.875 mg/mL) and the gram-negative bacteria, *Proteus hauseri* (30 mg/mL) [36]. The antimicrobial potential of five *G. lucidum* samples was tested against five fungal pathogens such as *Fusarium oxysporum*, *Aspergillus niger*, *A. flavus*, *Penicillium* sp., and *Alternaria alternata*. Significant inhibition was observed using higher concentrations of the methanolic extract (3% and 4%). Thus, methanolic extracts of *G. lucidum* had shown better antimicrobial activity against all plant fungal pathogens, when compared with the water extracts [37]. *G. boninense* isolated compounds have shown antibacterial activity, most potent against *S. aureus* ATCC 25923 and the multidrug-resistant *S. aureus* NCTC 11939. The compounds, further characterized by GC-MS, HPLC, and FTIR analysis, confirmed two antibacterial compounds, which were identified as 4, 4,14α-trimethylcholestane (m/z = 414.75; lanostane, $C_{30}H_{54}$) and ergosta-5,7,22-trien-3β-ol (m/z = 396.65; ergosterol, $C_{28}H_{44}O$) [38]. The antimicrobial activity of the essential oils from *S. verrucosum, C. infractus, H. capnoides*, and *H. fasciculare* were tested against the bacteria, including *E. coli, K. pneumoniae, P. aeruginosa, E. faecalis, S. aureus*, and *B. cereus*, respectively, by the agar dilution method, against seven bacteria and two yeast-like fungi. The essential oil of the fungus, *C. tropicalis*, in hexane has shown potential antibacterial and antifungal activities [11].

7.4.6 Antioxidant activities

In addition, the antioxidant activity of the oil of *G. pfeifferi* Bres. was determined using the DPPH assay. Moreover, the oil exhibited vigorous radical scavenging activity in the DPPH assay. This first report on the chemical composition and biological properties of the *G. pfeifferi* volatile oil makes its pharmaceutical uses rational and provides a basis for the biological and phytochemical investigations of the volatile oils of the Ganodermataceae species [12]. *Ganoderma* contains meroterpenoids, which are oxygen-containing heterocycles and have significant antioxidant activity. Thus, *Ganoderma* has been used as a medicine to treat diseases for more than 2,000 years, and we also reviewed its traditional uses [9]. The hot water extract (HWE) of *G. lucidum* was a high-quality medium for yeast and AAB growth; their Fourier transform infrared analysis revealed various compounds, including polysaccharides, phenols, proteins, and lipids. They are good sources of Total phenolic content. This is due to the potential and maximal effective concentrations for DPPH scavenging activity and reducing power, respectively [36, 39]. Protecting against cell damage by free radicals is one of the biological activities exerted by the *Cordyceps* species extracts. This activity corresponds to the polysaccharide fraction. *C. sinensis* has potent antioxidant and antiaging properties. Many studies elucidated the antioxidant effect of the extracts obtained from *C. militaris*. The fruiting bodies extract of *C. militaris* had a potent DPPH· radical scavenging activity, whereas the fermented mycelia extract had more vigorous total antioxidant activity and reducing ability [1].

7.5 Conclusion and future prospects

Medicinal and edible mushrooms are rich biosources for diverse bioactive compounds. They are classified based on their high dimension molecular structure. They are classified into various groups such as terpenes, phenolic compounds, fatty acids, glucans, polysaccharides, proteins, and many more. With the continuing development of science and technology and the need for a healthy and novel biomaterial development, the worldwide scientific community has attempted many unexplored mushroom species. Mushrooms' has well-known bioresources for edible and medicinal application. Thus, it has been considered a significant source of varieties of bioactive compounds. A number of findings were revealed in the past years, including bioactive compounds, essential oils, and other targeted specific components. The present chapter has presented the recent research on mushroom bioactive compounds and their extraction procedure. Many mushroom species are still unexplored, offering scope for unearthing new biomaterial for future health and other applications. At the same time, the extraction measurement process still needs to be developed for targeted validation, and will be jointly approached for *in silico* biology to optimize bioactive compounds.

References

[1] Elkhateeb WA, Daba GM. *Cordyceps* more than edible mushroom-a rich source of diverse bioactive metabolites with huge medicinal benefits. J Biomed Res Environ Sci 2022, 3(5), 566–574.

[2] Caglarimark N. The nutrients of exotic mushrooms *(Lentinula edodes* and *Pleurotus* species) and an estimated approach to the volatile compounds. Food Chem 2007, 105, 1188–1194.

[3] Yayli N, Yilmaz N, Ocak M, Sevim A, Sesli E, Yayli N. Essential oil compositions of four mushrooms: *Scleroderma verrucosum*, *Cortinarius infractus*, *Hypholamacapnoides* and *Hypholama fasciculare* from Turkey. Asian J Chem 2007, 19(5), 4102–4106.

[4] Kepler RM, Sung GH, Harada Y, Tanaka K, Tanaka E, Hosoya T, Bischoff JF, Spatafora JW. Host jumping onto close relatives and across kingdoms by Tyrannicordyceps (Clavicipitaceae) gen. nov. and Ustilaginoidea(Clavicipitaceae). Am J Bot 2012, 99(3), 552–61.

[5] Smiderle FR, Baggio CH, Borato DG, Santana-Filho AP, Sassaki GL, et al.. anti-inflammatory properties of the medicinal mushroom *Cordyceps militaris* might be related to its linear (1R3)-b-D-Glucan. PLoS ONE 2014, 9(10).

[6] Shashikant M, Bains A, Chawla P, Sharma M, Kaushik R, Kandi S, Kuhad RC. In-vitro antimicrobial and anti-inflammatory activity of modified solvent evaporated ethanolic extract of *Calocybe indica*: GCMS and HPLC characterization. Int J Food Microbiol 2022, 2(376), 109741.

[7] Grienke U, Kaserer T, Pfluger F, Mair CE, Langer T, Schuster D, Rollinger JM. Accessing biological actions of *Ganoderma* secondary metabolites by *in silico* profiling. Phytochemistry 2015, 114, 114–24.

[8] Su HG, Peng XR, Shi QQ, Huang YJ, Zhou L, Qiu MH. Lanostane triterpenoids with anti-inflammatory activities from *Ganoderma lucidum*. Phytochemistry 2020, 173, 112256.

[9] Wang L, Li JQ, Zhang J, Li ZM, Liu HG, Wang YZ. Traditional uses, chemical components and pharmacological activities of the genus *Ganoderma* P. Karst.: A review. RSC Adv 2020, 18, 10 (69), 42084–42097.

[10] Usami A, Nakaya S, Nakahashi H, Miyazawa M. Chemical composition and aroma evaluation of volatile oils from edible mushrooms (*Pleurotus salmoneostramineus* and *Pleurotus sajor-caju*). J Oleo Sci 2014, 63(12), 1323–1332.

[11] Yayli N, Yilmaz N, Ocak M, Sevim A, Sesli E, Yayli N. Essential oil compositions of four mushrooms: *Scleroderma verrucosum*, *Cortinarius infractus*, *Hypholamacapnoides* and *Hypholama fasciculare* from Turkey. Asian J Chem 2007, 19(5), 4102–4106.

[12] Al-Fatimi M, Wurster M, Lindequist U. Chemical composition, antimicrobial and antioxidant activities of the volatile oil of *Ganoderma pfeifferi* Bres. Medicines (Basel) 2016, 28, 3(2), 10.

[13] Aisala H, Sola J, Hopia A, Linderborg KM, Sandell M. Odor-contributing volatile compounds of wild edible Nordic mushrooms analyzed with HS-SPME-GC-MS and HS-SPME-GC-O/FID. Food Chem 2019, 15, 283, 566–578.

[14] Chen X, Huynh N, Cui H, Zhou P, Zhang X, Yang B. Correlating supercritical fluid extraction parameters with volatile compounds from Finnish wild mushrooms (*Craterellus tubaeformis*) and yield prediction by partial least squares regression analysis. RSC Adv 2018, 31, 8(10), 5233–5242.

[15] Cho IH, Kim SY, Choi HK, Kim YS. Characterization of aroma-active compounds in raw and cooked pine-mushrooms (*Tricholoma matsutake* Sing.). J Agric Food Chem 2006, 23, 54(17), 6332–5.

[16] Cho IH, Lee SM, Kim SY, Choi HK, Kim KO, Kim YS. Differentiation of aroma characteristics of pine-mushrooms (*Tricholoma matsutake* Sing.) of different grades using gas chromatography-olfactometry and sensory analysis. J Agric Food Chem 2007, 21, 55(6), 2323–8.

[17] Lenzi J, Costa TM, da Silva FHH, Alberton MD, Goulart JAG, Tavares LLB. Antiprotozoal activity of mycelial extracts of several medicinal Agaricomycetes mushrooms against *Giardia duodenalis*. Int J Med Mushrooms, doi: 10.1615/IntJMedMushrooms.2022045354.

[18] de Pinho PG, Ribeiro B, Gonçalves RF, Baptista P, Valentão P, Seabra RM, Andrade PB. Correlation between the pattern volatiles and the overall aroma of wild edible mushrooms. J Agric Food Chem 2008, 12, 56(5), 1704–12.

[19] Misharina TA, Mukhutdinova SM, Zharikova GG, Terenina MB, Krikunova NI. The composition of volatile components of cepe (*Boletus edulis*) and oyster mushrooms (*Pleurotus ostreatus*). Prikl Biokhim Mikrobiol 2009, 45(2), 207–13.

[20] Huang SZ, Ma QY, Kong FD, Guo ZK, Cai CH, Hu LL, Zhou LM, Wang Q, Dai HF, Mei WL, Zhao YX. Lanostane-type triterpenoids from the fruiting body of *Ganoderma calidophilum*. Phytochemistry 2017, 143, 104–110.

[21] Bryant JM, Bouchard M, Haque A. Anticancer activity of ganoderic acid DM: Current status and future perspective. J Clin Cell Immunol 2017, 8(6), 535.

[22] Chandrawanshi NK, Verma S. Recent Research and Development in Stem Cell Therapy for Cancer Treatment: Future Promising and Challenges. In: Kumar S, Rizvi MA, Verma (Eds), IGI Global, USA, 2020, 514–533.

[23] Deepali, Kosre A, Chandrawanshi NK. Mushrooms Bioactive Compounds: Potential Source for the Development of Antibacterial Nanoemulsion. In: Ramalingam K (Eds), IGI Global, USA, 2021, doi: 10.4018/978-1-7998-8378-4.ch010.

[24] Chandrawanshi NK, Kosre A, Deepali, Mahish PK. Mushroom Derived Bioactive based Nanoemulsion Current Status and Challenges for Cancer Therapy. In: Ramalingam K (Eds), IGI Global, USA, 2021, doi: 10.4018/978-1-7998-8378-4.ch016.

[25] Cor Andrejc D, Knez Z, Knez Marevci M. Antioxidant, antibacterial, antitumor, antifungal, antiviral, anti-inflammatory, and neuro-protective activity of *Ganoderma lucidum*: An overview. Front Pharmacol 2022, 13,934982.

[26] Darshani P, Sen Sarma S, Srivastava AK, Baishya R, Kumar D. Anti-viral triterpenes: A review. Phytochem Rev 2022, 1–82.

[27] Elhusseiny SM, El-Mahdy TS, Elleboudy NS, Yahia IS, Farag MMS, Ismail NSM, Yassien MA, Aboshanab KM. *In vitro* anti SARS-CoV-2 activity and docking analysis of *Pleurotus ostreatus*, *Lentinula edodes* and *Agaricus bisporus* edible mushrooms. Infect Drug Resist 2022, 15, 3459–3475.

[28] Kadhila NP Evaluation of indigenous Namibian mushrooms and plants for antimalarial properties. UNAM, University of Namibia, 2019. http://hdl.handle.net/11070/2539.

[29] Lin YX, Sun JT, Liao ZZ, Sun Y, Tian XG, Jin LL, Wang C, Leng AJ, Zhou J, Li DW. Triterpenoids from the fruiting bodies of *Ganoderma lucidum* and their inhibitory activity against FAAH. Fitoterapia 2022, 158, 105161.

[30] Luo D, Xie JZ, Zou LH, Qiu L, Huang DP, Xie YF, Xu HJ, Wu XD. Lanostane-type triterpenoids from *Ganoderma applanatum* and their inhibitory activities on NO production in LPS-induced BV-2 cells. Phytochemistry 2020, 177, 112453.

[31] Choi S, Nguyen VT, Tae N, Lee S, Ryoo S, Min BS, Lee JH. Anti-inflammatory and heme oxygenase-1 inducing activities of lanostane triterpenes isolated from mushroom *Ganoderma lucidum* in RAW264.7 cells. Toxicol Appl Pharmacol 2014, 280(3), 434–42.

[32] Mothana RA, Jansen R, Jülich WD, Lindequist U. Ganomycins A and B, new antimicrobial farnesyl hydroquinones from the basidiomycete *Ganoderma pfeifferi*. J Nat Prod 2000, 63(3), 416–8.

[33] Chan YS, Chong KP. Bioactive compounds of *Ganoderma boninense* inhibited methicillin-resistant *Staphylococcus aureus* growth by affecting their cell membrane permeability and integrity. Molecules 2022, 27(3), 838.

[34] Mikolasch A, Hahn V. Laccase-catalyzed derivatization of antibiotics with Sulfonamide or Sulfone structures. Microorganisms 2021, 9(11), 2199.

[35] Karwa AS, Rai MK. Naturally occurring medicinal mushroom-derived antimicrobials: A case-study using lingzhi or reishi *Ganoderma lucidum* (W. Curt.:Fr.)P. Karst. (higher Basidiomycetes). Int J Med Mushrooms 2012, 14(5), 481–90.

[36] Sknepnek A, Pantić M, Matijašević D, Miletic D, Lević S, Nedović V, Niksic M. Novel kombucha beverage from lingzhi or reishi medicinal mushroom, *Ganoderma lucidum*, with antibacterial and antioxidant effects. Int J Med Mushrooms 2018, 20(3), 243–258.

[37] Baig MN, Shahid AA, Ali M. *In vitro* assessment of extracts of the lingzhi or reishi medicinal mushroom, *Ganoderma lucidum* (higher Basidiomycetes) against different plant pathogenic fungi. Int J Med Mushrooms 2015, 17(4), 407–11.

[38] Abdullah S, Oh YS, Kwak MK, Chong K. Biophysical characterization of antibacterial compounds derived from pathogenic fungi *Ganoderma boninense*. J Microbiol 2021, 59(2), 164–174.

[39] Chandrawanshi NK, Tandia DK, Jadhav SK. Determination of antioxidant and antidiabetic activities of polar solvent extracts of *Daedaleopsis confragosa* (Bolton) J. Schröt. Research J Pharm Tech 2018, 11(12), 5623–5630.

V. Shanthi and Shubha Diwan*

Chapter 8
Application of essential oils in industries and daily usage

Abstract: In the present time, people are getting more inclined toward natural commodities be it food, health, or medicine. One of the most exploited resources for natural well-being is plants and their essential olis (EOs). EOs are volatile substances which have numerous applications by virtue of important biologically active compounds like terpenes and other oxygenated derivatives. The applicability of EOs ranges from food, textile, cosmetics to pharmaceutical industries. EOs directly or indirectly are becoming a part of daily routine ranging from its use in skin or hair care products to our food items as flavoring agents. These wide range of applications of EOs are a consequence of its diverse properties and attributes. The use of EOs in food sector is well known and validated. The antimicrobial, anti-inflammatory, antiparasitic, and analgesic properties of EOs enable them to be exploited by pharmaceutical industries. They are getting used as complementary medicine to enhance the permeability of certain drugs through skin. They are widely used even in cosmetic and textile industries for the fragrances it impart to the formulations apart from other benefits. This chapter is designed to summarize the recent advancement in the researches and applications on EOs which will facilitate stakeholders to reveal unexplored benefits of EOs.

Keywords: EOs, flavoring agent, natural preservative, medicinal applications, extraction

8.1 Introduction

Owing to the technological modernization of our society, lifestyle disorders have become the most prevailing diseases in the present era. One of the primary causes of these disorders is unhealthy food habits that include processed foods, artificial sweeteners, energy drinks, and fast foods [1]. Moreover, may it be fruits and vegetables, processed food items, cosmetic products, or household items, most of the human populations are prone to chemicals utilized in improving the quality or processing of these items. Strong consumer awareness thus has developed about augmenting natural constituents in day-to-day requirements. Consumers are more inclined toward purchase of

*Corresponding author: V. Shanthi and Shubha Diwan,** Department of Microbiology and Biotechnology, St. Thomas College, Bhilai, Chhattisgarh, India, e-mail: shubha2315@gmail.com

https://doi.org/10.1515/9783110791600-008

products that are natural and in case of food items, minimally processed eatables are preferred. Products with "chemical free" or "natural" taglines have become trending advertising techniques by manufactures and certainly catching more attention from consumers in recent years. Essential oils (EOs) are such natural compounds that are being used since long for its associated aromas and gained significant attention due to its other remarkable biological properties.

EOs are volatile odorous liquids obtained from different parts of plants like leaves, stem, bark, flowers, buds, and seeds. They are the secondary metabolites of aromatic plants having complex mixture of bioactive compounds which impart incredible biological properties to it like antimicrobial, antioxidant, anti-inflammatory, analgesic, and insecticidal [2]. EOs are mainly composed of complex mixtures of hydrocarbons and oxygenated compounds whose composition varies in each plant species. There are a number of methods to isolate these oils from plants like cold pressing, solvent extraction, steam distillation, and hydrodistillation. However, to improvise the quality of EOs and to reduce extraction time newer technologies like microwave-assisted extraction, supercritical fluid extraction, and ultrasound-assisted extraction have been standardized in various plant species [3].

The remarkable biological properties of EOs popularize its widespread application in industries. The global EO market size was valued at USD 10.3 billion in 2021 and is expected to grow at a compound annual growth rate of 9.3% from 2021 to 2026 [4]. It is used as flavoring agent and natural preservative in food product thus offering safety and increased shelf life to it. Moreover, due to their peculiar aroma, EOs are an important part of perfume and fragrances industries and also impart pleasant fragrances to cosmetic products, cleaning products, and industrial solvents. EOs are also utilized in pharmaceutical industries and in aromatherapy due to their positive effects on human health [5]. Out of 3,000 EOs identified from different plant species, about 150 are found to be commercially important and are sold in global market [6]. Thus, the review will give an insight on the recent developments in the EOs research and their potential application in daily life and industries.

8.2 Components of essential oil

EOs are primarily secondary metabolites with a complex chemical composition. The primary metabolites; carbohydrates, proteins, lipids, and nucleic acids contribute very little to the production of EOs. However, some EOs may be considered as degradative products of lipids. Various factors contribute in influencing the chemical composition of EOs, environmental factors being the most significant.

Every EO has its own complex chemical profile which decides its odor, taste, and ultimately its applicability. All the chemical components work synergistically to make it effective as an EO. Almost all EOs are basically composed of two major components: hydrocarbons and oxygenated compounds. Hydrocarbons constitute the most significant portion of all EOs and the most important hydrocarbon being the terpenoids which make up almost 80% of the EO composition. The other lesser significant component being oxygenated compounds.

8.2.1 Terpenes

One of the essential and important constituents of Eos are terpenes. They are made of 5-carbon units called isoprene. Some of the terpenes based on number of isoprenes are monoterpenes (C_{10}), sesquiterpenes (C_{15}), and diterpenes (C_{20}). Each EO has its own unique and significant fragrance, flavor, and piquant, all of which are contributed by the highly aromatic terpenes; the most abundant being monoterpenes and sesquiterpenes. Monoterpenes are the predominant terpenes in EOs. They are small, low molecular weight compounds. These highly volatile and non-viscous components impart strong aroma in EOs. Sesquiterpenes are big molecules with high molecular weight. Although their quantity is significantly less than monoterpenes, their increased stability is responsible for less volatility and enhanced aroma of EOs. Both monoterpenes and sesquiterpenes possess significant biological activity thereby making EOs of therapeutic importance.

8.2.2 Oxygenated compounds

All EOs in addition to terpenes also contain nonterpenic compounds called oxygenated compounds, whose composition varies on types of EOs. These compounds apart from possessing a variety of bioactive properties offer flavor and odor to EOs. They include esters, alcohols, aldehydes, esters, ethers, oxides, phenols, acids, acetals, and lactones.

EOs are estimated to contain about 300 different types of compounds with myriad biological properties making them potentially applicable and valuable in various domains of industries.

8.3 Essential oils in pharmaceutical industry

EOs are important in pharmaceutical industries because of their ability to demonstrate various properties like anti-inflammatory, antiseptic, antimicrobial, anticarcinogenic, and many more. All these properties enable them to be emphasized for development and discovery of novel drugs and antimicrobial agents. Many different EOs possess inhibitory action against a range of microbial etiological agents and therefore are being used for the treatment of various diseases since ancient times. However, recently pharmaceutical industries engaged in production of different antibiotics and drugs are facing the grave problem of multidrug resistance. Hence, the antimicrobial components of various EOs may offer a decent opportunity to overcome the multidrug resistance problem and can form an integral part of novel drug discovery program due to their effectiveness against broad spectrum of microbes.

8.3.1 Essential oils as antioxidants

Aging is a natural phenomenon and associated with aging and reportedly other degenerative diseases like cancer, diabetes, and cardiovascular diseases are the leading problem of oxidative stress. Moreover, in the recent years, humans are constantly at high risk due to their exposure to exogenous sources of free radicals. The problem of oxidative stress is usually dealt with antioxidant defense mechanisms of the body and of course defenses provided by dietary ingredients. Among the dietary ingredients is EOs whose terpene components possess antioxidant potential due to their free radical scavenging capacity. Many EOs sourced from aromatic species of Lamiaceae family are rich in antioxidants. Some EOs also contain phenols which are responsible for ameliorating the negative effects of free radicals. The active constituents in EOs are set to exhibit myriad capacities to manage the oxidative stress like reduction of metal ion, scavenging of free radicals, and preventing/delaying of lipid oxidation.

Strongest antioxidant activities are exhibited by EOs containing phenolic groups or metabolites carrying conjugated double bonds. For instance, cinnamon, clove, thyme, oregano, and basil EOs have superior antioxidant properties [7] due to the presence of phenolic groups. It is believed that phenolic structures are significant in decomposition of peroxides and neutralization of free radicals owing to their strong redox properties. Nevertheless, the antioxidant properties of EOs are also contributed by other constituents like alcohols, ketones, aldehydes, and ethers.

Many chronic diseases like cancer, diabetes, asthma, brain dysfunction, and atherosclerosis are a result of cellular damage caused by free radicals. Hence, the free radical scavenging property of EOs plays a pivotal role in pharmaceutical industries for prevention or treatment of such diseases. Some of the significant EOs with antioxidant properties are geranial, isomenthone, α-terpinene, menthone, citronellal, 1,8-cineole, α-terpinolene, and so on.

One of the chronic diseases which has been attracting the therapeutic potential of EOs is cancer. These pharmacologically important EOs constituents possess myriad anticancer or cancer protective attributes. EOs may display these properties by following different ways:
- Activating detoxifying enzymes in lungs [8]
- Inhibiting carcinogenesis by activating glutathione S-transferases [9]

- Inducing apoptosis of immature nerve cells which otherwise may cause neuroblastoma [10]
- Decreasing resistance of cancer cells to anticancer drugs [11]
- Lowering toxicity of anticancer agents to healthy cells [12]
- Exhibiting cytotoxic activity toward cancerous cells [13]

All these mechanisms of actions explain the therapeutic values of EOs as cancer protective agents. Although EOs alone cannot be considered for targeting cancerous cells, but their role in fighting cancer in combination with conventional therapeutic strategies however cannot be ignored.

8.3.2 Essential oils as skin penetration boosters

There are many methods of administering drugs into the human system. Some are delivered intravenously, some subcutaneously and others intramuscularly. However, there are several reasons for drugs to be delivered topically or through transdermal route. One of the reasons might be that the drugs need to avoid the initial passage through hepatic system, some drugs might need to maintain constant concentration in the blood plasma or some drugs probably might need to be discharged quickly after their metabolism through urine. Some drugs which are required to exhibit low metabolism are usually administered through transdermal route. However, the human skin acts as a natural barrier for many such drugs.

The special structure of skin layers especially the outermost layer functions in protecting the underlyingskin tissues from dehydration, infection, and other factors like stress. This ability of the skin epidermal layer restricts the effective transdermal penetration of many drugs and hence drug penetration enhancers are required which work by disrupting the skin lipid layer and/or by promoting partitioning. Skin penetration enhancers act in a temporary reversible manner thereby enhancing diffusion of drugs effectively and faster [14]. The most important factor which decides superiority of skin penetration enhancers is that it should exhibit both biocompatibility and also drug compatibility. It should be pharmacologically inert with good solvent properties.

There are many chemical compounds which have been detected to possess penetration enhancing properties. However, their poor skin permeability requires them to be used in higher doses to maintain their activity which no doubt reportedly cause severe side effects like skin irritation and cytotoxicity. The use of these less efficient chemical penetration enhancers is limited sighting safety issues.

EOs serve as good alternatives as skin penetration enhancers for both hydrophilic and lipophilic drugs. The constituents in EOs improve the permeation of drugs and some are reportedly known to enhance their activity too. They are considered GRAS (generally regarded as safe) thereby aiding as transdermal delivery systems for drugs.

Some EOs serve as successful penetration enhancers by decreasing the polarity of lipophilic drugs (Table 8.1). Some help in removing the lipid layer on the skin and others help in enhancing penetration by triggering denaturation of keratin layer thereby causing reversible skin protein composition change for easy diffusibility of drugs.

Table 8.1: Essential oils as skin permeation boosters.

S. no.	Plant name	Bioactive constituents	Permeation booster used for drug
1	*Aloe vera*	3,6-Octatriene, 3-cyclohexanol-1-methanol, bornylene	Losartan potassium (antihypertensive drug)
	Cuminum cyminum	β-Pinene, ocimene, γ-terpinene, safranal	
	Melaleuca alternifolia	γ-Terpinene, terpinene-4-01, α-pinene	
2	*Cinnamomum verum*	Cinnamaldehyde, coumaric acid, eugenol	Ibuprofen (NSID)
3	*Syzygium aromaticum*	Eugenol, caryophyllene, eugenyl acetate	
4	*Eucalyptus globulus*	1,8-Cineole, *p*-cymene, eucamalol, linalool, citronellol	Chlorhexidine digluconate (antiseptic) 2,3,5,6-tetramethylpyrazine (for treating cardiovascular disorders)
5	*Mentha piperita*	Menthol, menthone, menthyl acetate, β-pinene	Ketoconazole (antifungal)
6	*Citrus limon*	Terpenes, limonene, citral, geranyl acetate	Felodipine (antihypertensive)
7	*Boswellia serata*	α-Pinene, myrcene, sabinene, limonene, α-thujene	Chuanxiong
	Commiphora myrrha	β-Elemene, β-bourbonene	
8	*Pogostemon cablin*	γ-Guaiene, seychellene, α-patchoulene, α-bulnesene	Indomethacin (NSID)
9	*Salvia rosmarinus*	α-Pinene, 1,8-cineole, camphor	Diclofenac sodium (NSID)
10	*Sinapsis alba*	Allylisothiocyanate, thymol, limonene, octadecacene	5-Fluorouracil (antineoplastic), paeonol (anti-inflammatory)

Table 8.1 (continued)

S. no.	Plant name	Bioactive constituents	Permeation booster used for drug
11	*Guatheria procumbens*	Methyl salicylate, triacontane, formaldehyde, gawtheriline	Geniposide (hepatoprotective), (antidiabetic)
12	*Zanthoxylum bungeanum*	Limonene, linalool, linalyl anthranilate, 4-terpinenol	Indomethacin (NSID)

8.3.3 Essential oils as anti-inflammatory agents

Another property of EOs which makes their applicability for pharmaceutical use more appealing is there anti-inflammatory and analgesic property. Infection or injury leads to an inducible protective response called inflammation which helps combat external invaders of the body. A typical inflammatory response causes generation of free radicals called reactive nitrogen species (RNS) which are responsible for combating the invading organism. Nevertheless, overproduced RNS may also damage the inflammatory sites and trigger the action of a series of unwanted reactions.

There are currently many chemical agents especially nonsteroidal anti-inflammatory agents in use. Many diseases like arthritis, diabetes, and other autoimmune diseases demand regular use of these nonsteroidal agents and other corticosteroids triggering a number of serious side effects. Hypersensitivity, stomach ulcers, weight gain, and indigestion are a few to mention. Reports suggest that prolonged usage of corticosteroids lead to malfunctioning of various important glands and renal disorders. EOs act as noninvasive therapeutics and provide an excellent strategy for the treatment of chronic inflammatory responses. They are currently being considered to provide and edge over conventional chemical anti-inflammatory agents.

Infection caused by many bacterial agents especially gram-negative pathogens induces inflammation. This lipopolysaccharide-induced inflammation is the result of the synthesis of prostaglandins by blood monocytes. There are many isoenzymes of cyclooxygenases responsible for the constitutive synthesis of prostaglandins. However, cyclooxygenase-2 isoform is only induced upon inflammatory stimulation due to lipopolysaccharides of gram-negative pathogens indicating its role as inflammation-specific enzyme. Hence, treatment strategies for many inflammatory disorders are currently targeting cyclooxygenase-2 inhibitors. Nevertheless, the nonsteroidal anti-inflammatory drugs are proving unsuccessful due to reports of risk of myocardial infarction, stroke, and other cardiovascular side effects. Studies have revealed the role of EO constituents to drastically reduce the synthesis of prostaglandins, for instance; EOs from *Cha maecyparis obtusa* reportedly possessed ameliorating effects as an anti-inflammatory agent by affecting prostaglandins synthesis by blood monocytes. The terpenes, elemol, and sabinene present in *Chamaecyparis obtusa*, *Cryptomaria japonica*, and *Hyptis pectinata*, reportedly are said to be responsible for selectively targeting cyclooxygenase-2 enzyme and hence inhibiting prostaglandin synthesis. Also, these terpenes were also responsible for inhibiting the production of nitric oxide, cytokines, and interleukins (IL-1 and IL-6); all of which induce inflammation [15, 16]. Similarly EOs of lemon grass mediate its anti-inflammatory effect by suppressing cyclooxygenase-2 promoter thereby inhibiting cyclooxygenase-2 gene expression [17].

EOs function as anti-inflammatory agents by interacting with anti-inflammatory cascade response involving cytokines and other factors like nitric oxide and prostaglandins (Table 8.2). The anti-inflammatory attributes of EOs can be explained by their effect on arachidonic acid metabolism or on cytokine production or by modifying the pro-inflammatory gene expression patterns.

Table 8.2: Essential oils as anti-inflammatory agents.

S. no.	Plant name	Bioactive constituents	Use
1	*Salvia officinalis*	Linalool, linalyl acetate, germacrene D, 4 geranyl	Regulates menstrual problems, eases tension muscle cramps, and controls dry skin, acne, and wrinkles.
2	*Evcalyptus globulus*	Cineole, terpinene, cymene, phyllandrene, pinene	Regulates nervous system, headache, and used to treat respiratory diseases.
3	*Pelargonium graveolens*	Eugenol, geranic, citronellol, linalol, methone, sabinene	Best perfume, used in soaps, control emotions, anxiety, stress, and sedative.
4	*Lavandula officinalis*	Camphor, β-ocimene, 1-8-cineole, terpinene-4-01	Causes depression, sedative, cures sleep pattern, increases mental alertness, and suppresses aggression.
5	*Citrus limon*	Terpenes, limonene, phyllandrene	Astringent, detoxifying properties, causes blemishes, and boosts immune system.
6	*Mentha piperita*	Carvenol, menthol, carvone, menthone	Relieves pain spasms, arthritis.
7	*Anthemis mobilis*	Esters of angelic acid, tiglic acid, 2-Methyl butanoic acid	Calms emotions, anti-inflammatory, relieves anxiety, and stress.
8	*Rosmarinus officinalis*	Borneol, camphor, cineol, pinene, camphene	Relieves indigestion, pain, sleep disorders, and skin toxic.
9	*Melalevca alternifolia*	Terpinen-4-01, α-sabine, cineole	Cures insect bites, dandruff, and herpes.
10	*Cananga odorata*	Geranyl acetate, linalol, geraniol methyl chavicol	Retards heart beat, antidepressant, feeling of well-being.

8.3.4 EOs as antibacterial agents

Since historical times EOs are being used to treat various types of diseases which can be attributed to their antimicrobial property. But presently, man is not only tackling with new diseases but also is grappling with the issue of multidrug resistance. Many EOs contain constituents which possess antiviral, antiparasitic, antifungal, and antibacterial properties. They are being considered as potential source of antimicrobial compounds as they have the ability to address the problem of antibiotic resistance. EOs and their constituents for that matter are hydrophobic in nature, a property which enables them to partition with cell membrane and mitochondrial lipids of the invading organism making the membrane more permeable. This disturbance in membrane integrity renders loss in cell structure and eventually death. Some constituents are involved in modulating the efflux mechanism of organisms especially gram-negative bacteria affecting their drug resistance pattern [18]. EOs has an edge over antibiotics as antimicrobial agents as they have a complex composition which makes development of resistance tougher against them. Also, antibiotics in order to exhibit their antimicrobial activity need to make direct contact. But on the other hand, some EOs need not make direct contact and also since they can exist in highly active vapor phase can function as antifungal agents [19].

Another significant feature of EOs which makes them more applicable as antibacterial agents is their wide range of chemical compounds enabling them to have multiple targets. Some EOs act by disintegrating the outer membranes, and some act by depleting the intracellular ATP while some act deep inside the target cells by gaining access through the periplasm. It is believed that EOs abundant in phenols or aldehydes usually are better antibacterial agents. Thymol, eugenol, citral, and cinnamaldehyde are some phenol and aldehyde derivatives of EOs which possess highest antibacterial activities followed by EOs containing alcohol derivatives [20]. The high antibacterial activity of phenol/aldehyde containing EOs can be attributed to the presence of hydroxyl group which are the prime source of disrupting the periplasmic membrane and hence, cell death due to leakage of cell contents. Many important pathogens like *Pseudomonas*, *Bacillus*, *Staphylococcus*, *Listeria*, *Escherichia*, *Salmonella*, *Clostridium*, *Lactobacillus*, and *Helicobacter* are acted upon by phenolic/aldehyde-containing EOs suggesting that such EOs display antibacterial effect against wide range of bacteria which include both gram-positive and gram-negative species. However, presence of outer membrane and lipopolysaccharide structures in gram-negative bacteria makes the action of EOs less effective probably due to the restricted movement of hydrophobic compounds across the lipopolysaccharide.

8.4 Application of essential oils in food industry

Food is the basic necessity of mankind. Food commodities need to be prevented from microbial infestation during its postharvest storage. Moreover, with the modernization of society and technological advancements, people are inclined more toward precooked, processed food that can save both labor and time. Maintenance of food quality amid its processing and storage is the major challenge of food industry. Also, rising demand of good quality food product with natural ingredients has prompted the use of EOs in food products. EOs are mainly augmented in food products to mainly serve two purposes first to add flavors to the food and second as natural preservative due to its antimicrobial and antioxidant properties. EOs have already been used to enhance flavors of food for centuries. Due its strong aroma, EOs are used to mask the odd flavors generated due to processing of food, thereby increasing the palatability of product. The use of EOs as a food additive has been approved as substances generally recognized as safe by US Food and Drug Admistration [21].

Owing to the various other potential characteristic properties of EOs like antibacterial, antifungal, and antioxidant it has widely used in food industries and even myriad of researches are still revealing the potential of newer uncommon EOs to be utilized in food industries (Table 8.3). Antimicrobial agents prevent the natural spoilage of food and also controls the growth of invading microorganism thus offering both preservation and safety of food [22]. The efficacy of EOs as natural food preservative has been tested in many food products like bakery products, fruits and vegetables, and meat products. Mishra et al. [23] reviewed the literature reporting the potential of EOs in dairy products. It was found that EOs from oregano, clove, thyme, and orange had powerful inhibitory activity against bacteria and yeasts. Addition of EOs to dairy products both enhanced the aroma of product and imparted antimicrobial properties to it making it more consumer acceptable. Similarly addition of EOs to fruits and vegetables helps to extend its shelf life. The primary problem of associated with fruits and vegetables is its fungal spoilage. Researchers have come up showing reduction in microorganism count by the use of EOS in fruits. Viuda-Martos et al. [24] reported significant antifungal activities of clove, thyme, and oregano against *Aspergillus niger* and *Aspergillus flavus* which are related to food spoilage. Incorporation of EOs in fruit juices along with physical treatment like mild heat or pulsed electric field can help in preservation of fruit juices for longer time [25]. In addition to abovementioned categories of food products meat and its products also hold a significant share in processed food industries and are very vulnerable to spoilage. Shaltout and Koura [26] reported the use of thyme oil at concentration of 1% and cinnamon oil at 2% as preservative of minced meat to increase its shelf life. Antibacterial activity of basil EO was evaluated in Italian-type sausage by Gaio et al. [27]. In the study the activity of EOs was tested against 18 microorganisms and for all the microorganisms EOs was effective except *Pseudomonas aeruginosa*. The minimum inhibitory concentration was found between 0.25 and 1.00 mg/g.

EOs inspite of having an array of beneficial characteristics, its volatibilty and hydrophobic nature makes it difficult to be directly used in food systems. Nanotechnology has been reported as a solution for delivery system of EOs that could protect the active compounds against degradation and can enhance its efficacy in food system. Nanoencapsulation protects EOs against deteriorating factors like light, moisture, pH, and also during processing and storage. It also helps to solubilize lipophilic compounds in aqueous media and to release them in the target location [28]. Liao et al. [29] reviewed the research work reported in relation to the methods of nanoencapsulation of EOs for their improvised delivery into food products. The efficacy of various methods has been tested but spray-drying and freeze-drying are widely used in the food industry [30]. Delivery of EOs in fruit juices has been tested by Bento et al. [31], by chitosan-based nanotechnology where they encapsulated sweet orange EO using chitosan nanoemulsions as nanocarrier and evaluated its antimicrobial activity in combination with mild heat and found the strategy effective for bactericidal action in fruit juices.

Table 8.3: Potential uses of some essential oils in foods.

Essential oil	Scientific name	Parts used	Uses	References
Basil	*Ocimum basilicum*	Aerial part	Flavoring, antimicrobial, and antioxidant agent	Li and Chang [32]
Cinnamon	*Cinnamomum zeylanicum*	Bark and leaves	Flavoring, antimicrobial, and antioxidant agent	Cardoso-Ugarte et al. [33]
Clove	*Syzygium aromaticum* L.	Flowers	Flavoring, antimicrobial, and antioxidant agent	Haro-González et al. [34]
Cumin	*Nigella sativa*	Seeds	Flavoring and food preservative	Hassanien et al. [35]
Fennel	*Foeniculum vulgare* Mill.	Seeds	Flavoring, antimicrobial, and antioxidant agent	Diao et al. [36]
Garlic	*Allium sativum* L.	Cloves	Flavoring, antimicrobial, antioxidant agent	Khan et al. [37]
Lavender	*Lavandula* species	Flower	Antimicrobial	Wells et al. [38]
Lemon	*Citrus limon* L.	Fruit	Flavoring, preservative	Klimek-Szczykutowicz et al. [39]
Lemon Grass	*Cymbopogon* sp.	Leaves	Preservative	Faheem et al. [40]
Oregano	*Origanum* sp.	Aerial part	Flavoring, antimicrobial, antioxidant agent	Rodriguez-Garcia et al. [41]

Table 8.3 (continued)

Essential oil	Scientific name	Parts used	Uses	References
Peppermint	*Mentha piperita* L.	Leaves	Flavoring	Nayak et al. [42]
Rosemary	*Rosmarinus officinalis* L.	Leaves	Flavoring, antimicrobial, and antioxidant	Hernández et al. [43]
Thyme	*Thymus vulgaris* L.	Leaves	Flavoring, antimicrobial, and antioxidant	Stahl-Biskup and Venskutonis, [44]

8.5 Application of EOs in cosmetics

Cosmetics have become an obligatory part of modern life style. Its trending usage by both women and men has ignited the global cosmetic demand of 5.3% from 2021 to 2027. The market size valued in 2019 was $380.2 billion [45]. Cosmetic manufacturers are always innovative in advertising the product that can seek consumers attention. EOs have proved to be an effective promotion tool as being a natural compound it is easily acceptable while at the same time justifying its use by its incredible properties. In Cosmetic industry, EOs are majorly used for their pleasant fragrances and as natural preservative. Fragrances are important factor in selection of any cosmetic product and also it aids to overcome the unpleasant odor of other ingredients of cosmetic formulations. Some of the high-valued EOs used as fragrances are citrus, lavender, eucalyptus, tea tree, and other floral oils while linalool, geraniol, limonene, citronellol, and citral are much-appreciated fragrance components used in different cosmetics [46].

Due to the presence of pharmacologically important compounds in EOs it has antibacterial, antifungal, and antioxidant properties which impart various beneficial effects to skin and hair (Table 8.4). It can also be exploited in cosmetic products as natural preservative which can reduce the addition of synthetic preservative in the products. Muyima et al. [47] tested the potential of *Artemisia afra*, *Pteronia incana*, *Lavandula officinalis*, and *Rosmarinus officinalis* EOs as natural cosmetic preservatives in an aqueous cream formulation. The results showed the potential of plant EOs to be used as natural preservatives against *E. coli*, *Ps. aeruginosa*, *S. aureus*, *Ra. pickettii*, *C. albicans*, and *A. niger* in cosmetic products.

Even though EOs have always proved to have beneficial effects on skin and hair, new researches are always coming up that keep testing uncommon EO properties and manufactures can improvise and upgrade their product and its advertising strategy by adding newer ingredients to their product. Choi et al. [48] reported the antiwrinkle activity of *Chrysanthemum boreale* MAKINO EO (CBMEO). EO and juice from bergamot and sweet orange was found to improve *Acne Vulgaris*, prevalent dermatologic disease effecting 80% of young individuals [49]. EOs were found to decrease the growth rate of sebaceous gland spots and reduce the release of inflammatory cytokines thereby promoting apoptosis in the sebaceous gland. *Curcuma longa* EO proved to have antiaging effects on skin. It was found to reduce cutaneous photoaging in a UVB-irradiated nude mouse model suggesting its application in cosmetic products [50].

Table 8.4: Application of essential oils in skin and hair care.

Application	Essential oil	Functions	References
Skin care	Sandalwood	Conditions and brightens skin, and antiaging	Francois-Newton et al. [51]
	Chamomile	Antiaging and antiacne	Guzmán and Lucia [52]
	Rosemary	Skin hydration and antiaging	Montenegro et al., [53]
	Bergamot	Antiacne and repairs UV-induced damage	Forlot and Pevet [54]
	Sweet orange	Antiacne	Sun et al. [49]
Hair care	Peppermint oil	Hair growth	Oh et al. [55]
	Rosemary	Promotes hair growth and nourishes the hair	Panahi et al. [56]
	Bergamot	Antidandruff	Abelan et al. [57]
	Lavender	Hair growth and conditioning	Guzmán and Lucia [52]
	Tea tree	Antidandruff	Yadav et al. [58]
	Lemon grass	Strengthens hair follicles and promotes hair growth	Goyal et al. [59]

8.6 Essential oils in active and intelligent food packaging industries

The rise in living standards, increase in awareness, and consciousness about health is making people opt for packaged foods. Packaged food industry is an important global business. Packaging is intended to keep the food safe, fresh, and hygienic till it reaches the customer along the supply chain. Hence, packaging materials play crucial role in protecting the food product from external environment. Packaging industries were earlier dominated with synthetic polymers like polyvinylchloride polyethylene terephthalate and other petrochemical-based plastics. Nevertheless, health and environmental concerns paved way for ideas called active and intelligent packaging. The strategy of active packaging basically involves prolonging the shelf life of foods by arresting microbial growth, preventing food oxidation, removing moisture, and preventing other food safety risks. Whereas intelligent packaging encompasses materials with the ability to monitor the condition of the packaged food.

These advancements in active and intelligent packaging have led to the incorporation of various chemicals as additives in the packaging material. However, the presence of chemicals as an active agent in the packaging material such as ethylene absorbers to arrest ripening of fruits, moisture control agents to control water activity and thereafter microbial growth, oxygen scavengers to the decelerate oxidative reactions which otherwise may cause food deterioration and carbon dioxide absorbers or emitters to retard microbial growth in meat and poultry products, is proving a matter of concern. These chemicals can cause adverse health effects therefore leading to novel, yet sustainable alternatives such as incorporation of natural bioactive compounds. EOs are currently being considered as the promptest solution and alternative to this.

EOs are being considered as active ingredients in active food packaging due to their antimicrobial and antioxidant properties prolonging the shelf life of foods. EOs are recently being incorporated into active packaging as coatings which are applied on food surface or as films used as wrapping material or covers. Incorporating EOs into packaging material provides many benefits like

1. Improvement of tensile strength: This probably could be due to the role of phenols present in EOs in reorganizing and improving the cross-linking pattern of packaging matrix.
2. Improvement of barrier and permeation properties: These properties are important for maintaining increased product quality and is dependent on hydrophobicity and hydrophilicity ratio of the packaging material in order to control the vapor pressure. The hydrophobic nature of EO constituents when incorporated into hydrophilic packaging matrix enhances the barrier properties.
3. Improvement in optical properties: Optical properties of packaging material effect the protective function of food product and hence various optical properties like color, gloss, and transparency are significant. Incorporation of EOs into the packaging material not only impacts the color of the material but also functions as barriers against UV light.
4. Improvement of food quality: EOs possess antioxidative properties and when these EOs are incorporated into the packaging material the antioxidants present in the EOs prevent the food from spoilage due to oxidation. They also contain antibacterial agents which increase the shelf life of food product preventing microbial growth.

In meat and poultry industries, incorporating EOs into the packaging material with antioxidant property is said to release them into the product, preventing or delaying lipid oxidation. The antioxidants do so by binding or absorbing oxygen in and around the product. This use of EO in such industries is proving a bone and good cost-effective alternative to prevent foodborne illness and outbreaks. Many EOs derived from different plants are being used to be incorporated into the packaging material as they are considered as GRAS by FDA. For example, *Rosmarinus officinalis*, *Matricaria chamomilla*, and *Ocimum basilicum*.

EOs from bergamot and EOs containing linalool, which have a strong antibacterial effect, are being used in packaging material for food products like chicken and cabbage to prevent food spoilage by *E. coli*, *B. cereus*, and *Listeria monocytogenes*. Food packaging materials incorporated with cinnamon EO with antifungal properties are being used to prevent growth of important fungi like *Aspergillus*, *Penicillium*, and *Eurotium* which otherwise deteriorate food quality especially acidic foods.

Horticultural products are highly prone to fast deterioration and hence huge loss due to wastage. Food packing industries have worked out on this problem to minimize economic loss incurred due to spoilage and wastage of fruits and vegetables. Currently the packaging material or the corrugated cardboardsare widely being used as inexpensive and sustainable packaging material incorporated with encapsulated EOs. These encapsulated EOs are engaged in release of antimicrobial agents which are controlled by relative humidity and temperature within the packaging material thereby preventing these perishable products from spoilage. EO which are incorporated into foot packaging material have the ability to be released in a slow and control manner into the food to display their efficacy as an agent to suffice their shelf life and also in maintaining the organoleptic properties of the food [60].

8.7 Essential oils in textile industry

Functional textiles are the textiles materials developed to impart new functions to the product to meet the end use requirement apart from its basic use. The functional finishing decipher new properties to the fabric such as antimicrobial activity, antibiotic, antiwrinkle, UV resistance, photocatalytic activity, and flame retardency [61]. EOs have a great potency to be used in textile finishing due to the "green" and biodegradable nature which can possibly augment or replace the chemical agents used in textile industries. They are becoming increasingly popular because of their low toxicity and strong costumer approval. EOs are mainly reported to be used as functional agents for making fragrant textiles, antimicrobial textiles, biomedical textiles, and mosquito-repellent textiles.

8.7.1 Aroma-finished textiles

Fragrance finishing of textiles is often useful in aromatherapy. Perfumed textiles even provide greater acceptance by end consumer due its pleasant fragrances. Srivastava and Srivastava [62] impregnated citronella and lemongrass oil on wool and silk fabric and outcomes were assessed by olfactrometery analysis which reported long retention of aroma upto 20 wash cycles. Khanna et al. [63] infused EOs of lavender, eucalyptus, peppermint, jasmine, cedarwood, and clove to cotton fabric directly and with anchors as cyclodextrin both in native and modified form like monochlorotriazine-β-cyclodextrin. These aromated fabrics were studied for the retention time of oils in the fabrics and found that the presence of anchoring host provided better retention time and slower release of oils form fabric as compared to direct application of EOs.

8.7.2 Mosquito-repellent textiles

Various EOs have also been reported to have mosquito-repellent properties and used widely due its eco-friendly nature [64]. A number of researches have been reported with infusing EOs in fabrics to develop mosquito-repellent textiles and also found to be functional. Specos et al. [65] developed microcapsules containing citronella EO applied it to cotton textiles. Assessment of repellent activity of fabric was done by exposure of a human hand and arm covered with the treated textiles to *Aedes aegypti* mosquitoes. Tested fabric showed a long-lasting protection from insects. Litsea, lemon, and rosemary EOs were applied to polyester and cotton fabrics by Soroh et al. [66] to access their mosquito-repellent activity which was found to be 71.43%.

8.7.3 Antimicrobial textiles

External factors like high humidity, temperature, and insufficient air circulation enhance the growth of microorganisms in textiles containing natural fibers, for example, cotton and linen fabrics, flax, and hemp nonwovens [67]. EOs are becoming increasingly popular to develop antimicrobial textile because of being natural, safe, and eco-friendly perspective. Antimicrobial textiles are useful in apparels like jackets, sanitary pads, sportswear, and winter wear; health care like lab coats, bandage, mask, and protective kits and in households like curtains, covers, carpet, and towels. Walentowska and Foksowicz-Flaczyk [68] tested thyme EO on linen cotton-blended fabric and linen fabric to form antibacterial textile. Thyme EO applied as 8% concentration in methanol to linen cotton-blended fabric showed very high antibacterial and antifungal activity. Chitosan microencapsulation of sandalwood and eucalyptus EOs was reported by Javid et al. [61] to enhance the functional properties of cotton fabric. They carried out emulsive fabrication of chitosan microcapsules encapsulating EOs which was applied on a bleached, desized mercerized cotton fabric. The finished fabric retained antibacterial activity and also the activity increased with increasing concentration of EOs.

8.7.4 Biomedical textiles

Medical textiles are formulated to be used to provide wound healing and other medical benefits of EOs to the consumers. Their use may range from simple bandages to prostheses of body implants such as artificial heart, heart valve, blood vessel, and skin. Copaiba EO were added to cotton fabric by Aruda et al. [69], which was tested for its healing enhancement potential. Gong et al. [70] prepared thermosensitive hydrogel and micelles loaded with curcumin and was prepared and applied for cutaneous wound repair. Rafiq et al. [71] developed of sodium alginate/PVA antibacterial nanofibers by the incorporation of cinnamon, clove, and lavender EOs. Nanofibers with 1.5% cinnamon oil showed highest antimicrobial properties and the nanofibrous-coated cotton gauze validated its potential to be used for wound dressings due its considerable liquid absorption and antibacterial activity.

8.8 Conclusion

EOs are volatile substances of plant origin. They have plethora of applications in many areas ranging from medicine, pharmaceuticals, cosmetics, perfumes, food and food packaging, general health, therapeutics, and so on. EOs are composed of terpenes, alcohols aldehydes, esters, and many other bioactive components. These bioactive and helpful components of EOs are currently receiving enormous attention making them widely applicable in various industries. EOs are finding immense applicability in food industries either as an active ingredient of food product or as an important constituent of food packaging materials. EOs are characterized by their strong aroma and fragrance, and this very asset makes their applicability more prominent in cosmetic and perfumery industries. EOs and their distinctive bioactive compounds are being considered as effective and safe alternative therapeutics to conventional therapeutic strategies. The antioxidant, anti-inflammatory, antimicrobial, and antiparasitic attributes of EOs have extended the applicability to health care, personal care, food, and pharmaceutical industries. Therefore, EOs which are bestowed with natural, useful biological and medicinal properties are hence making consumers inclined toward them and finding permanent place in modern world. But still many more EOs still needs to be explored which can be utilized in industries as alternative to the synthetic chemicals and even in day-to-day life. Yet some researchers report a number of constraints in the use of EOs commercially. The researchers need to be more focused toward the toxicity profiling of components of EOs so health promoting benefits can be availed from EOs easily and safely.

References

[1]   National Health Portal, 2015. http://nhp.gov.in/lifestyle-disorder_mtl#:~:text=The%20most
        %20common%20lifestyle%20disorders,also%20fall%20under%20this%20category.
        (Accessed on 25 May 2022)
[2]   Ribeiro-Santos R, Andrade M, Sanches-Silva A, de Melo NR. Essential Oils for Food
        Application: Natural Substances with Established Biological Activities. Food Bioprocess
        Technol 11, 2018, 43–71.
[3]   Stratakos AC, Koidis A. Methods for Extracting Essential Oils. In: Essential Oils in Food
        Preservation, Flavor and Safety. Elsevier, London, UK, 2016, 31–38.
[4]   Markets and Markets, 2021, Published March 2021 https://www.marketsandmarkets.com/
        Market-Reports/essential-oil-market-119674487.html(accessed on 31 May 2022)
[5]   Burt S. Essential oils: their antibacterial properties and potential applications in foods—a
        review. Int J Food Microbiol 94, 2004, 223–253.
[6]   Barbieri C, Borsotto P. Essential Oils: Market and Legislation. In: El-Shemy HA (Ed), Potential
        Essent Oil IntechOpen; London, UK, 2018, 107–127.
[7]   Aruoma OI. Free radicals, oxidative stress, and antioxidants in human health and disease.
        J Am Oil Chem Soc 75, 1998, 199–212.
[8]   Wu CC, Sheen LY, Chen HW, Kuo WW, Tsai SJ, Lii CK. Differential effects of garlic oil and its
        three major organosulfur components on the hepatic detoxification system in rats. J Agric
        Food Chem 50, 2002, 378–383.
[9]   Ahmad H, Tijerina MT, Tobola AS. Preferential overexpression of a class MU glutathione
        S-transferase subunit in mouse liver by myristicin. Biochem Biophys Res Commun 236, 1997,
        825–828.
[10]   Zheng G, Kenny P, Lam L. Inhibition of benzo-pyrene-induced tumorigenesis by myristicin,
        a volatile aroma constituent of parsley leaf oil. Carcinogenesis 13, 1992, 1921–1923.
[11]   Lee BK, Kim JH, Jung JW, Choi JW, Han ES, Lee SH, Ko KH, Ryu JH. Myristicin
        inducedneurotoxicity in human neuroblastoma MSK-N-SH cells. Toxicol Lett 157, 2005, 49–56.
[12]   Carnesecchi S, Langley K, Exinger F, Gossé F, Raul F, Geraniol, a component of plant essential
        oils sensitizes human colonic cancer cells to 5-fluorouracil treatment. Pharmacol J. Exp Ther
        301, 2002, 625–630.
[13]   El Hadri A, Gómez Del Río MA, Sanz J, González Coloma A, Idaomar M, Ribas Ozonas B, Benedí
        González J, Sánchez Reus MI. Cytotoxic activity of humulene and transcaryophyllene from
        Salvia officinalis in animal and human tumor cells. An R Acad Nac Farm 76, 2010, 343–356.
[14]   Herman A, Herman AP, Essential oils and their constituents as skin penetration enhancer for
        transdermal drug delivery: A review. Pharm J. Pharmacol 67, 2015, 473–485.
[15]   Takeda K, Kaisho T, Akira S. Toll-like receptors. Annu Rev Immunol 21, 2003, 335–376.
[16]   Yoon WJ, Kim SS, Oh TH, Lee NH, Hyun CG. Cryptomeria japonica essential oil inhibits the
        growth of drug-resistant skin pathogens and LPS-induced nitric oxide and pro-inflammatory
        cytokine production. Pol J Microbiol 58, 2009, 61–68.

[17] Katsukawa M, Nakata R, Takizawa Y, Hori K, Takahashi S, Inoue H. Citral, a component of lemongrass oil, activates PPAR alpha and gamma and suppresses COX-2 expression. Biochim Biophys Acta Mol Cell Biol Lipids 1801, 2010, 1214–1220.

[18] Devi KP, Nisha SA, Sakthivel R, Pandian SK. Eugenol (an essential oil of clove) acts as an antibacterial agent against Salmonella typhi by disrupting the cellular membrane. J Ethanopharmacol 130, 2010, 107–115.

[19] Tyagi AK, Malik A. Antimicrobial potential and chemical composition of Eucalyptus globulus oil in liquid and vapour phase against food spoilage microorganisms. Food Chem 126, 2011, 228–235.

[20] Lambert RJW, Skandamis PN, Coote P, Nychas GJE. A study of the minimum inhibitory concentration and mode of action of oregano essential oil, thymol and carvacrol. J Appl Microbiol 91, 2001, 453–462.

[21] U.S. Food and Drug Administration, 2022. https://www.accessdata.fda.gov/scripts/cdrh/cfdocs/cfcfr/cfrsearch.cfm?fr=182.20 (Accessed on 25 May 2022)

[22] Tajkarimi MM, Ibrahim SA, Cliver DO. Antimicrobial herb and spice compounds in food. Food Control 2010, 21(9), 1199–1218.

[23] Mishra AP, Devkota HP, Nigam M, Adetunji CO, Srivastava N, Saklani S, Shukla I, Azmi L, Shariati MA, Melo Coutinho HD, Mousavi Khaneghah A. Combination of essential oils in dairy products: A review of their functions and potential benefits. Lwt 133, 2020, 110116.

[24] Viuda-Martos M, Ruiz-Navajas Y, Fernández-López J, Pérez-Álvarez JA. Antifungal activities of thyme, clove and oregano essential oils. J Food Saf 2007, 27(1), 91–101.

[25] Chueca B, Ramírez N, Arvizu-Medrano SM, García-Gonzalo D, Pagán R. Inactivation of spoiling microorganisms in apple juice by a combination of essential oils' constituents and physical treatments. Food Sci Technol Int 2016, 22(5), 389–398.

[26] Shaltout FA, Koura HA. Impact of some essential oils on the quality aspect and shelf life of meat. Bvmj 2017, 33(2), 351–364.

[27] Gaio I, Saggiorato AG, Treichel H, Cichoski AJ, Astolfi V, Cardoso RI, Toniazzo G, Valduga E, Paroul N, Cansian RL, Antibacterial activity of basil essential oil (Ocimum basilicum L.) in Italian-type sausage. Verbr J. Lebensm 10, 2015, 323–329.

[28] Bazana MT, Codevilla CF, de Menezes CR. Nanoencapsulation of bioactive compounds: Challenges and perspectives. Curr Opin Food Sci 26, 2019, 47–56.

[29] Liao W, Badri W, Dumas E, Ghnimi S, Elaïssari A, Saurel R, Gharsallaoui A. Nanoencapsulation of essential oils as natural food antimicrobial agents: an overview. Appl Sci 2021, 11(13), 5778.

[30] de Souza Simões L, Madalena DA, Pinheiro AC, Teixeira JA, Vicente AA, Ramos OL. Micro- and nano bio-based delivery systems for food applications: In vitro behavior. Adv Colloid Interface Sci 243, 2017, 23–45.

[31] Bento R, Pagán E, Berdejo D, de Carvalho RJ, García-Embid S, Maggi F, Magnani M, de Souza EL, García-Gonzalo D, Pagán R. Chitosan nanoemulsions of cold-pressed orange essential oil to preserve fruit juices. Int J Food Microbiol 331, 2020, 108786.

[32] Li QX, Chang CL. Basil (*Ocimum Basilicum* L.) Oils. In: Preedy VR (Eds), Essential Oils in Food Preservation, Flavor and Safety. Elsevier, London, UK, 2016, 231–238.

[33] Cardoso-Ugarte GA, López-Malo A, Sosa-Morales ME. Cinnamon (Cinnamomum Zeylanicum) Essential Oils. In: Preedy VR (Eds), Essential Oils in Food Preservation. Flavor and Safety, Elsevier, London, UK, 2016, 339–347.

[34] Haro-González JN, Castillo-Herrera GA, Martínez-Velázquez M, Espinosa-Andrews H. Clove Essential Oil (Syzygium aromaticum L. Myrtaceae): Extraction, Chemical Composition, Food Applications, and Essential Bioactivity for Human Health. Molecules 26, 2021, 6387.

[35] Hassanien MFR, Assiri AMA, Alzohairy AM, Oraby HF. Health-promoting value and food applications of black cumin essential oil: an overview. J Food Sci Technol 52, 2015, 6136–6142.

[36] Diao WR, Hu QP, Zhang H, Xu JG. Chemical composition, antibacterial activity and mechanism of action of essential oil from seeds of fennel (Foeniculum vulgare Mill.). Food Control 2014, 35(1), 109–116.

[37] Khan S, Das S, Malik N, Bhat SA. Antioxidant properties of garlic essential oil and its use as a natural preservative in processed food. Int J Chem Studies 2017, 5(6), 813–821.

[38] Wells R, Truong F, Adal AM, Sarker LS, Mahmoud SS. Lavandula Essential Oils: A Current Review of Applications in Medicinal, Food, and Cosmetic Industries of Lavender. Nat Prod Commun 2018, 13(10), 1403–1417.

[39] Klimek-Szczykutowicz M, Szopa A, Ekiert H. Citrus limon (Lemon) phenomenon—a review of the chemistry, pharmacological properties, applications in the modern pharmaceutical, food, and cosmetics industries, and biotechnological studies. Plants 2020, 9(1), 119.

[40] Faheem F, Liu ZW, Rabail R, Haq IU, Gul M, Bryła M, Roszko M, Kieliszek M, Din A, Aadil RM. Uncovering the Industrial Potentials of Lemongrass Essential Oil as a Food Preservative: A Review. Antioxidants 2022, 11(4), 720.

[41] Rodriguez-Garcia I, Silva-Espinoza BA, Ortega-Ramirez LA, Leyva JM, Siddiqui MW, Cruz-Valenzuela MR, Gonzalez-Aguilar GA, Ayala-Zavala JF. Oregano Essential Oil as an Antimicrobial and Antioxidant Additive in Food Products. Crit Rev Food Sci Nutr 56, 2016, 1717–1727.

[42] Nayak P, Kumar T, Gupta AK, Joshi NU. Peppermint a medicinal herb and treasure of health: A review. J Pharmacogn Phytochem 2020, 9(3), 1519–1528.

[43] Hernández MD, Sotomayor JA, Hernández A, Jordán MJ. Rosemary (*Rosmarinus Officinalis L.*) Oils. In: Preedy VR (Eds), Essential Oils in Food Preservation, Flavor and Safety. Elsevier, London, UK, 2016, 677–688.

[44] Stahl-Biskup E, Venskutonis RP. Thyme. In: Peter KV (Eds), Handbook of Herbs and Spices. Elsevier, London, UK, 2012, 499–525.

[45] Allied Market Research, 2021(accessed on 31 May 2022) https://www.alliedmarketresearch.com/cosmetics-market#:~:text=The%20global%20cosmetics%20market%20size,of%20modern%20lifestyle%20of%20individuals.

[46] Sharmeen JB, Mahomoodally FM, Zengin G, Maggi F. Essential Oils as Natural Sources of Fragrance Compounds for Cosmetics and Cosmeceuticals. Molecules 26, 2021, 666.

[47] Muyima NYO, Zulu G, Bhengu T, Popplewell D. The potential application of some novel essential oils as natural cosmetic preservatives in an aqueous cream formulation. Flavour Fragr J 17, 2002, 258–266.

[48] Choi IH, Hwang DI, Kim DY, Kim HB, Lee HM. A Study on the Anti-wrinkle Properties of Cosmetics Containing Essential Oil from Chrysanthemum boreale MAKINO. J Life Sci 29, 2019, 442–446.

[49] Sun P, Zhao L, Zhang N, Wang C, Wu W, Mehmood A, Zhang L, Ji B, Zhou F. Essential Oil and Juice from Bergamot and Sweet Orange Improve Acne Vulgaris Caused by Excessive Androgen Secretion. Mediat Inflamm 2020, 1–10, doi: https://doi.org/10.1155/2020/8868107.

[50] Zheng Y, Pan C, Zhang Z, Luo W, Liang X, Shi Y, Liang L, Zheng X, Zhang L, Du Z. Antiaging effect of Curcuma longa L. essential oil on ultraviolet-irradiated skin. Microchem J 2020, 154, 104608, https://doi.org/10.1016/j.microc.2020.104608.

[51] Francois-Newton V, Brown A, Andres P, Mandary MB, Weyers C, Latouche-Veerapen M, Hettiarachchi D. Antioxidant and Anti-Aging Potential of Indian Sandalwood Oil against Environmental Stressors In Vitro and Ex Vivo. Cosmetics 8, 2021, 53.

[52] Guzmán E, Lucia A. Essential Oils and Their Individual Components in Cosmetic Products. Cosmetics 8, 2021, 114.

[53] Montenegro L, Pasquinucci L, Zappalà A, Chiechio S, Turnaturi R, Parenti C, Rosemary Essential Oil-Loaded Lipid Nanoparticles: In Vivo Topical Activity from Gel Vehicles. 2017, 9, 48.

[54] Forlot P, Pevet P. Bergamot (Citrus bergamia Risso et Poiteau) essential oil: Biological properties, cosmetic and medical use. A review. J Essent Oil Res 24, 2012, 195–201.

[55] Oh JY, Park MA, Kim YC. Peppermint Oil Promotes Hair Growth without Toxic Signs. Toxicol Res 2014, 30, 297–304.

[56] Panahi Y, Taghizadeh M, Marzony ET, Sahebkar A. Rosemary oil vs minoxidil 2% for the treatment of androgenetic alopecia: a randomized comparative trial. Skinmed 2015, 13(1), 15–21.

[57] Abelan US, de Oliveira AC, Cacoci ESP, Martins TEA, Giacon VM, Velasco MVR, C.r.r.d.c. L. Potential use of essential oils in cosmetic and dermatological hair products: A review. J Cosmet Dermatol 2021.

[58] Yadav E, Kumar S, Mahant S, Khatkar S, Rao R. Tea tree oil: a promising essential oil. J Essent Oil Res 29, 2017, 201–213.

[59] Goyal A, Sharma A, Kaur J, Kumari S, Garg M, Sindhu RK, Rahman MH, Akhtar MF, Tagde P, Najda A, Banach-Albińska B, Masternak K, Alanazi IS, Mohamed HRH, El-kott AF, Shah M, Germoush MO, Al-malky HS, Abukhuwayjah SH, Altyar AE, Bungau SG, Abdel-Daim MM. Bioactive-Based Cosmeceuticals: An Update on Emerging Trends. Molecules 27, 2022, 828.

[60] Ribeiro-Santos R, Andrade M, de Melo NR, Sanches-Silva A. Use of essential oils in active food packaging: Recent advances and future trends. Trends Food Sci Technol 61, 2017, 132–140.

[61] Javid A, Raza ZA, Hussain T, Rehman A. Chitosan microencapsulation of various essential oils to enhance the functional properties of cotton fabric. J Microencapsul 31, 2014, 461–468.

[62] Srivastava S, Srivastava S, Essential Oil Impregnation on Wool Fabric for Aromatherapy. International Conference on Inter Disciplinary Research in Engineering and Technology, Published by ASDF International, Registered in London, United Kingdom, 2016, 59–62.

[63] Khanna S, Sharma S, Chakraborty JN. Performance assessment of fragrance finished cotton with cyclodextrin assisted anchoring hosts. Fash Text 2, 2015, 1–7.

[64] Prajapati V, Tripathi A, Aggarwal K, Khanuja S. Insecticidal, repellent and oviposition-deterrent activity of selected essential oils against Anopheles stephensi, Aedes aegypti and Culex quinquefasciatus. Bioresour Technol 96, 2005, 1749–1757.

[65] Specos MMM, García JJ, Tornesello J, Marino P, Vecchia MD, Tesoriero MVD, Hermida LG. Microencapsulated citronella oil for mosquito repellent finishing of cotton textiles. Trans R Soc Trop Med Hyg 104, 2010, 653–658.

[66] Soroh A, Owen L, Rahim N, Masania J, Abioye A, Qutachi O, Goodyer L, Shen J, Laird K. Microemulsification of essential oils for the development of antimicrobial and mosquito repellent functional coatings for textiles. J Appl Microbiol 131, 2021, 2808–2820.

[67] Tomsic B, Simoncic B, Orel B, Vilcnik A, Spreizer H. Biodegradability of cellulose fabric modified by imidazolidinone. Carbohydr Polym 2007, 69, 478–488.

[68] Walentowska J, Foksowicz-Flaczyk J. Thyme essential oil for antimicrobial protection of natural textiles. Int Biodeterior Biodegrad 84, 2013a, 407–411.

[69] Aruda LM, Teixeira M, Ribeiro AI, Souto AP, Cionek CA. Textiles, Identity and Innovation. In: Touch. Proceedings of the 2nd International Textile Design Conference (D_TEX 2019) Montagna G, Carvalho C (Eds), Taylor and Francis, CRC Press, London, 2020.

[70] Gong C, Wu Q, Wang Y, Zhang D, Luo F, Zhao X, Wei Y, Qian Z. A biodegradable hydrogel system containing curcumin encapsulated in micelles for cutaneous wound healing. Biomaterials 34, 2013, 6377–6387.

[71] Rafiq M, Hussain T, Abid S, Nazir A, Masood RJ. Development of sodium alginate/PVA antibacterial nanofibers by the incorporation of essential oils. Mater Res Express 2018, 5(3), 035007.

Joshua H. Santos and Mark Lloyd G. Dapar*

Chapter 9
Application of essential oils
in pharmaceutical industry

Abstract: Essential oils have been known for millennia for its therapeutic uses and properties and continually expanding worldwide due to its growing interest for a rediscovery of natural remedies. The sources, methods of extraction, diversity, uses, and applications of essential oils are variable in their application to pharmaceutics and pharmaceutical industry. The oily fragrance of essential oils from different plant parts can be extracted using different techniques mostly by steam distillation and hydrodistillation. Essential oils are widely used in massage therapy and aromatherapy for physical and psychological treatment. Most essential oils possessing bioactive compounds have potential therapeutic properties such as anti-inflammatory, antidiabetic, antivirals, antimicrobial, antioxidant, wound-healing, chemopreventive, chemotherapeutic, and anxiolytic activities. The essential oils of aromatic plants could be the active ingredients of their pharmaceutical properties. The use and application of essential oils in folk herbal remedies could support the therapeutic potential of most aromatic plants used in aromatherapy.

Keywords: flavor, food preservation, herbal medicine, pharmacy, plant biotechnology, textiles

9.1 Introduction

Since time immemorial, nature provided us with abundant compounds which can be applied in different aspects of our daily life, such as food, cosmetics, medicine, textiles, and many more. Essential oils are also known as volatile odoriferous oils that are aromatic in nature [1]. Essential oils can be produced constitutively or inducible, and as direct or indirect [2]. Constitutive essential oils are normally produced in plants,

---
*Corresponding author: Mark Lloyd G. Dapar, Department of Biology, College of Arts and Sciences, Central Mindanao University, University Town, Musuan, Bukidnon 8714, Philippines; Center for Biodiversity Research and Extension in Mindanao, Central Mindanao University, University Town, Musuan, Bukidnon 8714, Philippines; Microtechnique and Systematics Laboratory, Natural Science Research Center, Central Mindanao University, University Town, Musuan, Bukidnon 8714, Philippines, e-mail: f.marklloyd.dapar@cmu.edu.ph
Joshua H. Santos, Chemicals and Energy Division, Department of Science and Technology – Industrial Technology Development Institute, DOST compound, General Santos Avenue, Bicutan, Taguig City, Philippines, e-mail: jhsantos@dost.gov.ph

https://doi.org/10.1515/9783110791600-009

without any external stimuli. The main functions of this kind of essential oils are as coping mechanism for abiotic stress, allelopathy, to ward off herbivores, inter-plant signaling, protection against microbial pathogens, and to entice pollinators and seed dispersers [3]. In contrast with the constitutive essential oils, the production of inducible essential oils is mainly due to pathogenic or herbivore attacks. This type of essential oil acts also to reduce abiotic stress, to ward off herbivores, direct and indirect plant signaling mechanisms, and as protection against microbial pathogens [4], suggesting that essential oils are part of the immune system of the plant [5].

9.2 Sources of essential oils

Allelopathy has been demonstrated by *Salvia leucophylla* by secreting 1,8-cineole and camphor to prevent germination, thus taking advantage of the growing competition among other plants [6]. Essential oils have also taken part in the adaptation to abiotic stresses. Some compounds (isoprene and monoterpenes) impart a general adaptive maneuver in the plant by increasing the thermal tolerance in photosynthesis, protecting the photosynthetic apparatus, and maintaining photosynthetic activity, even during high-temperature stress [7].

Plant signaling can also occur from damaged plant parts to undamaged plant parts through the secretion of essential oils (mainly composed of 3-hexenyl acetate), thereby promoting the defense genes, priming of the plant tissues, and other responses, after the attack. Interplant signaling is also done through the release of essential oils, alerting both conspecific and heterospecific plants. The uptake of these essential oils from other plants induces the transcription of defense-related genes, and metabolites [3, 8]. Essential oils attract pollinators to reproduce and continue their life cycle, such as in the case of rose flowers (*Rosa damascene* and *Rosa centifolia*). Compounds detected in rose oil include citronellol, geraniol, nerol, linalool, phenyl ethyl alcohol, farnesol, stearoptene, α-terpinene, limonene, α-pinene, β-pinene, *p*-cymene, camphene, β-caryophyllene, neral, citronellyl acetate, geranyl acetate, neryl acetate, eugenol, methyl eugenol, rose oxides [(4*R*,2*S*)-(−)-*cis*-rose oxide, (4*S*,2*R*)-(−)-*trans*-rose oxide], α-damascenone, β-damascenone, benzaldehyde, benzyl alcohol, rhodinyl acetate, β-ionone, and phenyl ethyl formate [9]. Protection against herbivores and other pathogens is carried out by essential oils that are toxic and repellant to the attackers. In particular, they can be antinutritional agents, prevent digestibility of the plant material, and affect the growth and reproduction of the attackers. The toxicity of these essential oils is mainly due to their effect on the nervous system, digestive system, and endocrine system [4]. Indirect defense can also occur, when the essential oil attracts and favors other species (such as predators of the herbivores) to increase the plant fitness and survival [2, 10]. Some essential oils can be used as mosquito repellants, like in the case of the oils from *Zanthoxylum piperitum* and *Citronella* plants [11].

Essential oils are extracted in almost all parts of the plants such as leaves, peels, barks, flowers, and many more. Some plants producing essential oils with their corresponding plant parts are tabulated in Table 9.1 [1].

Table 9.1: Some plants and their plant parts producing essential oils.

Parts	Common name	Scientific name
Leaves	Basil	*Ocimum basilicum*
	Bay leaf	*Laurus nobilis*
	Cinnamon	*Cinnamomum verum*
	Common sage	*Salvia officinalis*
	Eucalyptus	*Eucalyptus odorata* *Eucalyptus globulus* *Eucalyptus urophylla* *Eucalyptus grandis* *Eucalyptus camaldulensis* *Eucalyptus citriodora*
	Lemon grass	*Cymbogon citratus*
	Citronella grass	*Cymbogon nardus*
	Melaleuca	*Melaleuca alternifolia*
	Patchouli	*Pogostemon cablin*
	Peppermint	*Mentha piperita* *Mentha balsamea*
	Pine	*Pinus palustris*
	Rosemary	*Salvia rosmarinus*
	Spearmint	*Mentha spicata*
	Oregano	*Origanum vulgare*
	Thyme	*Thymus vulgaris*
	Wintergreen	*Gaultheria procumbens*
	Kaffir lime	*Citrus hystrix*
	Cypress	*Cupressus sempervirens*
	Savory	*Satureja hortensis*
	Tarragon	*Artemisia dracunculus*
	Cajuput	*Melaleuca cajuputi*
	Lantana	*Lantana camara*

Table 9.1 (continued)

Parts	Common name	Scientific name
	Lemon myrtle	*Backhousia citriodora*
	Lemon tea tree	*Leptospermum petersonii*
	Niaouli	*Melaleuca quinquenervia*
	May Chang	*Litsea cubeba*
	Petitgrain	*Citrus aurantium*
Seeds	Almond	*Prunus dulcis*
	Anise	*Pimpinella anisum*
	Cardamom	*Elettaria cardamomum*
	Caraway	*Carum carvi*
	Celery	*Apium graveolens*
	Coriander	*Coriandrum sativum*
	Cumin	*Cuminum cyminum*
	Nutmeg	*Myristica fragrans*
	Parsley	*Petroselinum crispum*
	Fennel	*Foeniculum vulgare*
Wood	Torchwood	*Amyris elemifera*
	Atlas cedarwood	*Cedrus atlantica*
	Himalayan cedarwood	*Cedrus deodara*
	Camphor	*Cinnamomum camphora*
	Rosewood	*Aniba rosaeodora*
	Sandalwood	*Santalum album*
	True myrtle	*Myrtus communis*
	Guaiac wood	*Guaiacum officinale*
Bark	Golden shower	*Cassia fistula*
	Cinnamon	*Cinnamomum verum*
	Sassafras	*Sassafras albidum*
	Katrafay	*Cedrelopsis grevei*
Berries	Allspice	*Pimenta dioica*
	Juniper	*Juniperus communis*

Table 9.1 (continued)

Parts	Common name	Scientific name
Resin	Frankincense	*Boswellia sacra*
	Myrrh	*Commiphora myrrha*
Flowers	Blue tansy	*Tanacetum annuum*
	Chamomile	*MAtricaria chamomilla*
	Clary Sage/Europe sage	*Salvia sclarea*
	Clove	*Syzigium aromaticum*
	Cumin	*Cuminum cyminum*
	Geranium	*Pelargonium graveolens (sweet-scented geranium)* *Pelargonium peltatum (Ivy geranium)*
	Helichrysum hyssop	*Hyssopus officinalis*
	Jasmine	*Jasminum officinale* (common jasmine) *Jasminum polyanthum* (white jasmine) *Jasminum sambac* (Arabian jasmine)
	Lavender	*Lavandula angustifolia* (English lavender) *Lavandula stoechas* (French lavender)
	Manuka	*Leptospermum scoparium*
	Marjoram	*Origanum majorana*
	Orange	*Citrus sinensis*
	Rose	*Rosa damascene* (Damask rose)
	Baccharises	*Pseudobaccharis cabrera*
	Palmorosa/palm rose	*Cymbopogon martini*
	Patchouli	*Pogostemon cablin*
	Dwarf rhododendron	*Rhododendron anthopogon*
	White Champaca	*Magnolia alba*
	Ajowan/Ajwain	*Trachyspermum ammi*
	Ylang-ylang	*Cananga odorata*
	Marjoram	*Origanum majorana*
	Tarragon	*Artemisia dracunculus*
	Immortelle	*Helichrysum angustifolium*
	Neroli	*Citrus aurantium* var. amara

Table 9.1 (continued)

Parts	Common name	Scientific name
Peel	Bergamot	*Citrus bergamia*
	Grapefruit	*Citrus paradisi*
	Kaffir lime	*Citrus hystrix*
	Lemon	*Citrus limon*
	Lime	*Citrus aurantifolia*
	Orange	*Citrus sinensis*
	Tangerine	*Citrus reticulata*
	Calamansi	*Citrofortunella macrocarpa*
Root	Ginger	*Zingiber officinale*
	Plai	*Zingiber montanum*
	Turmeric	*Curcuma longa*
	Valerian	*Valeriana officinalis*
	Vetiver	*Chrysopogon zizanioides*
	Spikenard	*Nardostachys jatamansi*
	Female ginseng	*Angelica sinensis*
Fruits	Prickly ash/toothache tree	*Zanthoxylum americanum*
	Nutmeg	*Myristica fragrans*
	Black pepper	*Piper nigrum*
	Camboim-de-cachorro	*Neomitranthes obscura*

The main components of essential oils are terpenoid and phenylpropanoid derivatives. In most plants, around 80% of the compounds are of terpenoid derivatives. However, the presence of phenylpropanoid derivatives imparts flavor, odor, and piquant [12] taste. Some essential oils contain polyketide, and nitrogen- and sulfur-containing compounds [13]. Constitutive essential oils are mainly composed of terpenoids, shikimate derivatives, and polyketides [3], while inducible essential oils are mainly composed of mono- and sesquiterpenes, with trace amounts of green leaf volatile oil, and phytohormones such as ethylene, methyl salicylate, jasmonic acid, and others [2, 4, 8]. Terpenoids are a heterogeneous group of terpenes also known as compounds with double bonds and their corresponding derivatives or functional groups. Thus, terpenes are simple hydrocarbon, while terpenoids are terpenes with functional groups or oxygenated moiety [14]. These compounds consist of isoprene (a five-carbon building block with one double bond as illustrated in Figure 9.1a) [15].

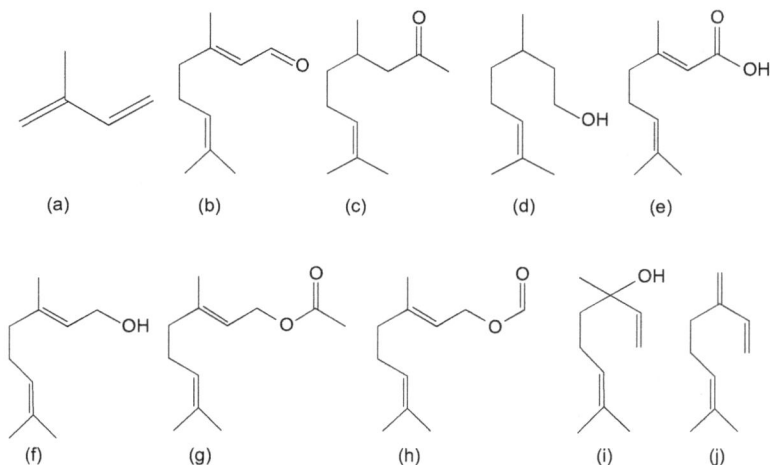

Figure 9.1: Structures of acyclic monoterpenes and monoterpenoids, namely (a) isoprene, (b) citral, (c) citronellal, (d) citronellol, (e) geranic acid, (f) geraniol, (g) geranyl acetate, (h) geranyl formate, (i) linalool, and (j) myrcene.

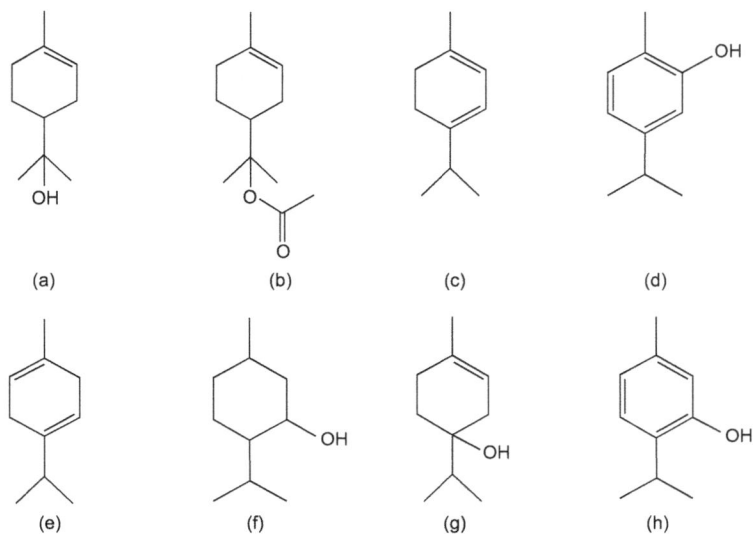

Figure 9.2: Structures of cyclic monoterpenes and monoterpenoids, namely (a) alpha-terpineol, (b) alpha-terpinyl acetate, (c) alpha-terpene, (d) carvacrol, (e) gamma-terpene, (f) menthol, (g) terpinen-4-ol, and (h) thymol.

Terpenes can be classified based on the number of isoprene units present in the structure. Monoterpene consists of 2 isoprene units (10 carbon atoms), sesquiterpene consists of 3 isoprene units (15 carbon atoms; Figure 9.3a–c), diterpene consists of 4 isoprene units (20 carbon atoms; Figure 9.3d–f), triterpene consists of 6 isoprene units (30 carbon atoms), and tetraterpene consists of 8 isoprene units (40 carbon atoms) [16]. Terpenes can be classified as acyclic or open ring (as illustrated in Figure 9.1b–j) while cyclic terpenes can be further classified as monocyclic (as illustrated in Figure 9.2a–h), bicyclic (Figure 9.3b), and tricyclic (Figure 9.3d–f) [17]. Most essential oils contain large amounts of monoterpenes and sesquiterpenes, while diterpenes, triterpenes, and tetraterpenes may be present at low levels in essential oils [18]. Terpenes can also be classified based on the presence of functional group, namely, hydrocarbons, oxygenated compounds, and sulfur and/or nitrogen containing compounds [19]. Some of the oxygenated compounds are phenols (thymol, eugenol, carvacrol, and chavicol), alcohols: monoterpene alcohols (borneol, isopulegol, lavanduol, and alpha terpineol), aldehydes (citral, myrtenal, cuminaldehyde, citronellal, cinnamaldehyde, and benzaldehyde), ketones (carvone, menthone, pulegone, fenchone, camphor, thujone, and verbenone), esters (bornyl acetate, linalyl acetate, citronellyl acetate, geraniol acetate, and geranyl acetate), oxides (1,8-cineole, bisabolone oxide, linalool oxide, and sclareol oxide), lactones (bergaptene, nepetalactone, psoralen, aesculatine, and citroptene), and ethers (1,8-cineol, anethole, elemicin, and myristicin).

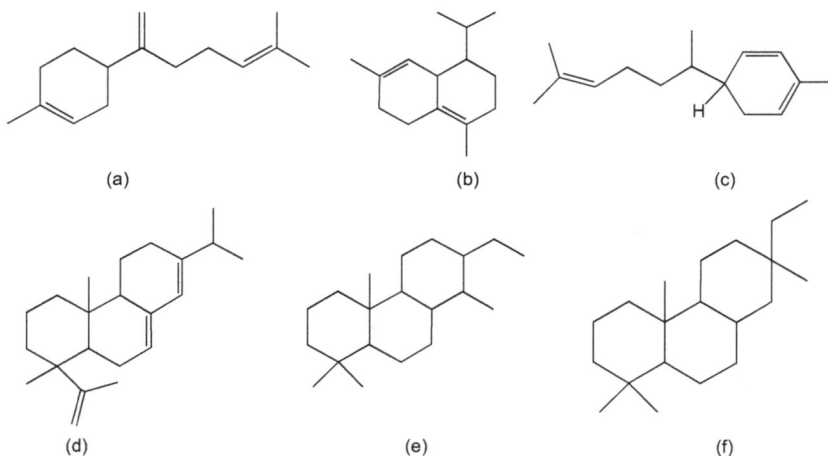

Figure 9.3: Structures of sesquiterpenes such as (a) bisabolene, (b) cadinene, and (c) zingiberene, and diterpenes such as (d) abietic acid, (e) cassane, and (f) pimarane.

Biosynthesis of terpenoids can occur in two pathways, in both cases leading to the production of isopentenyl diphosphate (IPP). First, the mevalonic acid pathway occurs in the cytosol, endoplasmic reticulum, and peroxisomes with the catalysis of multiple enzymes. Specifically, conversion of three units of acetyl-CoA leads to the production of 3-hydroxy-3-methylglutaryl CoA. This intermediate product is then converted to mevalonic acid, with the enzyme 3-hydroxy-3-methylglutaryl CoA reductase. IPP is produced via the process of phosphorylation and decarboxylation of mevalonic acid. The mevalonic acid undergoes a series of phosphorylation, and decarboxylation catalyzed by mevalonate kinase, phosphomevalonate kinase, and mevalonate diphosphode carboxylase. Conjugation of IPP with one dimethyl allyl pyrophosphate (DMAPP) results in formation of geranyl pyrophosphate (GPP) (Figure 9.4a), which is the precursor of monoterpenes.

Conjugation of one DMAPP with three units of IPP produces farnesyl diphosphate (FPP; a 15-carbon molecule; Figure 9.4b), with the aid of farnesyl diphosphate synthase. Farnesyl diphosphate is the major precursor in the synthesis of sesquiterpenes. Geranylgeranyl diphosphate (GGPP) (Figure 9.4c) is synthesized with one DMAPP and three units of IPP, with the enzyme Geranylgeranyl diphosphate synthase. GGPP is the main precursor for the synthesis of diterpenes. Triterpene synthesis, specifically, squalene, is formed through the condensation of two Farnesyl pyrophosphate (FPP) by squalene synthase. Squalene (Figure 9.4d) is then used as the precursor to other triterpenoids through oxidation (catalyzed by squalene epoxidase) producing 2,3-oxidosqualene. This intermediate product is converted to triterpene alcohols or aldehydes by oxidosqualene cyclase [16].

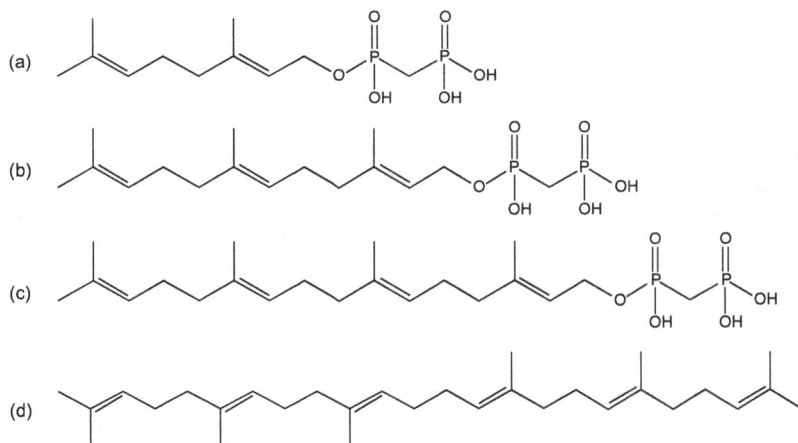

Figure 9.4: Structures of the intermediate products/precursor (a) geranyl pyrophosphate, (b) farnesyl pyrophosphate, (c) geranylgeranyl pyrophosphate, and (d) squalene.

Another pathway occurs in the plastid through the condensation of pyruvic acid and glyceraldehyde-3-phosphate, producing 1-deoxy-D-xylulose-5-phosphate with the aid of 1-deoxy-D-xylulose-5-phosphate synthase. This pathway is known as the DXP/MEP pathway. The product of the reaction is then catalyzed by 1-deoxy-D-xylulose-5-phosphate reducto isomerase into 2-C-methyl-D-erythritol-4-phosphate. Further reaction with 2-C-methyl-D-erythritol-4-phosphate and 4-cytidine-5-phosphate results in 4-cytidine-5-phosphate-2-C-methyl erythritol via 2-C-methyl-D-erythritol-4-phosphate cytidyltransferase. Downstream catalyzation of the product 2-C-methyl-D-erythritol-4-phosphate with the enzyme hydroxy methyl butenyl 4-diphosphate synthase forms the hydroxy methyl butenyl 4-diphosphate. This end product of the reaction is then converted to the isopentenyl diphosphate and dimethyl allyl diphosphate via IPP and DMAPP synthase [16].

Phenylpropanoids, also known as phenylpropenes or allylphenols or propenylphenols (Figure 9.5a), are a class of secondary compounds produced from primary metabolites such as phenylalanine or tyrosine [20]. Phenylpropanoid derivatives are divided into five major classes namely, flavonoids, monolignols, phenolic acids, stilbenes, and coumarins. In particular, some essential oils contain cinnamic acid (Figure 9.5b) and *p*-hydroxycinnamic acid (Figure 9.5c) but in some cases, these two phenolic acids act as intermediate for the synthesis of volatile phenylpropanoid [21].

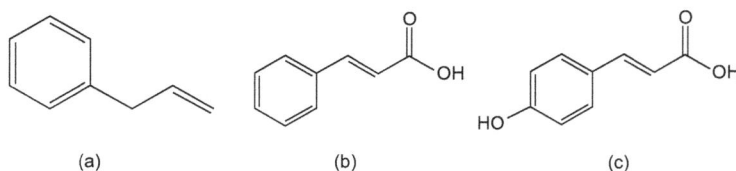

(a)   (b)   (c)

Figure 9.5: Structures of (a) phenylpropene, (b) cinnamic acid, and (c) *p*-hydroxycinnamic acid.

The essential oil of some *Piper* species demonstrated a high level of phenylpropanoid (safrole (Figure 9.6a), 3,4-methylenedioxypropiophenone (Figure 9.6b), anethole (Figure 9.6c), *p*-anisaldehyde (Figure 9.6d), isoosmorhizole (Figure 9.6e), asarone (Figure 9.6f), and dillapiole (Figure 9.6g). Phenylpropanoids are composed of an aromatic ring (six carbons) with a three-carbon propane side chain. Flavonoids, one of the members of the phenylpropanoids, are characterized to have a 15-carbon skeleton consisting of two aromatic phenyl rings (rings A and B), and a heterocyclic pyrane ring (ring C). Monolignols or hydroxycinnamyl alcohol monomers also have a C6–C3 skeleton like flavonoids, differing only in the degree of methoxylation at C3 and C5 positions of the aromatic ring. On the other hand, phenolic acids are aromatic acids, wherein the benzene ring is attached with one or more hydroxy- or methoxy- substituents. Coumarins are characterized by a fusion of benzene and α-pyrone ring (α-benzopyrone). Lastly, stilbenes are composed of 1,2-diphenylethylene structure (C6–C2–C6) [22].

Biosynthesis of phenylpropanoid starts with the either phenylalanine or tyrosine, which is known as the general phenylpropanoid pathway [23, 24]. Initially, phenylalanine, produced from the shikimate pathway, undergoes deamination with the aid of phenylalanine ammonia lyase, yielding cinnamic acid (Figure 9.5b). Cinnamic acid is hydroxylated to *para*-coumaric acid with the catalysis of cinnamate 4-hydroxylase. In some monocotyledon plants, fungi, and bacterial species, tyrosine is directly converted to *para*-coumaric acid, using either tyrosine ammonia lyase or bifunctional ammonia lyase [25, 26]. In this step, the hydroxylation of cinnamic acid is omitted.

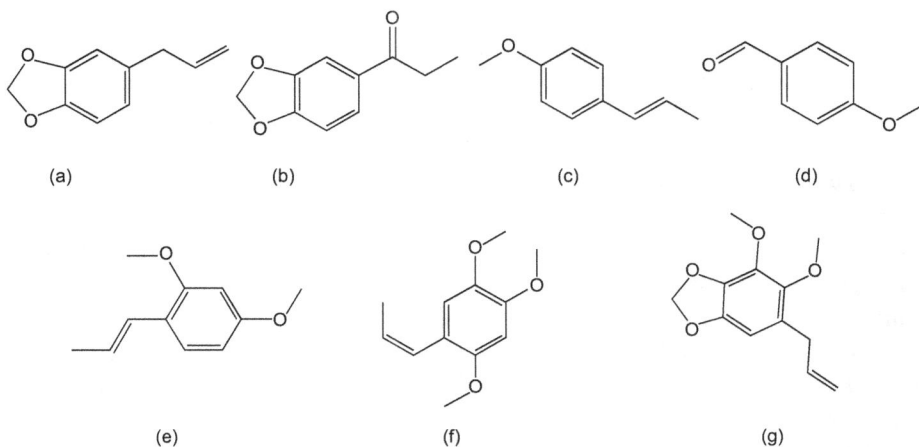

Figure 9.6: Structures of some phenylpropanoid derivatives, namely, (a) safrole, (b) 3,4-methylenedioxypropiophenone, (c) anethole, (d) *p*-anisaldehyde, (e) isoosmorhizole, (f) asarone, and (g) dillapiole.

From *para*-coumaric acid, another enzyme (4-coumaroyl CoA ligase) acts upon the conversion to *para*-coumaroyl-CoA. This intermediate product is then utilized in the synthesis of other phenylpropanoid compounds [27, 28]. Flavonoids, monolignol, and stilbene are synthesized from the *para*-coumaroyl-CoA, while most phenolic acids are synthesized from hydroxycinnamic acid [29–36]. Synthesis of coumarins is not only limited to the intermediate products of the general phenylpropanoid pathway (cinnamic acid and *para*-coumaric acid) but also includes other aromatic acids (caffeic acid and ferulic acid) [37].

Some families of plants develop a special secretory tissue where essential oils are secreted, such as schizogenous and lysigenous cavities or ducts, oil tubes or vittae, modified parenchymal cells, and glandular trichomes.

Plants belonging to the Pinaceae and Rutaceae secrete their essential oil through the schizogenous and lysigenous cavities or ducts. Schizogenous cavities (Figure 9.7a) are formed by the separation of the glandular cells, leading to a hollow space surrounded by secretory epithelial cells. In comparison, lysigenous cavities (Figure 9.7b) are developed through the autolysis (or cell death) of glandular cells. When these cavities degenerate, secretory cells release the products into an enlarging space (holocrine secretion). A combination of the two cavities is known as schizolysigenous cavities, wherein, the cavity is formed schizogenously. The epithelial lining then undergoes autolysis that further enlarges the cavity [38]. The formation of the cavity is done on an outward centrifugal direction. Essential oil synthesis precedes the cavity formation for both schizogenous and lysigenous cavities [39]. Apiaceae plants are well known for the presence of ridges or oil canals known as oil tubes or vittae (Figure 9.7 c–d) in the vegetative or reproductive parts of the plant [40]. However, most oils from the Apiaceae family is concentrated to their seeds, giving a characteristic odor, as in the case of anise (*Pimpinella anisum*), fennel (*Foeniculum vulgare*), and coriander (*Coriandrum sativum*) [40–43]. Production and storage of biologically active secondary metabolites (such as flavonoids and coumarins) also occur in the oil tubes. In *Conium maculatum*, presence of provittae is noted in the ovaries of the flower, and then transform to vittae upon fruit development and maturation [44]. In some plants of the Apiaceae family, irregular-shaped vittae are present rather than regular-shaped vittae. *Sanicula europaea* possesses branching vittae, while *Eryngium planum* possesses anastomosing vittae. Combination of the two irregular vittae is observed in *Steganotaenia araliacea* and *Actinolema macrolema* [45]. The parenchyma is a type of plant cells considered to be important in secondary xylem of most angiosperms. They act as storage and defense [46]. Parenchyma can be modified with the addition of some components such as chlorenchyma (with chloroplast), collenchyma (primary cell material), and others [47, 48]. In Piperaceae, such modified parenchyma exists as idioblasts or oil-secreting cells or oil cells (Figure 9.7e–f) [49–52]. Idioblasts are plant cells that are abnormal in shape, size, and content. Some idioblasts contain oils, resin, gums, crystals, and many more [53]. In *Piper aduncum*, idioblasts are distributed in the leaves. Thus, a large amount of essential oils are extracted from its leaves [54]. In *Piper callosum*, idioblasts are morphological, spherical to elliptical in shape, with the presence of lignified intermediary layer in the cell wall occurring as a trilamellar structure [51]. The presence of these secretory cells is not limited only to the Piperaceae but can also be seen in Poaceae (*Cymbogon citratus*) [55] and Saururaceae (*Saururus cernuus*) [56].

Glandular trichomes are specialized hair-like projections found on the surface of vascular plants. Trichomes are mainly found in the leaves and stems of the plants, but in some species, they can also be present in other parts. These hair-like projections are not connected to the vascular system of the plant, but are, rather, an extension of the epidermis [60]. These trichomes are specialized secretory cells of essential oils in the plants [61]. Glandular trichomes can produce, store, and secrete secondary metabolites [62]. Such secretions are placed on the surface of the plants and can serve as primary defense against herbivores and insects [63]. These secretory hair-like projections are mainly found in the Lamiaceae family, particularly, mint (*Mentha piperita*), basil (*Ocimum basilicum*), lavender (*Lavandula spica*), oregano (*Origanum vulgare*), and thyme (*Thymus vulgaris*) [62]. Other compounds secreted from trichomes are artemisin (*Artemisia annua*) and gossypol (*Gossypium hirsutum*) [64–66]. The morphological classification of glandular trichomes includes capitate and peltate trichomes. Both types are observed in the families of Asteraceae, Lamiaceae, and Solanaceae. Capitate trichomes (Figure 9.7g) are described to have one basal cell with one to several stalk cells and secretory cells at the tip of the stalk. They primarily secrete poorly to nonvolatile oils [67, 68]. On the other hand, peltate trichomes (Figure 9.7h) are described to have one short stalk cell attached to one basal cell. A characteristic head, which carries several secretory cells, is seen at the end of the stalk cell [69]. In some species in the family, both capitate and peltate are present.

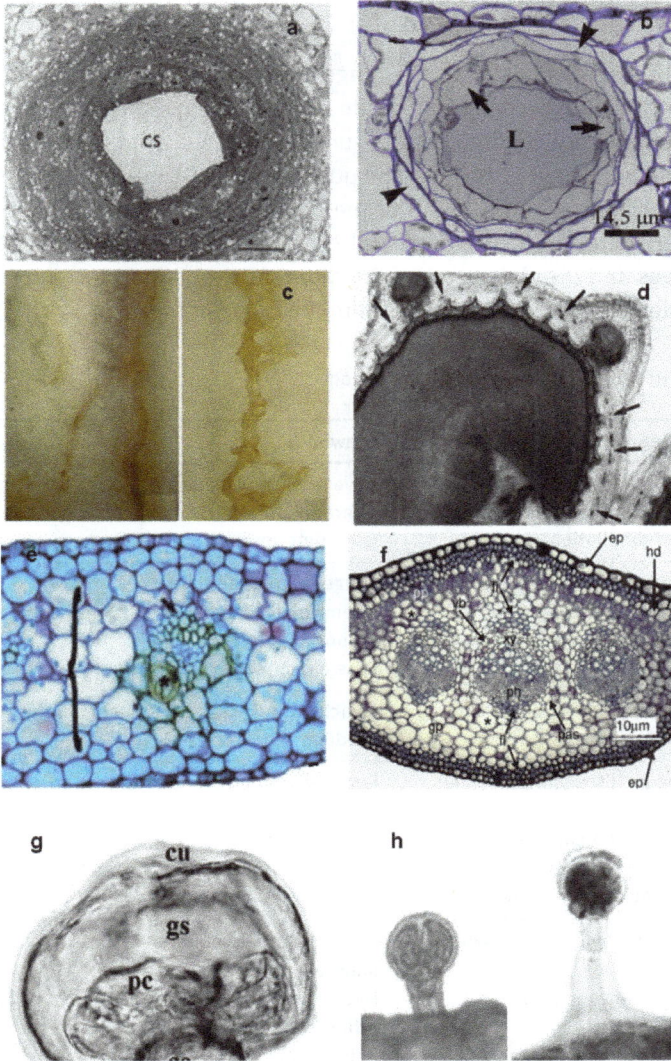

Figure 9.7: (a) Schizogenous cavity from the cross section of *Citrus limon* [38]; (b) Lysigenous cavity from the cross section of *Dictamnus dasycarpus*, indicating the secretory cells (arrow) and thick-walled sheath cells (arrowhead) [57]; (c) portion of the *Steganotaenia araliacea* mesocarp showing the oil tube or vittae [45]; (d) cross-sectional portion of *Conium maculatum* [44]; (e) pictograph of *Piper lepturum* var. *lepturum* depicting the compactly arranged parenchyma ((), vascular bundle (arrow), and oil-secreting cell (*) [58]; (f) cross-sectional portion of the leaf blade of *Piper callosum* depicting epidermis (ep), hypodermis (hd), palisade parenchyma (pp), fiber (fi), phloem (ph), xylem (xy), parenchymatic sheath (pas), ground parenchyma (gp), and oil-secreting cells (*) [51]; (g) cross section of the peltate glandular trichome of *Salvia smyrnea* depicting the cuticle (cc), glandular space (gs), periphery cells (pc), and central cells (cc); and (h) cross section of the capitate glandular trichome of *Salvia smyrnea* depicting the stalk and secretory cells [59].

9.3 Extraction of essential oils

Due to the abundance of essential oils in a wide range of plants with different plant parts, extraction of essential oils becomes one of the most widely used approach in using these oils for different applications. Factors such as plant material (state and form) and extraction method can greatly affect the quality of the essential oil, altering or deteriorating the chemical composition of the oils. In worst cases, discoloration, loss of characteristic odors, oxidation, and other physical changes might occur [1]. The advantages and disadvantages of the different extraction procedures are summarized in Table 9.2 and are also discussed in the following text.

Table 9.2: Advantages and drawbacks of essential oil extraction.

Methods	Advantages	Drawback
Steam distillation	Less artifacts produced Shorter extraction time compared to hydrodistillation	Several hours of heating Degradation of thermolabile compounds Odor deterioration
Hydrodistillation	Separation of essential oils and water are achieved easily Higher yield for some essential oils	Longer extraction time Chemical alteration due to prolonged boiling Loss of some polar compounds
Solvent extraction	Absences of alterations and chemical artifacts	Toxicity due to residual solvents in the product
Cold pressing	Pristine qualities are preserved Low cost No safety problems due to solvent residue Environmentally friendly	Low yield of nutraceutical extracted Pungent odor due to breakdown of glucosinolates
Supercritical fluid extraction	Higher quality of extracts with better activities Relatively low temperature Chemical inertness No safety problems due to solvent residue	High cost due to equipment and maintenance Need of high carbon dioxide purity Affinity of the carbon dioxide to low-polar and nonpolar compounds
Subcritical water extraction	Less extraction time Volatile and thermolabile compounds are preserved Low costs	Less yield of monoterpenes compounds High amount of plant material needed

Table 9.2 (continued)

Methods	Advantages	Drawback
Ultrasound-assisted extraction	High efficiency Low temperature Reduced solvent consumption Less energy input	Possibility of free radical formation due to sonolysis of the solvent Possible oxidation of the components
Microwave-assisted extraction	Less extraction time Environmentally friendly Solvent reduction Faster and more efficient extraction Better sensory properties	Frequent use of toxic organic solvents Use of high temperatures with formation of undesirable compounds

9.3.1 Distillation

Extraction of essential oils based on distillation is still considered the most common technique in extracting essential oils. This method is advantageous due to its flexibility, versatility, and prevention of essential oil degradation, and can be applied to small to large volumes [70]. However, some disadvantages can also occur, such as loss of volatile compounds, specially unsaturated compounds, either by hydrolysis or thermal degradation, long processing time, and high energy consumption [71–73].

Steam distillation is mostly applied for temperature-sensitive materials such as oils, resins, and hydrocarbons, which are insoluble in water and may deteriorate at their boiling point. This method has been utilized since the early 1980s. With the aid of steam, the essential oils are volatilized, as steam passes through the sample, without reaching the boiling point of the essential oil. The water vapor softens and breaks the plant material, allowing the volatilization of essential oils. The essential oil droplets mix with the water vapor and are then passed through a condenser, where the vapor transforms to liquid. Separation of essential oils and the water (called as hydrosol) occurs (Figure 9.8a). In most cases, the density of the essential oil is much less than that of water. The essential oil is collected at the top of the distillate, while the bottom layer is composed of the hydrosol [74]. The amount of essential oil that can be collected in this method ranges up to 93% [75]. Some chemical reaction can still occur during steam distillation, due to the diversity of the chemical composition of each essential oil. In particular, chamomile oil extraction leads to the formation of chamazulene [76]. Such products are known as artificial chemicals or artefacts. A different kind of steam distillation is known as the hydrodiffusion. This method is applied for dried plant material that is not damaged at boiling temperature. The difference of this method lies in the inlet of steam.Here, steam distillation involves entry of steam to the bottom of the plant material, while in hydrodiffusion, the steam enters at the top of the plant material. Another main difference is the temperature of the steam, which is at below 100 °C.

With all of this deviation from the steam distillation, more oil yield and less processing time are achieved [77].

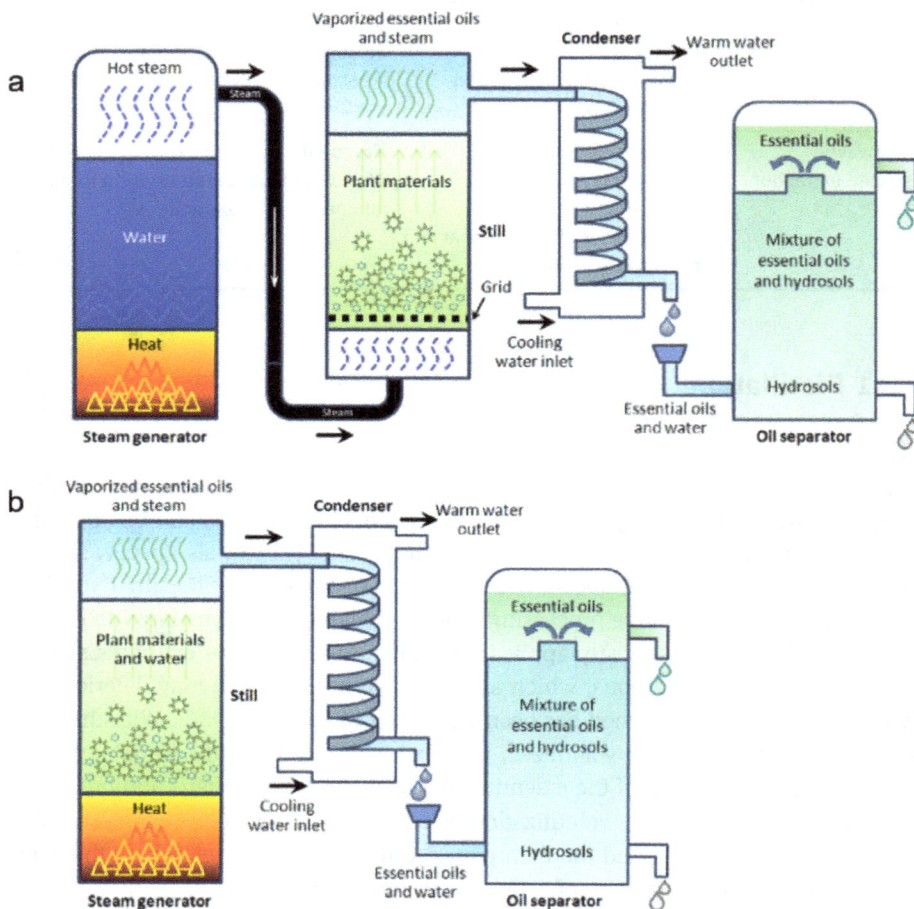

Figure 9.8: Illustration of (a) steam distillation, and (b) hydrodistillation setup (image taken from the journal article of Tongnuanchan and Benjakul [1]).

In comparison with the steam distillation, hydrodistillation is used for essential oil extraction from wood or flower. This method is highly recommended for non-aqueous components with high boiling point. The plant material is immersed in water, followed by boiling. The integrity of the oil is preserved, since water acts as barrier against overheating. The water vapor with the droplets of the essential oil will pass through a condenser, where the separation of oil and water occurs (Figure 9.8b) [78]. The advantage of this method compared to the steam distillation is that the extracting temperature is only limited up to 100 °C, since the mixture of essential oil and water leads to an azeotropic mixture [79]. Innovation in essential oil extraction occurs with the aid of microwave-assisted hydrodistillation [78]. This approach addresses the disadvantages of the conventional method such as loss of volatile components, longer extraction time, poor extraction outputs, and even residual toxic solvents in the end product.

9.3.2 Solvent extraction

Organic solvent extraction is done when the material is delicate such as flower petals or materials, and the essential oil is heat-labile. Different solvents are applied such as acetone, hexane, petroleum ether, methanol, or ethanol [80–82]. Conventionally, the plant material is submerged or macerated into the organic solvent, for a certain period of time. The supernatant portion is obtained either through filtration, spontaneous separation of the residue and solvent, and other techniques that can be applied for separation. The essential oil-rich liquid is concentrated by removing the solvent through spontaneous evaporation or using a rotary evaporator under pressure. The concentrate is composed resin or concrete. This concrete is a composed of wax, fragrance, and essential oil. Pure alcohol is added to the concrete to extract the fragrance and essential oil and distilled at low temperature [83]. The product of this process is known as absolute, a wax-free residue [84]. The main advantage of absolute is higher aromatic concentration than concretes. The main drawback of this method is a longer extraction time and solvent residue in the final product due to incomplete removal [83]. In a particular extraction, higher essential oil yield of 0.054 mL/g was obtained, compared to 0.048 mL/g from the conventional method [85]. Some techniques are applied to execute the solvent extraction method such as maceration, percolation, decoction, and reflux extraction. Maceration is the simplest technique applied for heat labile compounds but is limited due to a longer extraction time and low extraction efficiency. Percolation is considered as a more efficient method, due to the continuity of the extraction process. In this technique, the saturated solvents are drained and collected, and fresh solvent is applied to the remaining plant material. On the other hand, decoction involves the use of boiling water to extract the components; however, this method is only limited to heat-resistant compounds. In decoction, most of the components extracted are water-soluble compounds [86]. A more effective extraction procedure compared to maceration and percolation involves the use of the reflux extraction or the Soxhlet extraction method. This method is applied to extract a certain compound whose solubility is limited to a particular solvent. This method allows the recycling of the solvent to dissolve a larger amount of material, thus allowing the method to be executed in an unmonitored and unmanaged manner. Solid–liquid contact is observed in this method for the removal of one or more compounds from the plant material, by dissolving into the refluxing solvent. The repetitive contact of unsaturated solvent to the plant material allows maximum extraction of the components. This eliminates the inefficiency caused by solvent saturation. The near-boiling temperature of the solvent allows for more extraction kinetic of the system. However, longer extraction time is needed, and the components can be thermally degraded [86, 87].

9.3.3 Cold-press method

Cold-press method, also known as the scarification method, is a method of essential oil extraction done to obtain seeds, flowers, lemon, and tangerine oils. In this method, the resulting oils (carrier oil and essential oils) are of highest purity, while retaining almost all properties of the plant material. The method does not need other substance for extraction but rather uses mechanical forces in the form of crushing and pressing. The method is done at low temperature and pressure, thus bearing the name, cold press. The whole plant or plant material is squeezed to release the essential oil from the pouches. The plant residue and oil are separated by means of gravitational or centrifugal forces. Examples of cold-press machines are gear, rack, and level, hydraulic press, power screws, and two-step power screws [88]. Higher essential oil yields are also obtained using this method, as compared to the hydrodistillation method. However, the amount of limonene extracted from *Citrus paradisi* and *Citrus grandis* is more favorable in distillation method compared to the cold-press method (*C. paradisi* – 92.83% and 96.06%; *C. grandis* – 32.63% and 55.74% for cold press and distillation methods, respectively). In addition, oils obtained from the distillation method showed greater free radical scavenging activity compared to the cold-press method [89].

Some essential oils are not capable of being distilled, and the quality of the oil will deteriorate. Thus, other techniques involving mechanical processing are more viable. One of the techniques involves the use of the ecuelle, a piquer process. The rinds of the plant materials, usually from the citrus family, are mechanically ruptured with the use of numerous fine needles, which induces the release of the essential oils. A simpler method involves cutting the material and soaking the material in water, after which, the material is pressed against a sponge, where the oils are absorbed, and collection is done by squeezing the sponge [90].

9.3.4 Enfleurage

Enfleurage is a type of cold-press technique using a cold fat as the adsorbent material or lipophilic carrier applied with pressure to the flower petals. In this method, the cold fat is prepared by mixing an equal portion of vegetable and animal fat and heated up to 60 °C with mixing. The fat mixture is placed into a chassis, and the flower petals are placed on top. The essential oils from the flower are then absorbed by the fat and extracted using alcohol. The alcoholic mixture of the essential oil is vacuum-distilled to separate the oil from the alcohol. Other techniques involve the use of hexane and petroleum ether to extract the oil from the cold fat. This method is applied when the oil component can be damaged due to hydrolysis and polymerization process [91]. Components with high boiling temperature cannot be carried together with the water vapor, thus suggesting that heat-inclusive methods are not viable [90, 92].

9.3.5 Innovative approach

9.3.5.1 Ultrasound-assisted extraction

Ultrasound-assisted extraction, also known as ultrasonic extraction or sonication, uses ultrasonic energy. This energy creates cavitation, which enhances the permeation of the solvent to the plant material, allowing more dissolved components into the solvent, which enhances the extraction efficiency [87]. In addition, ultrasound increases the surface wetness evaporation and causes oscillating velocities at the interfaces [93]. This method is advantageous due to the low solvent and energy consumption as well as shorter extraction time and lower extraction temperature. This method is applied for thermolabile and unstable compounds [94, 95]. Moreover, this method is more selective and intensifies essential oil extraction, when combined with other methods such as solvent extraction and hydrodistillation.

Supercritical carbon dioxide extraction offers numerous advantages over conventional methods of extraction of essential oils. This includes shorter preparation time, no to fewer amounts of organic solvent used, high extraction efficiency, prevention of unnecessary degradation, and no toxic solvent residues. In this method, carbon dioxide is used as the supercritical fluid, due to its modest critical points, 31.1 °C and 73 atm [97–99]. In addition, supercritical carbon dioxide is selective and capable of extracting heat-sensitive compounds, low polarity compounds, and small molecules [100, 101]. The use of carbon dioxide provides advantages such as high diffusivity, safety as regards human health and the environment, along with nontoxic, nonreactive, nonflammable, noncorrosive, and reusable [102] properties. The liquid carbon dioxide at high pressure is used as the solvent for extraction but offers the advantage of no residual concentration in the oils, since carbon dioxide reverts to gas at normal atmospheric pressure (Figure 9.9b). Problems like slow extraction rate do occur in using carbon dioxide. Thus, application of 15 min static extraction with methylene chloride, followed by 15 min dynamic extraction with pure carbon dioxide, produces a higher oil recovery. The extraction using this technique is comparable with that of the hydrodistillation technique, wherein more than 90% of the monoterpenes are extracted. More aromatic compounds are reported to be extracted using supercritical carbon dioxide than the conventional methods [103, 104]. Overall, the use of carbon dioxide supercritical extraction proves to be more economically viable than other methods [105].

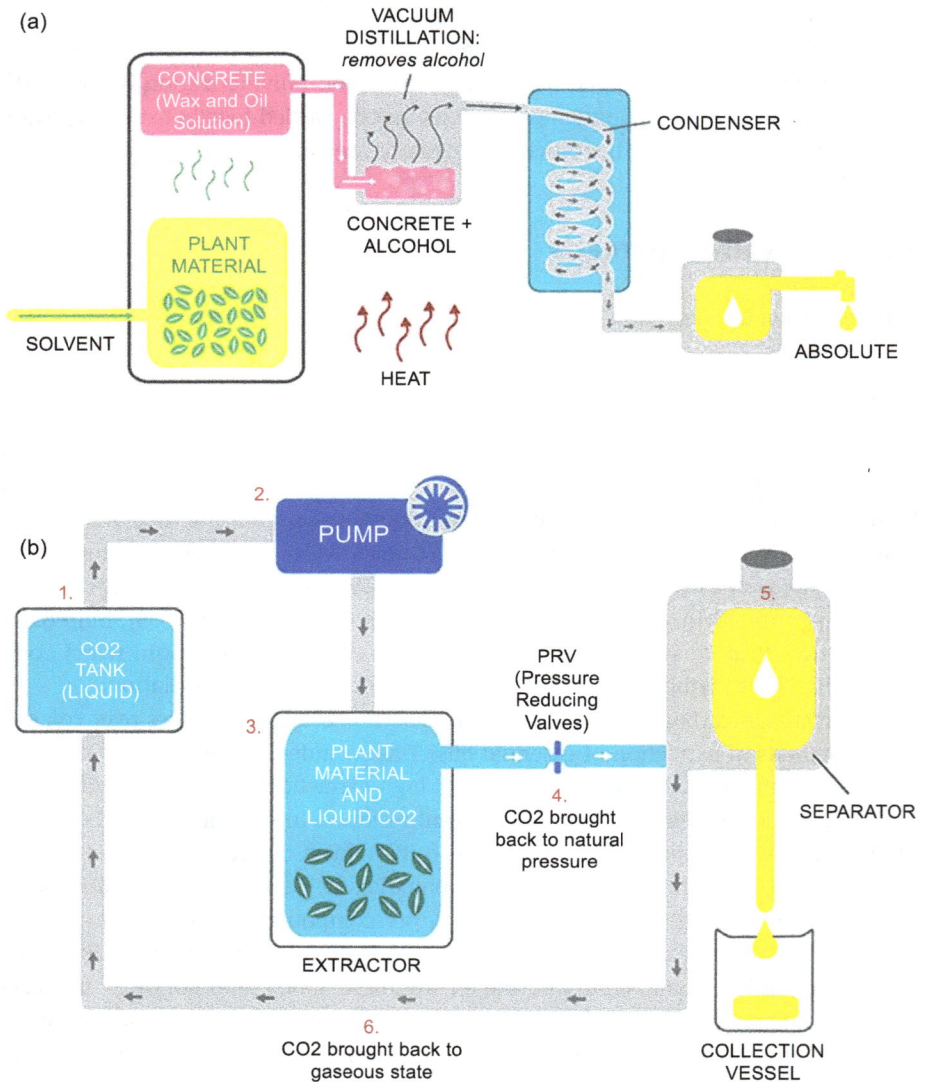

Figure 9.9: Illustration of the (a) solvent extraction forming the concrete and absolute, (b) supercritical carbon dioxide extraction (images taken from the website of New Direction Aromatics [96]).

Subcritical water extraction involves the use of pressurized hot water under dynamic conditions. It means that water is preserved at high pressure and temperature ranging from 100 to 374 °C, while still maintaining the liquid state of water. With this method, the extraction rate is noted to be 5.1 times greater than in hydrodistillation. In addition, this method is quicker, and capable of extracting more valuable essential oils such as oxygenated compound, with little or no trace of terpenes. Due to the efficiency in extraction, a small amount of energy and plant material is used, compared to other methods. As an example, even the use of 100 °C for 2 h or 175 °C for 20 min resulted in a greater amount (40–60%) of extraction compared to the conventional Soxhlet apparatus (6 h extraction) [98, 106].

Some plant materials extracted with the different extraction process as summarized and reported by Tongnuanchan and Benjakul [1], Ozdemir and Gozel [107], and Ashrafur Rahman and coworkers [108] are tabulated in Table 9.3.

Table 9.3: Examples of plants extracted by different extraction methods.

Method	Plants
Solvent extraction	Benzoin (*Styrax benzoin*)
	Sage (*Salvia officinalis*)
	Apiaceae (*Ptychotis verticillate*)
	Chasteberry (*Vitex agnus-castus*)
	Lemon (*Citrus limon*)
Supercritical carbon dioxide extraction	Rosemary (*Rosmarinus officinalis*)
	Fennel (*Foeniculum vulgare*)
	Anise (*Pimpinella anisum*)
	Cumin seed (*Cuminum cyminum*)
	Sage (*Salvia officinalis*)
	Lemon (*Citrus limon*)
	Carrot fruit (*Daucus carrota*)
	Marjoram (*Majorana hortensis*)
	Catnip (*Nepeta cataria* L.)
	Oregano (*Origanum vulgare*)
	Lavender (*Lavandula angustifolia*)
	Thyme (*Thymus vulgaris*)
	Hyssop (*Hyssopus officinalis*)
	Anise hyssop (*Lophantus anisatus*)
	Patchouli (*Pogostemon cablin*)
	Cumin (*Cuminum cyminum*)
	Clove (*Eugenia caryophyllata*)
	Coriander (*Coriandrum sativum*)
	Chamomile (*Matricaria chamomilla*)
	Baccharises (*Baccharis uncinella, Baccharis anomala*, and *Baccharis dentata*)
Subcritical water extraction	*Fructus amomi*
	Marjoram (*Origanum majorana*)
	Olive (*Olea europaea*)
	Coriander seeds (*Coriandrum sativum* L.)

Table 9.3 (continued)

Method	Plants
Steam distillation	Black pepper (*Piper nigrum*)
	Rose-scented geranium (*Pelargonium* sp.)
	Thyme (*Thymus kotschyanus*)
	Germander *(Teucriumorientale)*
	Rosemary (*Rosmarinus officinalis*)
	Fennel (*Foeniculum vulgare*)
	Field mint (*Mentha arvensis*)
	Anise (*Pimpinella anisum*)
	Eucalyptus (*Eucalyptus citriodora*)
	Basil (*Ocimum basilicum*)
	Lavender (*Lavandula dentata*)
	Lemon-scented gum (*Eucalyptus citriodora*)
	Patchouli (*Pogostemon cablin*)
	Ginger (*Zingiber officinalis*)
	Clove (*Eugenia caryophyllata*)
	Orange (*Citrus sinensis*)
	Wild thyme (*Thymus serpyllum*)
	Wormwood (*Artemisia absinthium*)
	Bay laurel (*Laurus nobilis*)
	Muskrat root (*Acorus calamus*)
	Chamomile (*Matricaria chamomilla*)
	Camphor (*Cinnamomum camphora*)
	Caraway (*Carum carvi*)
	Tea tree (*Melaleuca alternifolia*)
	Sandalwood (*Santalbum album*)
	Rosewood (*Dalbergia sissoo*)
	Roseberry (*Fragaria xananassa*)
	Rose (*Rosa damascena*)
	Pine (*Pinus palustris*)
	Nutmeg (*Cyperus rotundus*)
	Lemongrass (*Cymbopogon citratus*)

Table 9.3 (continued)

Method	Plants
Hydrodistillation	Rose-scented geranium (*Pelargonium* sp.) Germander (*Teucrium orientale*) Rosemary (*Rosmarinus officinalis*) Lemon (*Citrus limon*) Oregano (*Origanum vulgare L.)* Marjoram (*Majorana hortensis*) Catnip (*Nepeta cataria*) Lavender (*Lavandula angustifolia*) Hyssop (*Hyssopus officinalis*) Anise hyssop (*Lophantus anisatus*) Sage (*Salvia officinalis*) Cumin (*Cuminum cyminum*) Clove (*Eugenia caryophyllata*) Caraway (*Carum carvi*) Thyme (*Thymus vulgaris*) Basil (*Ocimum basilicum*) Garden mint (*Mentha crispa*) Clove (*Syzygium aromaticum*)
Hydrodiffusion—microwave-assisted extraction	Orange (*Citrus sinensis*) Rosemary leaves (*Rosmarinus officinalis*)
Solvent and steam extraction	Cumin (*Cuminum cyminum*) Tobacco (*Nicotiana tabacum*) Jasmine (*Jasminum polyanthum*)
Cold-press extraction	Bergamot (*Citrus bergamia*) St. John's Wort (*Hypericum perforatum*) Grapefruit (*Citrus paradoso*) Lemon (*Citrus limon*) Lime (*Citrus aurantifolia*) Mandarin (*Citrus reticulata*) Orange (*Citrus sinensis*)

9.4 Essential oils in the pharmaceutical industry

The diversity in the chemical composition of different essential oils allows them to be used in the pharmaceutical industry for different purposes. Such applications are in aromatherapy and as antioxidant, antimicrobial, antifungal, pain relievers, to treat anxiety or depression, and in perfumery [109]. The United States Food and Drug administration has recognized these essential oils as safe for human application (as reported by Cimino and coworkers [110].

Table 9.4: US FDA approved essential oil sources as safe for human consumption.

Common name	Botanical name of plant source
Alfalfa herb and seed	*Medicago sativa* L.
Allspice	*Pimentaofficinalis* Lindl.
Ambrette seed	*Hibiscus abelmoschus* L.
Angelica	*Angelica archangelica* L. or other spp. of Angelica
Angelica root	Do
Angelica seed	Do
Angostura (cusparia bark)	*Galipea officinalis* Hancock.
Anise	*Pimpinella anisum* L.
Anise, star	*Illicium verum Hook.* f.
Balm (lemon balm)	*Melissa officinalis* L.
Basil, bush	*Ocimum minimum* L.
Basil, sweet	*Ocimum basilicum* L.
Bay	*Laurus nobilis* L.
Calendula	*Calendula officinalis* L.
Camomile (chamomile), English or Roman	*Anthemis nobilis* L.
Camomile (chamomile), German or Hungarian	*Matricaria chamomilla* L.
Capers	*Capparis spinosa* L.
Capsicum	*Capsicum frutescens* L. or *Capsicum annuum* L.
Caraway	*Carum carvi* L.
Caraway, black (black cumin)	*Nigella sativa* L.
Cardamom (cardamon	*Elettaria cardamomum* Maton.
Cassia, Chinese	*Cinnamomum cassia* Blume.
Cassia, Padang or Batavia	*Cinnamomum burmanni* Blume.
Cassia, Saigon	*Cinnamomum loureirii* Nees.
Cayenne pepper	*Capsicum frutescens* L. or *Capsicum annuum* L.
Celery seed	*Apium graveolens* L.
Chervil	*Anthriscus cerefolium* (L.) Hoffm.
Chives	*Allium schoenoprasum* L.

Table 9.4 (continued)

Common name	Botanical name of plant source
Cinnamon, Ceylon	*Cinnamomum zeylanicum* Nees.
Cinnamon, Chinese	*Cinnamomum cassia* Blume.
Cinnamon, Saigon	*Cinnamomum loureirii* Nees.
Clary (clary sage)	*Salvia sclarea* L.
Clover	*Trifolium* spp.
Cloves	*Eugenia caryophyllata* Thunb.
Coriander	*Coriandrum sativum* L.
Cumin (cummin)	*Cuminum cyminum* L.
Cumin, black (black caraway)	*Nigella sativa* L.
Dill	*Anethum graveolens* L.
Elder flowers	*Sambucus canadensis* L.
Fennel, common	*Foeniculum vulgare* Mill.
Fennel, sweet (finocchio, Florence fennel)	*Foeniculum vulgare* Mill. var. duice (DC.) Alex.
Fenugreek	*Trigonella foenum-graecum* L.
Galanga (galangal)	*Alpina officinarum* Hance.
Garlic	*Allium sativum* L.
Geranium	*Pelargonium* spp.
Ginger	*Zingiber officinale* Rosc.
Glycyrrhiza	*Glycyrrhiza glabra* L. and other spp. of *Glycyrrhiza*
Grains of paradise	*Amomum melegueta* Rosc.
Horehound (hoarhound)	*Marrubium vulgare* L.
Horseradish	*Armoracia lapathifolia* Gilib.
Hyssop	*Hyssopus officinalis* L.
Lavender	*Lavandula officinalis* Chaix.
Licorice	*Glycyrrhiza glabra* L. and other spp. of *Glycyrrhiza*
Linden flowers	*Tilia* spp.
Mace	*Myristica fragrans* Houtt.

Table 9.4 (continued)

Common name	Botanical name of plant source
Marigold, pot	*Calendula officinalis* L.
Marjoram, pot	*Majorana onites* (L.) Benth.
Marjoram, sweet	*Majorana hortensis* Moench.
Mustard, black, or brown	*Brassica nigra* (L.) Koch.
Mustard, brown	*Brassica juncea* (L.) Coss.
Mustard, white or yellow	*Brassica hirta* Moench.
Nutmeg	*Myristica fragrans* Houtt.
Oregano (oreganum, Mexican oregano, Mexican sage, origan)	*Lippia* spp.
Paprika	*Capsicum annuum* L.
Parsley	*Petroselinum crispum* (Mill.) Mansf.
Pepper, black	*Piper nigrum* L.
Pepper, cayenne	*Capsicum frutescens* L. or *Capsicum annuum* L.
Pepper, red	Do.
Pepper, white	*Piper nigrum* L.
Peppermint	*Mentha piperita* L.
Poppy seed	*Papaver somniferum* L.
Pot marigold	*Calendula officinalis* L.
Pot marjoram	*Majorana onites* (L.) Benth.
Rosemary	*Rosmarinus officinalis* L.
Rue	*Ruta graveolens* L.
Saffron	*Crocus sativus* L.
Sage	*Salvia officinalis* L.
Sage, Greek	*Salvia triloba* L.
Savory, summer	*Satureia hortensis* L. (Satureja).
Savory, winter	*Satureiamontana* L. (Satureja).
Sesame	*Sesamum indicum* L.
Spearmint	*Mentha spicata* L.
Star anise	*Illicium verum* Hook. f.

Table 9.4 (continued)

Common name	Botanical name of plant source
Tarragon	*Artemisia dracunculus* L.
Thyme	*Thymus vulgaris* L.
Thyme, wild, or creeping	*Thymus serpyllum* L.
Turmeric	*Curcuma longa* L.
Vanilla	*Vanilla planifolia* Andr. or *Vanilla tahitensis* J. W. Moore.
Zedoary	*Curcuma zedoaria* Rosc.

Source: https://www.accessdata.fda.gov/scripts/cdrh/cfdocs/cfcfr/CFRSearch.cfm?fr=182.20.

9.5 Uses of essential oils in pharmaceutics

Essential oils are composed of monoterpenes and sesquiterpenes, aromatic and aliphatic compounds, including other compounds like terpenoids, alcohols, ethers, esters, ketones, and aldehydes [110, 111]. The pharmaceutical potential of essential oils is sourced mostly from aromatic plants [112]. A majority of essential oils are composed of around 20–60 components up to more than 100 single substances in variable concentrations (110). Differences in concentration pose significant heterogeneity in the products used in various studies, making them more complex for comparison [113, 114]. The classification and qualification of essential oils to variable groups can be based on the characteristic physical properties like color and water solubility [115]. While essential oils are naturally complex with multicomponent systems primarily of terpenes and other non-terpene components [112], Essential oils have limited use, due to several factors such as hydrophobicity, instability, and high volatility [116, 117]. Hence, researchers, nowadays, focus on innovative formulation for essential oil encapsulation [110].

Most studies on the applications of essential oils as pure compounds are usually for external applications, due to its safety in use like in mouthwashes and inhalation. However, the use of essential oils for the UV-sensitive skin could result in irritation or darkening skin reactions [110, 112]. Therefore, it is vital to regulate the use of administered dose of essential oils, particularly on broken skin, to avoid significant systemic absorption and other serious side effects. Also, the frequent and rapid inhalation of essential oils is not be recommended as it causes eye irritation. While essential oils are considered generally safe for oral administration, they can be diluted using essential oils, but this is very rare. Nonetheless, the primary limitations to their use are their hydrophobicity, instability, high volatility, and risk of toxicity [112].

9.6 Some essential oils and therapeutic applications

Essential oils have been known since time immemorial for their applications to medicinal and ritual purposes, due to their therapeutic properties [110–112]. A variety of methods and applications were discovered over the centuries with various stereochemical structure, type, and quantity of the essential oil molecules. Most essential oils with promising therapeutic applications possess bioactive compounds that have potential anti-inflammatory, antidiabetic, antivirals, antimicrobial, antioxidant, wound-healing, chemopreventive, chemotherapeutic, and anxiolytic properties [110–112, 115].

Essential oils are also widely used for aromatherapy and massage therapy [110, 112]. The word "aromatherapy" was derived from aroma (a fragrance or sweet smell) and therapy (a treatment) [112]. Aroma and massage therapy are commonly practiced using essential oils for physical and psychological treatment, through massage and inhalation therapy. While the term "aromatherapy" which refers to the application of odorous substances, may be confusing, particularly to nonspecialists, massage therapy using essential oils cannot be considered as aromatherapy [118–120]. Aroma delivery through inhalation that induces psychological or physical effects can be considered as aromatherapy [119]. Even so, the clinical use and interest of essential oils, including the discovery of their volatile constituents used in aromatherapy, have expanded worldwide.

References

[1]   Tongnuanchan P, Benjakul S. Essential oils: Extraction, bioactivities, and their uses for food preservation: Bioactivities and applications of essential oils. J Food Sci 2014, 79, R1231–49, doi: https://doi.org/10.1111/1750-3841.12492.

[2]   Holopainen JK, Blande JD. Where do herbivore-induced plant volatiles go?. Front Plant Sci 2013, 4, doi: https://doi.org/10.3389/fpls.2013.00185.

[3]   Fürstenberg-Hägg J, Zagrobelny M, Bak S. Plant defense against insect herbivores. IJMS 2013, 14, 10242–10297, doi: https://doi.org/10.3390/ijms140510242.

[4]   War AR, Paulraj MG, Ahmad T, Buhroo AA, Hussain B, Ignacimuthu S, et al. Mechanisms of plant defense against insect herbivores. Plant Signal Behav 2012, 7, 1306–1320, doi: https://doi.org/10.4161/psb.21663.

[5]   Butnariu M, Sarac I. Essential oils from plants. JBBS 2018, 1, 35–43, doi: https://doi.org/10.14302/issn.2576-6694.jbbs-18-2489.

[6]   Sakai A, Yoshimura H. Monoterpenes of Salvia leucophylla. CBC 2012, 8, 90–100, doi: https://doi.org/10.2174/157340712799828205.

[7]   Pichersky E, Gershenzon J. The formation and function of plant volatiles: Perfumes for pollinator attraction and defense. Curr Opin Plant Biol 2002, 5, 237–243, doi: https://doi.org/10.1016/S1369-5266(02)00251-0.

[8]   War AR, Sharma HC, Paulraj MG, War MY, Ignacimuthu S. Herbivore induced plant volatiles: Their role in plant defense for pest management. Plant Signal Behav 2011, 6, 1973–1978, doi: https://doi.org/10.4161/psb.6.12.18053.

[9]   Cseke LJ, Kaufman PB, Kirakosyan A. The biology of essential oils in the pollination of flowers. Nat Prod Commun 2007, 2, 1934578X0700201, doi: https://doi.org/10.1177/1934578X0700201225.

[10]   Hare JD. Ecological role of volatiles produced by plants in response to damage by herbivorous insects. Annu Rev Entomol 2011, 56, 161–180, doi: https://doi.org/10.1146/annurev-ento-120709-144753.

[11]   Lee MY. Essential oils as repellents against arthropods. Biomed Res Int 2018, 2018, 1–9, doi: https://doi.org/10.1155/2018/6860271.

[12]   Sangwan NS, Farooqi AHA, Shabih F, Sangwan RS. Regulation of essential oil production in plants. Plant Growth Regul 2001, 34, 3–21, doi: https://doi.org/10.1023/A.

[13]   Sharifi-Rad J, Sureda A, Tenore G, Daglia M, Sharifi-Rad M, Valussi M, et al. Biological activities of essential oils: From plant chemoecology to traditional healing systems. Molecules 2017, 22, 70, doi: https://doi.org/10.3390/molecules22010070.

[14]   Hanuš LO, Hod Y. Terpenes/Terpenoids in Cannabis: Are they important?. Med Cannabis Cannabinoids 2020, 3, 25–60, doi: https://doi.org/10.1159/000509733.

[15]   Baptiste Hzounda Fokou J, Michel Jazet Dongmo P, Fekam Boyom F. Essential oil's chemical composition and pharmacological properties. In: El-Shemy H (Ed), Essential Oils – Oils of Nature, IntechOpen, 2020, doi: https://doi.org/10.5772/intechopen.86573.

[16]   Singh B, Sharma RA. Plant terpenes: Defense responses, phylogenetic analysis, regulation and clinical applications. 3 Biotech 2015, 5, 129–151, doi: https://doi.org/10.1007/s13205-014-0220-2.

[17]   Buckle J. Basic Plant Taxonomy, Basic Essential Oil Chemistry, Extraction, Biosynthesis, and Analysis. Clinical Aromatherapy, Elsevier, 2015, 37–72, doi: https://doi.org/10.1016/B978-0-7020-5440-2.00003-6.

[18]   Bakkali F, Averbeck S, Averbeck D, Idaomar M. Biological effects of essential oils – A review. Food Chem Toxicol 2008, 46, 446–475, doi: https://doi.org/10.1016/j.fct.2007.09.106.

[19]   Burčul F, Blažević I, Radan M, Politeo O. Terpenes, phenylpropanoids, sulfur and other essential oil constituents as inhibitors of cholinesterases. CMC 2020, 27, 4297–4343, doi: https://doi.org/10.2174/0929867325666180330092607.

[20]   Gang DR, Wang J, Dudareva N, Nam KH, Simon JE, Lewinsohn E, et al. An investigation of the storage and biosynthesis of phenylpropenes in sweet basil. Plant Physiol 2001, 125, 539–555, doi: https://doi.org/10.1104/pp.125.2.539.

[21]   Friedrich H. Phenylpropanoid constituents of essential oils. Lloydia 1976, 39, 1–7.

[22]   Deng Y, Lu S. Biosynthesis and regulation of phenylpropanoids in plants. CRC Crit Rev Plant Sci 2017, 36, 257–290, doi: https://doi.org/10.1080/07352689.2017.1402852.

[23]   Le Roy J, Huss B, Creach A, Hawkins S, Neutelings G. Glycosylation is a major regulator of phenylpropanoid availability and biological activity in plants. Front Plant Sci 2016, 7, doi: https://doi.org/10.3389/fpls.2016.00735.

[24]   Cho M-H, Lee S-W. Phenolic phytoalexins in rice: Biological functions and biosynthesis. IJMS 2015, 16, 29120–29133, doi: https://doi.org/10.3390/ijms161226152.

[25]   Jendresen CB, Stahlhut SG, Li M, Gaspar P, Siedler S, Förster J, et al. Highly active and specific tyrosine ammonia-lyases from diverse origins enable enhanced production of aromatic compounds in bacteria and saccharomyces cerevisiae. Appl Environ Microbiol 2015, 81, 4458–4476, doi: https://doi.org/10.1128/AEM.00405-15.

[26] Barros J, Serrani-Yarce JC, Chen F, Baxter D, Venables BJ, Dixon RA. Role of bifunctional ammonia-lyase in grass cell wall biosynthesis. Nat Plants 2016, 2, 16050, doi: https://doi.org/10.1038/nplants.2016.50.

[27] Vogt T. Phenylpropanoid biosynthesis. Mol Plant 2010, 3, 2–20, doi: https://doi.org/10.1093/mp/ssp106.

[28] Liu J, Osbourn A, Ma P. MYB transcription factors as regulators of phenylpropanoid metabolism in plants. Mol Plant 2015, 8, 689–708, doi: https://doi.org/10.1016/j.molp.2015.03.012.

[29] Falcone Ferreyra ML, Rius SP, Casati P. Flavonoids: Biosynthesis, biological functions, and biotechnological applications. Front Plant Sci 2012, 3, doi: https://doi.org/10.3389/fpls.2012.00222.

[30] Liu Q, Luo L, Zheng L. Lignins: Biosynthesis and biological functions in plants. IJMS 2018, 19, 335, doi: https://doi.org/10.3390/ijms19020335.

[31] Vanholme R, Storme V, Vanholme B, Sundin L, Christensen JH, Goeminne G, et al. A systems biology view of responses to lignin biosynthesis perturbations in *Arabidopsis*. Plant Cell 2012, 24, 3506–3529, doi: https://doi.org/10.1105/tpc.112.102574.

[32] Zhao Q, Zhang Y, Wang G, Hill L, Weng J-K, Chen X-Y, et al. A specialized flavone biosynthetic pathway has evolved in the medicinal plant, *Scutellaria baicalensis*. Sci Adv 2016, 2, e1501780, doi: https://doi.org/10.1126/sciadv.1501780.

[33] Widhalm JR, Dudareva N, Familiar A. Ring to it: Biosynthesis of plant benzoic acids. Mol Plant 2015, 8, 83–97, doi: https://doi.org/10.1016/j.molp.2014.12.001.

[34] Chen H-C, Li Q, Shuford CM, Liu J, Muddiman DC, Sederoff RR, et al. Membrane protein complexes catalyze both 4- and 3-hydroxylation of cinnamic acid derivatives in monolignol biosynthesis. Proc Natl Acad Sci USA 2011, 108, 21253–21258, doi: https://doi.org/10.1073/pnas.1116416109.

[35] Chong J, Poutaraud A, Hugueney P. Metabolism and roles of stilbenes in plants. Plant Sci 2009, 177, 143–155, doi: https://doi.org/10.1016/j.plantsci.2009.05.012.

[36] Jeandet P, Delaunois B, Conreux A, Donnez D, Nuzzo V, Cordelier S, et al. Biosynthesis, metabolism, molecular engineering, and biological functions of stilbene phytoalexins in plants. BioFactors 2010, 36, 331–341, doi: https://doi.org/10.1002/biof.108.

[37] Shimizu B-I. 2-Oxoglutarate-dependent dioxygenases in the biosynthesis of simple coumarins. Front Plant Sci 2014, 5, doi: https://doi.org/10.3389/fpls.2014.00549.

[38] Turner GW, Berry AM, Gifford EM. Schizogenous secretory cavities of Citrus limon (L.) Burm. F. and A reevaluation of the lysigenous gland concept. Int J Plant Sci 1998, 159, 75–88, doi: https://doi.org/10.1086/297523.

[39] Bennici A, Tani C. Anatomical and ultrastructural study of the secretory cavity development ofCitrus sinensis and Citrus limon: Evaluation of schizolysigenous ontogeny. Flora Morphol Distrib Funct Ecol Plants 2004, 199, 464–475, doi: https://doi.org/10.1078/0367-2530-00174.

[40] Sayed-Ahmad B, Talou T, Saad Z, Hijazi A, Merah O. The Apiaceae: Ethnomedicinal family as source for industrial uses. Ind Crops Prod 2017, 109, 661–671, doi: https://doi.org/10.1016/j.indcrop.2017.09.027.

[41] Anastasopoulou E, Graikou K, Ganos C, Calapai G, Chinou I. Pimpinella anisum seeds essential oil from Lesvos island: Effect of hydrodistillation time, comparison of its aromatic profile with other samples of the Greek market safe use. Food Chem Toxicol 2020, 135, 110875, doi: https://doi.org/10.1016/j.fct.2019.110875.

[42] Barros L, Carvalho AM, Ferreira ICFR. The nutritional composition of fennel (Foeniculum vulgare): Shoots, leaves, stems and inflorescences. LWT Food Sci Technol 2010, 43, 814–818, doi: https://doi.org/10.1016/j.lwt.2010.01.010.

[43] Bahmani K, Darbandi AI, Ramshini HA, Moradi N, Akbari A. Agro-morphological and phytochemical diversity of various Iranian fennel landraces. Ind Crops Prod 2015, 77, 282–294, doi: https://doi.org/10.1016/j.indcrop.2015.08.059.

[44] Corsi G. Secretory structures and localization of alkaloids in conium maculatum L. (Apiaceae). Ann Bot 1998, 81, 157–162, doi: https://doi.org/10.1006/anbo.1997.0547.

[45] Liu M, Van Wyk B-E, Tilney PM. Irregular vittae and druse crystals in Steganotaenia fruits support a taxonomic affinity with the subfamily Saniculoideae (Apiaceae). S Afr J Bot 2007, 73, 252–255, doi: https://doi.org/10.1016/j.sajb.2006.10.003.

[46] Morris H, Plavcová L, Cvecko P, Fichtler E, Gillingham MAF, Martínez-Cabrera HI, et al. A global analysis of parenchyma tissue fractions in secondary xylem of seed plants. New Phytol 2016, 209, 1553–1565, doi: https://doi.org/10.1111/nph.13737.

[47] Leroux O. Collenchyma: A versatile mechanical tissue with dynamic cell walls. Ann Bot 2012, 110, 1083–1098, doi: https://doi.org/10.1093/aob/mcs186.

[48] Sañudo-Barajas JA, Lipan L, Cano-Lamadrid M, de la Rocha RV, Noguera-Artiaga L, Sánchez-Rodríguez L, et al. Texture. Postharvest Physiology and Biochemistry of Fruits and Vegetables, Elsevier, 2019, 293–314, doi: https://doi.org/10.1016/B978-0-12-813278-4.00014-2.

[49] Marinho CR, Zacaro AA, Ventrella MC. Secretory cells in Piper umbellatum (Piperaceae) leaves: A new example for the development of idioblasts. Flora Morphol Distrib Funct Ecol Plants 2011, 206, 1052–1062, doi: https://doi.org/10.1016/j.flora.2011.07.011.

[50] Periyanayagam K, Jagadeesan M, Kavimani S, Vetriselvan T. Pharmacognostical and phyto-physicochemical profile of the leaves of Piper betle L. var Pachaikodi (Piperaceae) – valuable assessment of its quality. Asian Pac J Trop Biomed 2012, 2, S506–10, doi: https://doi.org/10.1016/S2221-1691(12)60262-7.

[51] Silva RJF, de Aguiar-dias AC, Faial KDC, Mendonça MSD. Morphoanatomical and physicochemical profile of Piper callosum: Valuable assessment for its quality control. Rev Bras Farmacogn 2017, 27, 20–33, doi: https://doi.org/10.1016/j.bjp.2016.07.006.

[52] Gogosz AM, Boeger MRT, Negrelle RRB, Bergo C. Anatomia foliar comparativa de nove espécies do gênero Piper (Piperaceae). Rodriguésia 2012, 63, 405–417, doi: https://doi.org/10.1590/S2175-78602012000200013.

[53] Shirakawa M, Tanida M, Ito T. The cell differentiation of idioblast myrosin cells: Similarities with vascular and guard cells. Front Plant Sci 2022, 12, 829541, doi: https://doi.org/10.3389/fpls.2021.829541.

[54] Jacinto ACP, de Souza LP, Nakamura AT, Carvalho FJ, Simão E, Zocoler JL, et al. Idioblasts formation and essential oil production in irrigated Piper aduncum. Pesqui Agropecu Trop 2018, 48, 447–452, doi: https://doi.org/10.1590/1983-40632018v4853165.

[55] Lewinsohn E, Dudai N, Tadmor Y, Katzir I, Ravid U, Putievsky E, et al. Histochemical localization of citral accumulation in lemongrass leaves (Cymbopogon citratus(DC.) Stapf., Poaceae). Ann Bot 1998, 81, 35–39, doi: https://doi.org/10.1006/anbo.1997.0525.

[56] Tucker S. Intrusive growth of secretory oil cells in Saururus cernuus. Bot Gaz 1976, 137, 341–347.

[57] Zhou Y-F, Mao S-L, Li S-F, Ni X-L, Li B, Liu W-Z. Programmed cell death: A mechanism for the lysigenous formation of secretory cavities in leaves of Dictamnus dasycarpus. Plant Sci 2014, 225, 147–160, doi: https://doi.org/10.1016/j.plantsci.2014.06.007.

[58] Machado NSDO, Pereira FG, Santos PRDD, Costa CG, Guimarães EF. Comparative anatomy of the leaves of Piper lepturum(Kunth) C.DC. var. lepturum and Piper lepturum var. angustifolium (C.DC.) Yunck. Hoehnea 2015, 42, 1–8, doi: https://doi.org/10.1590/2236-8906-21/2013.

[59] Baran P, Aktas K, Ozdemir C. Structural investigation of the glandular trichomes of endemic Salvia smyrnea L. S Afr J Bot 2010, 76, doi: https://doi.org/10.1016/j.sajb.2010.04.011.

[60] Wagner GJ. New approaches for studying and exploiting an old protuberance, the plant trichome. Ann Bot 2004, 93, 3–11, doi: https://doi.org/10.1093/aob/mch011.

[61] Glas J, Schimmel B, Alba J, Escobar-Bravo R, Schuurink R, Kant M. Plant glandular trichomes as targets for breeding or engineering of resistance to herbivores. IJMS 2012, 13, 17077–17103, doi: https://doi.org/10.3390/ijms131217077.

[62] Schilmiller AL, Last RL, Pichersky E. Harnessing plant trichome biochemistry for the production of useful compounds. Plant J 2008, 54, 702–711, doi: https://doi.org/10.1111/j.1365-313X.2008.03432.x.

[63] Wagner GJ. Secreting glandular trichomes: More than just hairs. Plant Physiol 1991, 96, 675–679, doi: https://doi.org/10.1104/pp.96.3.675.

[64] Weathers PJ, Arsenault PR, Covello PS, McMickle A, Teoh KH, Reed DW. Artemisinin production in Artemisia annua: Studies in planta and results of a novel delivery method for treating malaria and other neglected diseases. Phytochem Rev 2011, 10, 173–183, doi: https://doi.org/10.1007/s11101-010-9166-0.

[65] Mellon JE, Zelaya CA, Dowd MK, Beltz SB, Klich MA. Inhibitory effects of gossypol, gossypolone, and apogossypolone on a collection of economically important filamentous fungi. J Agric Food Chem 2012, 60, 2740–2745, doi: https://doi.org/10.1021/jf2044394.

[66] Dayan FE, Duke SO. Trichomes and root hairs: Natural pesticide factories. Pest Outlook 2003, 14, 175, doi: https://doi.org/10.1039/b308491b.

[67] Maffei ME. Sites of synthesis, biochemistry and functional role of plant volatiles. S Afr J Bot 2010, 76, 612–631, doi: https://doi.org/10.1016/j.sajb.2010.03.003.

[68] Tissier A. Glandular trichomes: What comes after expressed sequence tags?: Glandular trichomes: What comes after ESTs?. Plant J 2012, 70, 51–68, doi: https://doi.org/10.1111/j.1365-313X.2012.04913.x.

[69] Turner GW, Gershenzon J, Croteau RB. Distribution of peltate glandular trichomes on developing leaves of peppermint. Plant Physiol 2000, 124, 655–664, doi: https://doi.org/10.1104/pp.124.2.655.

[70] Amenaghawon NA, Okhueleigbe KE, Ogbeide SE, Okieimen CO. Modelling the kinetics of steam distillation of essential oils from lemon grass (Cymbopogon Spp.). Int J Appl Sci Eng 2014, 12, 107–115, doi: https://doi.org/10.6703/IJASE.2014.12(2).107.

[71] Gavahian M, Farhoosh R, Farahnaky A, Javidnia K, Shahidi F. Comparison of extraction parameters and extracted essential oils from Mentha piperita L. using hydrodistillation and steam distillation. Int Food Res J 2015, 22, 283–288.

[72] Zeng Q-H, Zhao J-B, Wang -J-J, Zhang X-W, Jiang J-G. Comparative extraction processes, volatile compounds analysis and antioxidant activities of essential oils from Cirsium japonicum Fisch. ex DC and Cirsium setosum (Willd.) M. Bieb. LWT Food Sci Technol 2016, 68, 595–605, doi: https://doi.org/10.1016/j.lwt.2016.01.017.

[73] Memarzadeh SM, Ghasemi Pirbalouti A, AdibNejad M. Chemical composition and yield of essential oils from Bakhtiari savory (Satureja bachtiarica Bunge.) under different extraction methods. Ind Crops Prod 2015, 76, 809–816, doi: https://doi.org/10.1016/j.indcrop.2015.07.068.

[74] Immaroh NZ, Kuliahsari DE, Nugraheni SD. Review: Eucalyptus globulus essential oil extraction method. IOP Conf Ser Earth Environ Sci 2021, 733, 012103, doi: https://doi.org/10.1088/1755-1315/733/1/012103.

[75] Masango P. Cleaner production of essential oils by steam distillation. J Clean Prod 2005, 13, 833–839, doi: https://doi.org/10.1016/j.jclepro.2004.02.039.

[76] Povh NP, Marques MOM, Meireles MAA. Supercritical CO2 extraction of essential oil and oleoresin from chamomile (Chamomilla recutita [L.] Rauschert). J Supercrit Fluids 2001, 21, 245–256, doi: https://doi.org/10.1016/S0896-8446(01)00096-1.

[77] Vian MA, Fernandez X, Visinoni F, Chemat F. Microwave hydrodiffusion and gravity, a new technique for extraction of essential oils. J Chromatogr A 2008, 1190, 14–17, doi: https://doi.org/10.1016/j.chroma.2008.02.086.

[78] Moradi S, Fazlali A, Hamedi H. Microwave-assisted hydro-distillation of essential oil from rosemary: Comparison with traditional distillation. Avicenna J Med Biotechnol 2018, 10, 22–28.

[79] El Kharraf S, Farah A, Miguel MG, El-Guendouz S, El Hadrami EM. Two extraction methods of essential oils: Conventional and non-conventional hydrodistillation. J Essent Oil Bear Plants 2020, 23, 870–889, doi: https://doi.org/10.1080/0972060X.2020.1843546.

[80] Areias F, Valentão P, Andrade PB, Ferreres F, Seabra RM. Flavonoids and phenolic acids of sage: Influence of some agricultural factors. J Agric Food Chem 2000, 48, 6081–6084, doi: https://doi.org/10.1021/jf000440+.

[81] Pizzale L, Bortolomeazzi R, Vichi S, Überegger E, Conte LS. Antioxidant activity of sage (*Salvia officinalis* and *S fruticosa*) and oregano (*Origanum onites* and *O. indercedens*) extracts related to their phenolic compound content: Antioxidant activity of sage and oregano extracts. J Sci Food Agric 2002, 82, 1645–1651, doi: https://doi.org/10.1002/jsfa.1240.

[82] Koşar M, Dorman HJD, Hiltunen R. Effect of an acid treatment on the phytochemical and antioxidant characteristics of extracts from selected Lamiaceae species. Food Chem 2005, 91, 525–533, doi: https://doi.org/10.1016/j.foodchem.2004.06.029.

[83] Ferhat MA, Tigrine-Kordjani N, Chemat S, Meklati BY, Chemat F. Rapid extraction of volatile compounds using a new simultaneous microwave distillation: Solvent extraction device. Chroma 2007, 65, 217–222, doi: https://doi.org/10.1365/s10337-006-0130-5.

[84] Baydar H, Kineci S. Scent composition of essential oil, concrete, absolute and hydrosol from Lavandin (*Lavandula × intermedia* Emeric ex Loisel.). J Essent Oil Bear Plants 2009, 12, 131–136, doi: https://doi.org/10.1080/0972060X.2009.10643702.

[85] Bayramoglu B, Sahin S, Sumnu G. Solvent-free microwave extraction of essential oil from oregano. J Food Eng 2008, 88, 535–540, doi: https://doi.org/10.1016/j.jfoodeng.2008.03.015.

[86] Zhang Q-W, Lin L-G, Ye W-C. Techniques for extraction and isolation of natural products: A comprehensive review. Chin Med 2018, 13, 20, doi: https://doi.org/10.1186/s13020-018-0177-x.

[87] Rassem H, Nour A, Yunus R. Techniques for extraction of essential oils from plants: A review. Aust J Basic Appl Sci 2016, 10, 117–127.

[88] Geramitcioski T, Mitrevski V, Mijakovski V. Design of a small press for extracting essential oil according VDI 2221. IOP Conf Ser Mater Sci Eng 2018, 393, 012131, doi: https://doi.org/10.1088/1757-899X/393/1/012131.

[89] Ou M-C, Liu Y-H, Sun Y-W, Chan C-F. The composition, antioxidant and antibacterial activities of cold-pressed and distilled essential oils of *Citrus paradisi* and *Citrus grandis* (L.) Osbeck. Evid Complement Altern Med 2015, 2015, 1–9, doi: https://doi.org/10.1155/2015/804091.

[90] Soe'eib S, Asri NP, DiahAgustina P. Enfleurage essential oil from jasmine and rose using cold fat adsorbent. Jurnal Ilmiah Widya Teknik 2016, 15, doi: https://doi.org/10.33508/wt.v15i1.1525.

[91] Nugrahini AD, Baskara CD, Ainuri M. Effectiveness of the usage of various solvent in essential oil production from tuberose flower waste (Polianthes Tuberose L.) by Enfleurage Method. KLS 2016, 3, 104, doi: https://doi.org/10.18502/kls.v3i3.388.

[92] Eltz T, Zimmermann Y, Haftmann J, Twele R, Francke W, Quezada-Euan JJG, et al. Enfleurage, lipid recycling and the origin of perfume collection in orchid bees. Proc R Soc B 2007, 274, 2843–2848, doi: https://doi.org/10.1098/rspb.2007.0727.

[93] García-Pérez JV, Cárcel JA, de la Fuente-blanco S, Riera-franco de Sarabia E. Ultrasonic drying of foodstuff in a fluidized bed: Parametric study. Ultrasonics 2006, 44, e539–43, doi: https://doi.org/10.1016/j.ultras.2006.06.059.

[94] Chemat F, Rombaut N, Sicaire A-G, Meullemiestre A, Fabiano-Tixier A-S, Abert-Vian M. Ultrasound assisted extraction of food and natural products. Mechanisms, techniques, combinations, protocols and applications. A review. Ultrason Sonochem 2017, 34, 540–560, doi: https://doi.org/10.1016/j.ultsonch.2016.06.035.

[95] Barba FJ, Zhu Z, Koubaa M, Sant'Ana AS, Orlien V. Green alternative methods for the extraction of antioxidant bioactive compounds from winery wastes and by-products: A review. Trends Food Sci Technol 2016, 49, 96–109, doi: https://doi.org/10.1016/j.tifs.2016.01.006.

[96] New Direction Aromatics, Inc. A Comprehensive Guide to Essential Oil Extraction Methods 2017. https://www.newdirectionsaromatics.com/blog/articles/how-essential-oils-are-made.html (accessed March 25, 2022).

[97] Señoráns FJ, Ibañez E, Cavero S, Tabera J, Reglero G. Liquid chromatographic–mass spectrometric analysis of supercritical-fluid extracts of rosemary plants. J Chromatogr A 2000, 870, 491–499, doi: https://doi.org/10.1016/S0021-9673(99)00941-3.

[98] Jiménez-Carmona MM, Ubera JL, Luque de Castro MD. Comparison of continuous subcritical water extraction and hydrodistillation of marjoram essential oil. J Chromatogr A 1999, 855, 625–632, doi: https://doi.org/10.1016/S0021-9673(99)00703-7.

[99] Kessler JC, Vieira VA, Martins IM, Manrique YA, Afonso A, Ferreira P, et al. Obtaining aromatic extracts from Portuguese Thymus mastichina L. by hydrodistillation and supercritical fluid extraction with CO2 as potential flavouring additives for food applications. Molecules 2022, 27, 694, doi: https://doi.org/10.3390/molecules27030694.

[100] Wrona O, Rafińska K, Możeński C, Buszewski B. Supercritical fluid extraction of bioactive compounds from plant materials. J AOAC Int 2017, 100, 1624–1635, doi: https://doi.org/10.5740/jaoacint.17-0232.

[101] Azmir J, Zaidul ISM, Rahman MM, Sharif KM, Mohamed A, Sahena F, et al. Techniques for extraction of bioactive compounds from plant materials: A review. J Food Eng 2013, 117, 426–436, doi: https://doi.org/10.1016/j.jfoodeng.2013.01.014.

[102] Herrero M, Mendiola JA, Cifuentes A, Ibáñez E. Supercritical fluid extraction: Recent advances and applications. J Chromatogr A 2010, 1217, 2495–2511, doi: https://doi.org/10.1016/j.chroma.2009.12.019.

[103] Suetsugu T, Tanaka M, Iwai H, Matsubara T, Kawamoto Y, Saito C, et al. Supercritical CO2 extraction of essential oil from Kabosu (Citrus sphaerocarpa Tanaka) peel. Flavor 2013, 2, 18, doi: https://doi.org/10.1186/2044-7248-2-18.

[104] Uwineza PA, Waśkiewicz A. Recent advances in supercritical fluid extraction of natural bioactive compounds from natural plant materials. Molecules 2020, 25, 3847, doi: https://doi.org/10.3390/molecules25173847.

[105] Pereira CG, Meireles MAA. Economic analysis of rosemary, fennel and anise essential oils obtained by supercritical fluid extraction. Flavour Fragr J 2007, 22, 407–413, doi: https://doi.org/10.1002/ffj.1813.

[106] Kubátová A, Miller DJ, Hawthorne SB. Comparison of subcritical water and organic solvents for extracting kava lactones from kava root. J Chromatogr A 2001, 923, 187–194, doi: https://doi.org/10.1016/S0021-9673(01)00979-7.

[107] Ozdemir E, Gozel U. Nematicidal activities of essential oils against meloidogyne incognita on tomato plant. Fresenius Environ Bull 2018, 27, 4511–4517.

[108] Ashrafur Rahman SM, Rainey TJ, Ristovski ZD, Dowell A, Islam MA, Nabi MN, et al. Review on the use of essential oils in compression ignition engines. In: Agarwal AK, Gautam A, Sharma N, Singh AP (Eds), Methanol and the Alternate Fuel Economy, Springer Singapore, Singapore, 2019, 157–182, doi: https://doi.org/10.1007/978-981-13-3287-6_8.

[109] Alasmary AF, Assirey EA, El-Meligy RM, Awaad AS, El-sawaf LA, Allah MM, et al. Analysis of Alpina officinarum Hance, chemically and biologically. Saudi Pharm J 2019, 27, 1107–1112, doi: https://doi.org/10.1016/j.jsps.2019.09.007.

[110] Cimino C, Maurel OM, Musumeci T, Bonaccorso A, Drago F, Souto EMB, et al. Essential oils: Pharmaceutical applications and encapsulation strategies into lipid-based delivery systems. Pharmaceutics 2021, 13, 327, doi: https://doi.org/10.3390/pharmaceutics13030327.

[111] Raut JS, Karuppayil SM. A status review on the medicinal properties of essential oils. Ind Crops Prod 2014, 62, 250–264.

[112] Edris AE. Pharmaceutical and therapeutic potentials of essential oils and their individual volatile constituents: A review. Phytother Res 2007, 21, 308–323.

[113] Malcolm BJ, Tallian K. Essential oil of lavender in anxiety disorders: Ready for prime time?. Ment Health Clin 2017, 7, 147–155.

[114] Bilia AR, Guccione C, Isacchi B, Righeschi C, Firenzuoli F, Bergonzi MC. Essential oils loaded in nanosystems: A developing strategy for a successful therapeutic approach. Evid Complement Altern Med 2014, 651593.

[115] Soetjipto H. Antibacterial properties of essential oil in some Indonesian herbs. In: Potential Essentials Oils, IntechOpen, London, UK, 2018, vol. 41.

[116] Hosseini SF, Zandi M, Rezaei M, Farahmandghavi F. Two-step method for encapsulation of oregano essential oil in chitosan nanoparticles: Preparation, characterization and *in vitro* release study. Carbohydr Polym 2013, 95, 50–56.

[117] Trinetta V, Morgan MT, Coupland JN, Yuce L. Essential oils against pathogen and spoilage microorganisms of fruit juices: Use of versatile antimicrobial delivery systems. J Food Sci 2017, 82, 471–476.

[118] Buchbauer G, Jirovetz L. Aromatherapie-definition und discussionuber den stand der forschung (German). Arstezeitschriftfur Naturheilverfahren 1993, 34, 259–264.

[119] Buchbauer G, Jirovetz L. Aromatherapy use of fragrance and essential oils as medicament. Flavour Fragr J 1994, 9, 217–222.

[120] Buchbauer G, Jirovetz L, Jager W, Plank C, Dietrich H. Fragrance compounds and essential oils with sedative effects upon inhalation. J Pharm Sci 1993, 82, 660–664.

Siham Abdulrazzaq Salim
Chapter 10
Application of essential oils in agriculture and veterinary

Abstract: Although essential oils offer an alternative to conventional pesticides in controlling pathogenic agents, agricultural pests, and weeds, their uncontrolled usage has had a detrimental impact on the environment and living things. However, additional research is still required before botanical products may be fully integrated and used safely in agriculture. Additionally, as antimicrobial resistance poses a serious danger to contemporary veterinary care, excessive use of antibiotics has been linked to the establishment and spread of resistant bacteria, rendering the treatment of infectious diseases in animals ineffective. Because plant essential oils provide a safe alternative to growth stimulants and industrial pharmaceuticals, the majority of current studies have focused on their safe and effective application in animal nutrition and illness treatment.

Keywords: Essential oil, Microorganism, Antibacterial, Environmental friendly

10.1 Introduction

Essential oils are a mixture of organic compounds found in different parts of plants and are the cause of the distinctive smell of plants. Essential oils evaporate when exposed to air at normal temperature, and they volatilize with water vapor. These oils have several names, all bearing the same content, namely, essential oils, volatile oils, and ethereal oils. In general, what distinguishes essential oils from fatty or fixed oils is the volatile characteristic of essential oils [1–3].

Essential oils are found in different places in aromatic plants. They may be found in flowers, as in roses, jasmine, lavender, thyme, and lily, or in leaves, as in mint and basil, or in the bark of some trees (stem bark), as in cinnamon, or in wood such as sandalwood, or found in unopened flower buds like carnations (clove). They may be found in peels of fruits, as in lemons and oranges, or in fruits such as cumin, caraway, anise, and vanilla, and may be found in the roots and rhizomes, as in ginger and turmeric plant [2, 4].

In general, essential oils are formed in the plant during the metabolism process as byproducts (secondary metabolites) that are stored in special structures inside plant organs, including glandular hairs, oil glands, or oil ducts. It is to be noted that these structures protect the plant itself from the damages of essential oils that affect

Siham Abdulrazzaq Salim, Department of Biology, College of Education, Al-Iraqia University, Iraq

https://doi.org/10.1515/9783110791600-010

it [1]. Indeed, essential oils are assumed as viable environmental-friendly alternative agents in agricultural fields due to their proven capacity as antimicrobial [5, 6], antifungal [7], insecticidal [8], nematicidal [9, 10], or herbicidal, and insect repellent capabilities [11]. On the other hand, essential oils and plants extracts have been widely used in veterinary medicine for controlling the diseases of viral, bacterial, and parasitic origin, which shows that these oils have future importance as natural remedies that can override chemical treatments [12–16].

10.1.1 Applications of essential oils in agriculture

It may be difficult for some plant growers to stay away from the use of chemical pesticides because they are effective in killing agricultural pests such as insects, fungi, bacteria, and viruses. In addition, chemical control is a losing war due to the fact that agricultural pests are quick to adapt and have a high ability to generate immunity against pesticides, which leads farmers and agricultural guides into a vicious and deadly cycle, which is known as the pests vortex. Also, the excessive use of chemical pesticides has led to more environmental complex problems, causing disruption to the ecosystem and the death of many beneficial and necessary organisms for the soil and plants. The use of pesticides on crops around the world has been estimated at about 2.5 million tons each year, causing the worldwide damages to reach $100 billion annually.

The major reasons for this are the high toxicity of pesticides and their lack of biodegradation ability, in addition to the fact that their residues in soil, water resources, and crops affect the general health of humans and mammals in the future. On the other hand, the matter becomes difficult for organic farming, which increases the importance of the need to use new highly selective pesticides with biodegradation ability as an alternate to synthetic pesticides as a means to solve the problems of the negative impacts on human health, environment, and for the protection of crops, especially as most of the farmers do not have effective solutions for agricultural pests management as well as for developing techniques that can be used to minimize the use of pesticides while maintaining crop yields.

The recent trend is toward green chemistry and green pesticides through the use of plant products, including essential oils, as an alternative and safe means in the biological control of agricultural pests in a way that ensures the safe production of agricultural crops and the preservation of environmental safety, as compared to synthetic pesticides [17].

10.1.1.1 Essential oils as herbicides

Most essential oils have been investigated for their suspected inhibitory effect on one or both of the processes of seed germination, and shoot growth and development

[18]. This leads to the use of these essential oils as herbicides for biocontrol of weeds due to the presence of allelochemical compounds through allelopathy, which refers to plant–plant interaction (Table 10.1).

Table 10.1: Herbicidal properties of some essential oils [18].

Essential oils extracted from	Tested plants
Achillea gypsicola	*Amaranthus retroflexus* *Chenopodium album* *Cirsium arvense* *Lactuca serriola* *Rumex crispus*
Achillea biebersteinii	*Amaranthus retroflexus* *Chenopodium album* *Cirsium arvense* *Lactuca serriola* *Rumex crispus*
Angelica glauca	*Lemna minor*
Citrus aurantifolia	*Avena fatua* *Echinochloa crus-galli* *Phalaris minor*
Citrus × limon L.	*Portulaca oleracea*
Coriandrum sativum L.	*Amaranthus retroflexus* *Chenopodium album* *Echinochloa crus-galli*
Eucalyptus spp.	Annual ryegrass *Echinochloa crus-galli* *Lolium multiflorum* *Nicotiana glauca* *Phalaris minor* *Phalaris canariensis* *Parthenium hysterophorus* *Portulaca oleracea* *Sinapis arvensis* *Solanum elaeagnifolium*
Lavandula spp.	*Lolium rigidum*
12 Mediterranean species	*Lactuca sativa* *Lepidium sativum* *Raphanus sativus*
Origanum acutidens	*Amaranthus retroflexus* *Chenopodium album* *Rumex crispus*

Table 10.1 (continued)

Essential oils extracted from	Tested plants
Origanum vulgare L.	*Hordeum vulgare* *Lepidium sativum* *Matricaria chamomilla* L. *Sinapis alba* *Triticum aestivum*
Peumus boldus	*Portulaca oleracea*
Pinus nigra	*Phalaris canariensis* *Trifolium campestre* *Sinapis arvensis*
Pinus pinea	*Lolium rigidum* *Raphanus raphanistrum* *Sinapis arvensis*
Plectranthus rugosus	*Lemna minor*
Rosmarinus officinalis	*Amaranthus retroflexus* *Matricaria chamomilla* L. *Phalaris minor* *Rhaphanus sativus* *Silybum marianum* *Trifolium incarnatum*
Syzygium aromaticum	*Redwood pigweed*
Tagetes erecta	*Echinochloa crus-galli* L.
Tanacetum aucheranum	*Amaranthus retroflexus* *Chenopodium album* *Rumex crispus*
Tanacetum chiliophyllum	*Amaranthus retroflexus* *Chenopodium album* *Rumex crispus*
Tetraclinis articulata	*Phalaris canariensis* *Sinapis arvensis*
25 various plants	*Taraxacum officinale*
Valeriana wallichii	*Lemna minor*

10.1.1.2 Essential oils as antipathogenic on different plant pathogens

The use of essential oils could mean the appearance of new plant protection products to control the microbial pathogens and prevent their propagation and resistance [19]. Most pathogenic bacteria and fungi in crops can cause serious problems and symptoms in different parts of plants, such as cankers, rots, spots, wilting, among others [20]. Different studies have shown that essential oils offer antibacterial and antifungal properties against pathogens, which can cause post-harvest diseases in vegetables and fruits. In fact, the use of plants, herbs, or spices (for medical, preservatives, and/or pest control) has been reported in the ancient civilizations of Egypt, Greece, Rome, among others [21] and their properties have been systematically analyzed in laboratories since the beginning of the last century, and even earlier. Besides, they have demonstrated their efficacy against multidrug-resistant/antibiotic-resistant bacteria and fungi.

Essential oils have different components and it is possible that their actions work on different targets in the bacterial and fungal cells (Table 10.2). Their action

Table 10.2: Antipathogenic properties of some essential oils.

Essential oils extracted from		Target pathogens	Diseases caused by the pathogens
Achillea biebersteinii *Achillea millefolium*	**Bacteria**	*Clavibacter michiganensis*	Ring rot disease
Ocimum ciliatum		*Rhodococcus fascians*	Leafy gall disease
Origanum heracleoticum *Origanum onites* *Salmea scandens* *Satureja hortensis* *Satureja spicigera* *Solidago canadensis* L. *Tanacetum aucheranum* *Thymus fallax*		*Clavibacter michiganensis*	Ring rot disease
Achillea biebersteinii		*Pseudomonas* spp. *Xanthomonas* spp.	Bacterial canker Bacterial spots and blights
Achillea millefolium		*Pseudomonas* spp. *Xanthomonas* spp.	Bacterial canker Bacterial spots and blights
Citrus aurantium L.		*Agrobacterium tumefaciens* *Dickeya solani* *Erwinia amylovora*	Crown gall Black legands of trot Fire blight
Citrus reticulata		*Pseudomonas aeruginosa*	Soft rot

Table 10.2 (continued)

Essential oils extracted from		Target pathogens	Diseases caused by the pathogens
Cleistocalyx operculatus	**Bacteria**	*Xanthomonas* spp.	Bacterial spots and blights
Cynara scolymus(stems)		*Erwinia amylovora*	Fire blight
		Erwinia carotovora	Soft rot
		Pseudomonas syringae	Bacterial canker
		Xanthomonas vesicatoria	Bacterial leaf spot
Eriocephalus africanus L.		*Agrobacterium tumefaciens*	Crown gall
		Dickeya solani	Black leg and soft rot
		Erwinia amylovora	Fire blight
		Pseudomonas cichorii	Leaf blight and spots
Juglans regia L.(shells)		*Erwinia amylovora*	Fire blight
		Erwinia carotovora	Soft rot
		Pseudomonas syringae	Bacterial canker
		Xanthomonas vesicatoria	Bacterial leaf spot
Metasequoia glyptostroboides		*Xanthomonas* spp.	Bacterial spots and blights
Ocimum ciliatum		*Agrobacterium vitis*	Crown gall
		Brenneria nigrifluens	Cankers
		Pantoea stewartii	Stewart's wilt and leaf
		Pseudomonas spp.	blight
		Ralstonia solanacearum	Bacterial canker
		Xanthomonas spp.	Bacterial wilt
			Bacterial spots and blights
Ocimum basilicum		*Pseudomonas aeruginosa*	Soft rot
Origanum heracleoticum		*Pseudomonas* spp.	Bacterial canker
		Xanthomonas sp.	Bacterial spots and blights
Origanum majorana		*Pseudomonas* spp.	Bacterial canker
		Xanthomonas sp.	Bacterial spots and blights
Origanum onites		*Pseudomonas* spp.	Bacterial canker
		Xanthomonas spp.	Bacterial spots and blights
Origanum vulgare		*Erwinia amylovora*	Fire blight
		Erwinia carotovora	Soft rot
		Pseudomonas syringae	Bacterial canker
		Xanthomonas vesicatoria	Bacterial leaf spot
		Pseudomonas syringae	Bacterial canker
		Pseudomonas spp.	Bacterial canker

Table 10.2 (continued)

Essential oils extracted from		Target pathogens	Diseases caused by the pathogens
Piper sarmentosum	**Bacteria**	Xanthomonas oryzae pv. oryzae	Bacterial blight
Salmea scandens		Pseudomonas syringae	Bacterial canker
		Erwinia carotovora	Soft rot
Satureja hortensis		Pseudomonas spp.	Bacterial spots and blights
		Xanthomonas spp.	Bacterial spots and blights
Satureja spicigera		Pseudomonas spp.	Bacterial canker
		Xanthomonas spp.	Bacterial spots and blights
Solidago canadensis L.		Pseudomonas spp.	Bacterial canker
		Xanthomonas sp.	Bacterial spots and blights
Syzygium aromaticum		Erwinia amylovora	Fire blight
		Erwinia carotovora	Soft rot
		Pseudomonas syringae	Bacterial canker
		Xanthomonas vesicatoria	Bacterial leaf spot
Tanacetum aucheranum		Pseudomonas spp.	Bacterial canker
		Xanthomonas spp.	Bacterial spots and blights
Tanacetum chiliophyllum		Agrobacterium tumefaciens	Crown gall
		Pseudomonas spp.	Bacterial canker
		Xanthomonas spp.	Bacterial spots and blights
Thymus fallax		Pseudomonas spp.	Bacterial canker
		Xanthomonas spp.	Bacterial spots and blights
Thymus vulgaris		Pseudomonas syringae	Bacterial canker
Vetiveria zizanioides		Pseudomonas aeruginosa	Soft rot
Zataria multiflora		Xanthomonas campestris	Black rot and leaf spot
Carum carvi L. Carum opticum L. Foeniculum vulgare L. Echinophora platyloba (seed) Lavandula angustifolia Laurus nobilis Origanum vulgare L Salvia sclarea Thymus vulgaris L. Thymus zygiis	**Fungi**	Alternaria alternata	Leaf spot, alternariose
Asarum heterotropoides		Alternaria humicola	Alternariose

Table 10.2 (continued)

Essential oils extracted from		Target pathogens	Diseases caused by the pathogens
Angelica archangelica	**Fungi**	*Alternaria solani*	Early blight
Genista quadriflora *Pinus pinea* *Pulicaria mauritanica* *Warionia saharae*		*Alternaria* spp.	Alternariose
Citrus × limon L.		*Aspergillus carbonarius*	Ochra toxin producer
Angelica glauca *Artemisia nilagirica* *Curcuma longa* *Eucalyptus* spp. *Ferula galbaniflua* *Mentha × piperita* *Mentha spicata* *Michelia alba* *Ocimum basilicum* *Origanum* spp. *Plectranthus rugosus* *Rosmarinus officinalis* L *Schinusmole* L *Syzygium aromaticum* *Tagetes minuta* L *Thymus capitatus* *Valeriana wallichii* *Vetiveria zizanioides*		*Aspergillus flavus*	Rot and mold, Afla toxins production, aspergillosis
Cinnamomum zeylanicum *Cymbopogon citratus* *Origanum vulgare* L. *Syzygium aromaticum* L. *Thymus vulgaris* L *Zingiber officinale* Rosc		*Aspergillus* spp.	Rot and mold, Afla toxins production, aspergillosis
Piper sarmentosum		*Bipolaris oryzae*	Brown spot
Eucalyptus erythrocorys *Pinus pinea*		*Bipolaris sorokiniana*	Leaf blight/spot

Table 10.2 (continued)

Essential oils extracted from		Target pathogens	Diseases caused by the pathogens
Angelica archangelica *Carum carvi* L *Cestrum nocturnum* *Cinnamomum cassia* *Eucalyptus erythrocorys* *Foeniculum vulgare* L *Mentha pulegium* *Melissa officinalis* *Origanum heracleoticum* *Origanum majorana* *Solidago canadensis* L. *Tetraclinis articulate* *Thymus* spp.		*Botrytis cinerea*	Gray mold
Cinnamomum camphora *Syzygium cumini*		*Choanephora cucurbitarum*	Fruit and blossom rot
Citrus × limon L. *Eugenia caryophillata* L. *Lavandula angustifolia* *Origanum vulgare* L *Salvia sclarea* *Thuja plicata* *Thymus vulgaris* L.		*Cladosporium cladosporioides*	Rot
Cestrum nocturnum *Metasequoia glyptostroboides* *Piper chaba*		*Colletotrichum capsici*	Leaf spot
Asarum heterotropoides *Cymbopogon* sp.		*Colletotrichum gloeosporioides*	Leaf spot
Echinophora platyloba (seed)	**Fungi**	*Colletotrichum trichellum* *Curvularia fallax* *Cytospora sacchari*	Leaf spot Black sheath spot Stem canker
Citrus × limon L.		*Eurotium herbariorum*	Mold
Eucalyptus erythrocorys		*Fusarium avenaceum*	Ear blight and root rot on cereals

Table 10.2 (continued)

Essential oils extracted from		Target pathogens	Diseases caused by the pathogens
Echinophora platyloba (seed) Eucalyptus erythrocorys Genista quadriflora Mentha × piperita Metasequoia glyptostroboides Mikania scandens Piper chaba Salmea scandens Syzygium aromaticum		Fusarium oxysporum	Fusarium wilt
Zanthoxylum bungeanum		Fusarium sulphureum	Dry rot
Curcuma longa	**Fungi**	Fusarium verticillioides	Ear rot on maize
Angelica glauca Asarum heterotropoides Eucalyptus erythrocorys Marrubium vulgare Metasequoia glyptostroboides Piper chaba Plectranthus rugosus Valeriana wallichii		Fusarium solani	Root rot, soft rot of plant tissues
Thymus spp.		Geotrichum citri-aurantii	Sour rot (post-harvest)
Myrcialundiana		Lasiodiplodia theobromae	Rot and die back (forest species)
Echinophora platyloba (seed) Mentha × piperita Ocimum basilicum		Macrophomina phaseolina	Damping-off, seedling blight, rot
Pinus pinea		Microdochium nivale	Patch lawn disease
Mentha pulegium Solidago canadensis L.		Monilinia fructicola	Brown rot

Table 10.2 (continued)

Essential oils extracted from		Target pathogens	Diseases caused by the pathogens
Carum carvi L. *Carum opticum* L. *Foeniculum vulgare* L. *Marrubium vulgare* *Ocimum basilicum* L *Satureja hortensis* L. *Thymus* spp.		*Penicillium digitatum*	Green mold (post-harvest)
Melissa officinalis *Pulicaria mauritanica* *Solidago canadensis* L. *Warionia saharae*		*Penicillium expansum*	Postharvest mold
Thymus spp. *Rosmarinus officinalis*		*Penicillium italicum*	Blue mold
Allium sativum L. *Mentha × piperita* *Origanum onites* L. *Salvia officinalis* L.		*Penicillium verrucosum*	Ochra toxin producer
Syzygium aromaticum *Zanthoxylum xanthoxyloides*	**Fungi**	*Phytophthora megakarya*	Black pod disease
Mikania scandens *Thymus* spp.		*Pythium* spp.	Root rot
Angelica archangelica *Asarum heterotropoides* *Cestrum nocturnum* *Bunium persicum* *Foeniculum vulgare* *Juniperus polycarpus* *Mentha* spp. *Metasequoia glyptostroboides* *Mikania scandens* *Ocimum basilicum* *Piper chaba* *Piper sarmentosum* *Syzygium cumini* *Thymus vulgaris* *Zingiber officinale*		*Rhizoctonia solani*	Damping-off, root and stem rot

Table 10.2 (continued)

Essential oils extracted from	Target pathogens	Diseases caused by the pathogens
Ocimum basilicum *Vetiveria zizanioides*	*Rhizopus microsporus*	Rice seedling blight, various head, grain, and ear rots
Foeniculum vulgare Mill. *Melissa officinalis* *Ocimum basilicum* L. *Pulicaria mauritanica* *Satureja hortensis* L. *Thymus vulgaris* L. *Warionia saharae*	*Rhizopus stolonifer*	Storage(post-harvest rot)
Cestrum nocturnum *Metasequoi aglyptostroboides* *Ziziphora clinopodioides*	*Sclerotinia sclerotiorum*	White mold
35 plant's botanical species	*Verticillium dahliae*	*Verticillium* wilt
18 plant's botanical species	*Villosiclava virens*	Rice false smut

mechanisms are unknown, but they can be linked to their capacity to dissolve/disrupt the integrity of the bacterial cell walls, fungi membranes, and the cell walls; the antifungal properties are based on the inhibition of the fungal mycelial. Also, It can be said that their antimicrobial activity is due to the solubility of essential oils in the bilayer of these pathogens.

The use of essential oils (or their respective plants of origin) to combat these types of organisms is not a minor issue as the microbial pathogens cause different diseases in plants, which can have a notable economic impact. Due to this impact, in recent years, numerous studies have been carried out on the antibacterial and antifungal properties of essential oils against the different pathogens that can attack plants and food [18, 22].

10.1.1.3 Essential oils as green pesticides

Most previous studies and researches indicate that essential oils are used as pesticides for agricultural pests rather than as growth stimulators for agricultural crops, possibly because of their ability to volatile when sprayed on plants; so the focus is only on their effects on agricultural pests that cause significant losses in crop production [17].

There are many examples of essential oils that are used as pesticides for agricultural pests. For example, clove (*Eugenia caryophyllus*), lemon grass (*Cimbopogon winteriana*), *Eucalyptus globulus*, rosemary (*Rosemarinus officinalis*), thyme (*Thymus vulgaris*), and vetiver (*Vetiveria zizanioides*) are known for their pest control properties. On the other hand, penny royal (*Mentha pulegium*) wards off ants, fleas, mosquitoes, and moths; peppermint (*Mentha piperita*) repels ants, flies, and moths.

Basil (*Ocimum basilicum*) and spearmint (*Mentha spicata*) are also effective in warding off flies, while, essential oil from plants like *Artemesia vulgaris, Juniperus virginiana, Lavandula angustifolia, Melaleuca leucadendron, Mentha piperita*, and *Pelargonium roseum* are also effective against various insects. Essential oil from *Abelmoschus moschatus, Acorus calamus, Bunium persicum, Cedrus* spp., *Cinnamomum zeylanicum, Cuminum cymium, Cymbopogon citratus, Eugenia caryophyllata, Foeniculum vulgare, Gaultheria procumbens, Lavandula angustifolia* syn. *officinalis, Ocimum* spp., *Piper* spp., *Rabdosia melissoides, Tanacetum vulgare*, and *Trachyspermum ammi* are also known for their varied pest control properties [23].

Essential oils are also used as anti-feedants, either as repellents (without making direct contact with insects) or as suppressants, or as deterrents from feeding (once contact has been made with the insects). Essential oil of *Eucalyptus camaldulensis* var. *obtusa* and *Luvanga scandens* are effective as feeding deterrent against tobacco cutworm (*S. litura*)larvae. Feeding deterrence activities of *Curcuma longa* leaf's essential oil against adult and larvae of grain borer, *R.domestica*; rice weevil, *S.oryzae,* and red flour beetle were also tested [24].

10.1.1.4 Essential oils as food additives for honey bees

Honey is a natural food. It has been used since ancient times for nutrition and for its pharmacological properties. A study was carried out on the use of essential oils of basil, cinnamon, cloves, juniper, mint, oregano, rosemary and thyme in the supplementary feeding of bee families in spring season for three weeks to understand the effect of essential oils on the chemical components and antibacterial activity of the resulting honey [25]. Results of the study showed that there was a positive effect on the chemical composition of honey, through the determination of its acidity, flavonoid content (FC), humidity, impurities, pH, reducing sugar content, total phenolic content (TPC) and antioxidant activity, as well as the antibacterial activity when all of the essential oils were used, as compared with the batch that was fed only with sugar syrup. This opens up hope in the future to rely on essential oils instead of artificial nutrients to feed bee families to improve the good quality of honey produced, since honey bees feed naturally on the nectar of different plant flowers.

10.1.2 Applications of essential oils in veterinary

As the world's population grows, intensifying and manufacturing animal production to meet the growing demand for food involves food safety opportunities and challenges. These challenges place greater responsibility on food producers and their growers to ensure food safety. Over the past decade, all countries around the world have witnessed the spread of serious diseases transmitted by food of animal origin, as unsafe food poses a global threat to everyone's health. Governments should therefore make food safety a public health priority because they play a central role in the development of policies and regulatory frameworks, and in the establishment and implementation of effective food safety systems.

Veterinarians and veterinary researchers provide food security to all animals and treat them from various diseases to ensure their safety and prevent transmission of diseases from these animals to humans in case of contact with them or if food is consumed from these animals. For the treatment of animals, veterinarians use antibiotics to treat infections caused by microorganisms. However, the overuse of antibiotics in veterinary medicine has been associated with the emergence and spread of resistant microbes, making the treatment of infectious diseases ineffective in animals, as antimicrobial resistance is a major threat to modern animal medicine. For this sake, most recent studies focus on the safe and optimal use of plant essential oils in animal nutrition and treatment of diseases as it is a safe alternative to growth promoters and industrial drugs.

10.1.2.1 Essential oils as animal nutrition additives

Essential oils were used as feed additives in animal nutrition for improving food flavor, palatability and to increase the performance of animals. Most of the advantages of essential oils are associated with their role on many metabolic pathways, like stimulation of digestive enzyme secretion and activity, including on lipid metabolism and enhanced gut integrity of animals. Numerous feeding trials have been performed with such additives, but most results are reduced to growth-promoting parameters like feed intake, weight gain, and feed conversion rate [26–29].

In poultry nutrition, most of the achieved studies have shown no significant changes in feed intake parameters caused by essential oils additives, although growth was often enhanced and the feed conversion rate improved. In broiler chickens, when using a mixture of essential oils of anise, citrus, laurel, oregano, and sage in food additives; positive effects were detected on increasing feed intake, growth, and body weight gain, in a linear pattern [30]. On the other hand, feeding a mixture of essential oils or garlic oil to the diet of a laying hen led to increased egg production, increased egg weight, and reduced percentage of broken/cracked eggs [30].

In ruminants, there are less known reports about the effects of essential oils on feed intake and palatability, when the terpenes in essential oils affect the feed intake in sheep, which is quite important when grazing in Mediterranean farms where the potential of essential oils as manipulators of rumen metabolism is significant. This was better for altering feed managements [31].

10.1.2.2 Essential oils as anti-pathogenic in veterinary medicine

In the last few years, essential oils have been introduced in veterinary medicine. There is an increase in the number of studies related to understanding the characteristics of essential oils as anti-pathogens of bacterial and fungal diseases in animals due to the emergence of resistant strains of these pathogens for antibiotics and chemical drugs. On the other hand, the presence of these drugs' residues in animal food origin can cause health damage to humans in the future as a result of the consumption of these foods [32]. Hence, the trend in recent studies is on evaluating essential oils as an alternative to the treatment of animal infections and diseases in veterinary medicine.

Most studies have been conducted on the effect of essential oils on bacterial and fungal strains in vitro, but studies on their effect in vivo against animal clinical bacterial and fungal isolates are rare. On the other hand, few or rare studies have been conducted on the effect of essential oils on algae, specially *Prototheca* spp., which is a parasitic algae that lacks chlorophyll, and which can cause certain diseases (protothecosis) to economically important animals [32]. The use of essential oils in the therapy of farm and companion animals should follow careful studies on the side effects or toxicity of these oils in relation to animal species and the route of their administration. Moreover, considering the different behavior of these oils in relation to both the species and the pathogen strain, before starting a therapy, an aromatogram should be executed to choose the oil with the best antimicrobial activity. The effect of essential oils as antibacterial or antifungal can be observed in the results presented in Table 10.3.

10.1.2.3 Essential oils against ectoparasites in veterinary medicine

Most ectoparasites that infect different species of animals cause huge economic losses in livestock, which are resistant to industrial drugs. There is also the problem of the presence of residues of these drugs in food and the environment, leading to pollution problems. Therefore, the use of essential oils as alternative protocols have played remarkable role in controlling different ectoparasites (flies, lice, mites, and ticks) that infect economically important animals. This importance calls for further research as studies in this area are still in their early stage [33, 34]. In studies on the

Table 10.3: Antipathogenic properties of some essential oils in veterinary.

Essential oils extracted from	Target pathogens	Animal source
Zingiber officinale	*Mycobacterium* spp. (bacteria)	Not specified
Cinnamomum verum *Syzigium aromaticum*	*Streptococcus iniae* (bacteria)	Fish
Backhousia citriodora *Leptospermum* sp. *Melaleuca alternifolia*	*Campylobacter jejuni* (bacteria)	Poultry
Cinnamomum zeylanicum *Ocimum basilicum* *Syzygium aromaticum*	*Escherichia coli* (bacteria)	Poultry
Cinnamomum zeylanicum *Cymbopogon citratus* *Litsea cubeba* *Mentha piperita* *Ocimum basilicum* *Origanum vulgare* *Pelargonium graveolens* *Syzygium aromaticum* *Thymus vulgaris*	*Escherichia coli* (bacteria)	Dog
Cinnamomum zeylanicum *Syzygium aromaticum*	*Salmonella enterica* (bacteria)	Poultry
Ocimum basilicum *Rosmarinum officinalis* *Salvia sclarea*	*Pseudomonas aeruginosa* (bacteria)	Dog
Abies alba *Carum carvi* *Citrus aurantium* var. Dulce *Cinnamomum zeylanicum* *Foeniculum vulgare* *Lavandula angustifolia* *Mentha piperita* *Origanum vulgare* *Pinus montana* *Salvia officinalis*	*Pseudomonas* spp. (bacteria)	Fish
Lavandula angustifolia *Thymus vulgaris*	*Staphylococcus* spp. (bacteria)	Cattle
Thymus vulgaris	*Streptococcus suis* (bacteria)	Swine
Thymus vulgaris	*Enterococcus* spp. (bacteria)	Dog/cat
Cinnamomum sp. *Origanum vulgare*	*Mycobacterium avium* (bacteria)	Cattle

Table 10.3 (continued)

Essential oils extracted from	Target pathogens	Animal source
Commiphora myrrha Helichrysum trilineatum Leptospermum scoparium Melissa officinalis Origanum vulgare Satureja montana Thymus vulgaris	Staphylococcus spp. (bacteria)	Dog
Coriandrum sativum Cymbopogon nardus Helichrysum pandurifolium Litsea cubeba Origanum vulgare Satureja montana Thymus vulgaris	Microsporum canis (fungi)	Cat, dog, horse
Clinopodium nepeta Litsea cubeba Thymus numidicus	Trichophyton mentagrophytes (fungi)	Dog
Thymus serpyllum Litsea cubeba	Trichophyton mentagrophytes (fungi)	Cat
Mixture of Origanum vulgare, Rosmarinus officinalis, Thymus serphyllum	Trichophyton mentagrophytes (fungi)	Bovine
Melaleuca alternifolia	Trichophyton equinum (fungi)	Horse
Aloysia triphylla Chamaecyparis obtusa Cymbopogon spp. Melaleuca alternifolia Mentha piperita Origanum vulgare Satureja montana Thymus spp. Zataria multiflora	Malassezia pachydermatis (fungi)	Dog
Aloysia triphylla Cymbopogon citratus	Aspergillus fumigatus (fungi)	Poultry
Leptospermum petersonii Origanum syriacum	Aspergillus fumigatus (fungi)	Birds
Illicium verum Litsea cubeba Origanum vulgare Rosmarinus officinalis	Aspergillus fumigatus (fungi)	Dog

Table 10.3 (continued)

Essential oils extracted from	Target pathogens	Animal source
Origanum majorana *Origanum vulgare*	*Sporothrix brasiliensis* (fungi)	Cat/dog
Japanese cypress oil *Cymbopogon martinii*	*Microsporum gypseum* (fungi)	Stock culture strains
Cinnamomum sp.	*Pseudogymnoascus destructans* (fungi)	Bat
Cymbopogon flexuosus *Litsea cubeba* *Syzygium aromaticum*	*Ascosphaera apis* (fungi)	Bee chalk brood
Cryptocarya alba	*Nosema ceranae* (fungi)	Bee
Anethum graveolens *Mentha suaveolens* *Santolina chamaecyparissus*	*Candida* spp. (yeasts)	Mice
Origanum vulgare *Thymus vulgaris*	*Candida krusei* (yeasts)	Poultry
Origanum floribundum	*Candida albicans* (yeasts)	Bovine
Anethum graveolens *Geranium* sp. *Mentha suaveolens* *Santolina chamaecyparissus*	*Candida albicans* (yeasts)	Stock culture strains
Origanum vulgare *Rosmarinus officinalis* *Salvia sclarea*	*Candida tropicalis* (yeasts)	Dog
Atractyloides lancea *Cymbopogon flexuosus* *Eucalyptus camaldulensi* *Laureliopsis philippiana* *Origanum majorana* *Thymus vulgaris* *Zataria multiflora*	*Saprolegnia* spp. (oomycetes)	Fish
Citrus bergamia	*Saprolegnia parasitica* (oomycetes)	Fish
Melaleuca alternifolia	*Pythium insidiosum* (oomycetes)	Horse and dog
Mentha piperita *Origanum majorana* *Origanum vulgare*	*Pythium insidiosum* (oomycetes)	Horse

Table 10.3 (continued)

Essential oils extracted from	Target pathogens	Animal source
Citrus bergamia Litsea cubeba Origanum vulgare Thymus vulgaris	Prototheca blaschkeae (algae)	Bovine
Citrus bergamia	Prototheca wickerhamii (algae)	Bovine
Cinnamomum zeylanicum Litsea cubeba Mentha piperita Origanum vulgare Satureja hortensis Syzygium aromaticum Thymus vulgaris	Prototheca zopfii (algae)	Bovine

effects of essential oils on flies, it has been shown that the oils extracted from *Mentha piperita* plant were effective against the larvae of the *Muscadomestica* [35]. Essential oils of *Cinnamomum camphora*, *Matricaria chamomilla*, and *Mentha piperita* have also been tested for reducing and repelling nuisance flies (*Hippobosca equina*, *Musca domestica*, and *Stomoxys calcitrans*) on cattle [36]. Antirepellent effect is also found when using essential oil of *Nepeta cataria* against *Stomoxys calcitrans* [37]. On the other hand, essential oils may be of limited use against fleas when *Citrus sinensis* essential oil was tested against *Ctenocephalidesfelis* (cat flea) [38].

Significant effects of essential oils were observed against different types of lice that affect animals. Tea tree (*Melaleuca alternifolia*) essential oil was more effective against chewing lice (*Bovicola ocellatus*) [39]. Lethal effect of *Allium cepa*, *Cinnamomum camphora*, and *Mentha piperita* essential oils was shown against *Haematopinus tuberculatus* (cattle-sucking louse) and its eggs [35].

Other effects of essential oils have been significantly tested on mites that affect different animals [40]. Furthermore, good results were obtained when the *Leptospermum scoparium* essential oil was used against *Dermanyssus gallinae* mites in closed and open chambers. The results showed a 9.9% mortality rate in open chambers and a high percentage (80% mortality rate) in closed chambers. Also, a higher mortality was observed when these mites were treated with *Thymus vulgaris* essential oil in the closed chamber, as compared to open chambers [41].

However, the chemical composition of essential oils can also vary according to various important factors such as water availability, and soil conditions, including its type, season before or after flowering and the genetic composition of the plant, which is in accordance with the plant variety. All such factors, including genetic and epigenetic factors, affect the biochemical synthesis of essential oils in a particular plant. So, the same species of plant with a different chemical composition of essential oil may produce different biological and therapeutic effects. Differences in the composition of essential oils of two species of *Laurus* plant were used against the *Psoroptes cuniculi* mites. A reasonable effect of mortality of these mites (73%) was observed with the essential oil of *Laurus nobilis*, while a greater acaricidal effect rate (100%) was obtained when the essential oil of the second species *Laurus novocanariensis* was sprayed on these mites [42].

Many studies of acaricidal and repellent effects of essential oils were achieved on ticks, especially the species of the Ixodidae family. The use of *Ageratum houstonianum* essential oil against ticks biting goats led to a remarkable effect (94.9% of mortality rate of these ticks) [43]. The potent effect of essential oils against ticks' larvae varied among plant species. A high mortality of 100% of *Rhipicephalus microplus* larvae was achieved when treated with essential oils derived from *Dorystoechas hastata*, *Mentha longifolia*, and *Thymus vulgaris* [44]. On the other hand, the use of *Cuminum cyminum* and *Pimenta dioica* essential oils were more effective against the larvae of *Rhipicephalus microplus*, which showed significant mortality rate (99%), whereas the essential oil of *Ocimum basilicum* had no larvicidal properties [45].

The toxicity of essential oils of *Origanum minutiflorum* and *Satureja thymbra* demonstrated valuable acaricidal efficacy against *Hyalomma marginatum*, *Ixodes ricinus*, and *Rhipicephalus turanicus* [46]. The acaricidal effect of essential oils depended on the susceptibility of different types of ticks to these oils. The essential oil of *Tagetes minuta* was found to be effective for controlling different tick species, including *Amblyomma cajennense*, *Argas miniatus*, *Rhipicephalus microplus*, and *Rhipicephalus sanguineus* The efficacy of essential oil was assessed by adult immersion and larval packet tests [47, 48]. The repellent effect of essential oils on ticks has also been tested when essential oils of *Rhododendron tomentosum* and *Dianthus caryophyllus* plants were used against *Ixodes ricinus* ticks, which repelled 100% of the ticks, upon treatment [49, 50].

References

[1] Sharifi-Rad J, Sureda A, Tenore GC. et al., Biological activities of essential oils: From plant chemoecology to traditional healing systems. Molecules 2007, 22, 70.
[2] Dijilani A, Dicko A. The Therapeutic Benefits of Essential Oils. In: Bouayed J, Bohn T (Eds.), Nutrition, well-being and health. IntechOpen, London, UK, 2012, 156–178.
[3] Turek C, Stintzing FC. Stability of essential oils: A review. Comp Rev Food Sci Food Saf 2013, 12, 40–53.
[4] De Colmenares NG, Dellacassa E, Hasegawa M. et al., Los Aceitesesenciales. In: Bandoni AL (ed), Los recursos vegetales aromáticos en latinoamérica. su aprovechamiento industrial para la producción de aromas y sabores. Cyted, Buenos Aires, Argentina, 2003.
[5] Wińska K, Ma̧czka W, Łyczko J, Grabarczyk M, Czubaszek A, Szumny A. Essential oils as antimicrobial agents – Myth or real alternative?. Molecules 2019, 24, 2130.

[6]   Lang G, Buchbauer G. A review on recent research results (2008–2010) on essential oils as
        antimicrobials and antifungals. A review. Flavour Fragr J 2012, 27, 13–39.
[7]   Nazzaro F, Fratianni F, Coppola R, DeFeo V. Essential oils and antifungal activity.
        Pharmaceuticals 2017, 10, 86.
[8]   Yang Y, Isman MB, Tak JH. Insecticidal activity of 28 essential oils and a commercial product
        containing Cinnamomum cassia bark essential oil against sitophilus zeamais Motschulsky.
        Insects 2020, 11, 474.
[9]   Andrés MF, González-Coloma A, Sanz J, Burillo J, Sainz P. Nematicidal activity of essential
        oils: A review. Phytochem Rev 2012, 11, 371–390.
[10]   Barbosa P, Lima AS, Vieira P. et al, Nematicidal activity of essential oils and volatiles derived
        from Portuguese aromatic flora against the pinewood nematode, Bursaphelenchusxylophilus.
        J Nematol 2010, 42, 8–16.
[11]   Isman MB. Bioinsecticides based on plant essential oils: A short overview. Z Naturforsch
        2020, 75, 179–182.
[12]   Ntalli NG, Ferrari F, Giannakou I, Menkissoglu-Spiroudi U. Phytochemistry and nematicidal
        activity of the essential oils from 8 Greek lamiaceae aromatic plants and 13 terpene
        components. J Agric Food Chem 2010, 58, 7856–7863.
[13]   Awaad MHH, Afify MAA, Zoulfekar SA, Mohammed FF, Elmenawy MA, Hafez HM. Modulating
        effect of peppermint and eucalyptus essential oils on vVND infected chickens. Pak Vet J 2016,
        36, 350–355.
[14]   Fang F, Candy K, Melloul E. et al., *In vitro* activity of ten essential oils against
        Sarcoptesscabiei. Parasit Vectors 2016, 9, 1–8.
[15]   Radsetoulalova I, Hubert J, Lichovnikova M. Acaricidal activity of plant essential oils against
        poultry red mite(Dermanyssusgallinae). Mendel Net 2017, 24, 260–265.
[16]   Sharifi-Rad J, Sureda A, Tenore GC. et al., Biologicalactivitiesofessentialoils:
        fromplantschemoecologytotraditionalhealingsystems. Molecules 2017, 22, 70.
[17]   Koul O, Walia S, Dhaliwal GS. Essential oils as green pesticides: Potential and constraints.
        Biopestic Int 2008, 4, 63–84.
[18]   Raveau R, Fontaine J, Sahraoui ALH. Essential oils as potential alternative biocontrol products
        against plant pathogens and weeds: A review. Foods 2020, 9, 1–31.
[19]   Alonso-Gato M, Astray G, Mejuto JC, Simal-Gandara J. Essential oils as antimicrobials in crop
        protection: Review. Antibiotics 2021, 10, 34.
[20]   Buttimer C, McAuliffe O, Colin-Hill RPR, O'Mahony J, Coffey A. Bacteriophages and bacterial
        plant diseases. Front Microbiol 2017, 8, 34.
[21]   DeCarvalho CCCR, Caramujo MJ. Ancient procedures for the high-tech world: Health benefits
        and antimicrobial compounds from the Mediterranean empires. Open Biotechnol J 2008, 2,
        235–246.
[22]   Chang Y, Harmon PF, Treadwell DD, Carrillo D, Sarkhosh A, Brecht JK. Biocontrol potential of
        essential oils in organic horticulture systems: From farm to fork. Front Nutr 2022, 8, 1–26.
[23]   Kordali S, Cakir A, Mavi A, Kilic H, Yildirim A. Screening of chemical composition and
        antifungal activity of essential oils from three Turkish *Artemisia* species. J Agric Food Chem
        2005, 53, 1408–1416.
[24]   Hummelbrunner AL, Isman MB. Acute, sublethal, antifeedant and synergistic effects of
        monoterpenoid essential oil compounds on the tobacco cutworm (Lepidoptera: Noctuidae). J
        Agric Food Chem 2001, 49, 715–720.
[25]   Lazăr RN, Alexa E, Obis,tioiu D, Cocan I, Pătruică S. The effect of the use of essential oils in
        the feed of bee families on honey chemical composition and antimicrobial activity. Appl Sci
        2022, 12, 1–22.

[26] Zeng Z, Zhang S, Wang H, Piao X. Essential oil and aromatic plants as feed additives in non-ruminant nutrition: a review. J Anim Sci Biotechnol 2015, 7.

[27] Kim SJ, Lee KW, Kang CW, An BK. Growth performance, relative meat and organ weights, cecal microflora, and blood characteristics in broiler chickens fed diets containing different nutrient density with or without essential oils. Asian-Australian J Anim Sci 2016, 29, 549–554.

[28] Pirgozliev V, Mansbrige SC, Rose SP. et al., Dietary essential oils improve feed efficiency and hepatic antioxidant content of broiler chickens. Animal 2019, 13, 502–508.

[29] Irawan A, Hidayat C, Jayanegara A, Ratriyanto A. Essential oils as growth-promoting additives on performance, nutrient digestibility, cecal microbes, and serum metabolites of broiler chickens: Ameta-analysis. Anim Biosci 2021, 00, 1–15.

[30] Krishan G, Narang A. Use of essential oils in poultry nutrition: a new approach. J Adv Vet Anim Res 2014, 1, 156–162.

[31] Franz C, Baser KHC, Windisch W. Essential oils and aromatic plants in animal feeding- a European perspective: A review. FlavourFragr J 2010, 25, 327–340.

[32] Ebani VV, Mancianti F. Use of essential oils in veterinary medicine to combat bacterial and fungal infections. Vet Sci 2020, 7, 2–35.

[33] Abbas A, Abbas RZ, Masood S. et al., Acaricidal and insecticidal effects of essential oils against ectoparasites of veterinary importance. Bol Latinoam Caribe Plant Med Aromat 2018, 17, 441–452.

[34] Ellse L, Wall R. The use of essential oils in veterinary ectoparasite control: A review. Med Veter Entom 2014, 28, 233–43.

[35] Khater HF, Ramadan MY, El-Madawy RS. Lousicidal, ovicidal and repellent efficacy of some essential oils against lice and flies infesting water buffaloes in Egypt. Vet Parasitol 2009, 164, 257–266.

[36] Morey RA, Khandagle AJ. Bioefficacy of essential oils of medicinal plants against housefly, *Musca domestica* L. Parasitol Res 2012, 111, 1799–1805.

[37] Zhu JJ, Berkebile DR, Dunlap CA. et al., Nepetalactones from essential oil of *Nepetacatariare* present a stable flyfeeding and oviposition repellent. Med Vet Entomol 2012, 26, 131–138.

[38] Collart MG, Hink WF. Sublethal effects of D-limonene on the cat flea, *Ctenocephalidesfelis*. Entomol Exp Appl 1986, 42, 225–229.

[39] Talbert R, Wall R. Toxicity of essential and non-essential oils against the chewing louse, *Bovicola*(Werneckiella) *ocellatus*. Res Vet Sci 2012, 93, 831–835.

[40] Kim SI, Yi JH, Tak J, Ahn YJ. Acaricidal activity of plant essential oils against *Dermanyssus gallinae (Acari: Dermanyssidae)*. Vet Parasitol 2004, 120, 297–304.

[41] George DR, Sparagano OAE, Port G, Okello E, Shiel RS, Guy JH. Repellence of plant essential oils to *Dermanyssusgallinae* and toxicity to the non-target invertebrate *Tenebrio molitor*. Vet Parasitol 2009, 162, 129–134.

[42] Macchioni F, Perrucci S, Cioni P, Morelli L, Castilho P, Cecchi F. Composition and acaricidal activity of *Laurusnovocanariensis* and *Laurus nobilis* essential oils against *Psoroptescuniculi*. J Ess Oil Res 2006, 18, 111–14.

[43] Pamo ET, Tendon Keng F, Kana JR. etal, A study of the acaricidal properties of an essential oil extracted from the leaves of *Ageratum houstonianum*. Vet Parasitol 2005, 128, 319–323.

[44] Koc S, Oz E, Cinbilgel I, Aydin L, Cetin H. Acaricidal activity of *Origanum bilgeri* P.H.Davis (Lamiaceae) essential oil and its major component, carvacrol, against adult *Rhipicephalusturanicus* (Acari:Ixodidae). Vet Parasitol 2013, 193, 316–319.

[45] Martinez-Velazquez M, Castillo-Herrera GA, Rosario-Cruz R. et al., Acaricidal effect and chemical composition of essential oils extracted from *Cuminum cyminum*, *Pimentadioica* and *Ocimum basilicum* against the cattle tick *Rhipicephalus* (*Boophilus*) *microplus* (Acari: Ixodidae). Parasitol Res 2011, 108, 481–487.

[46] Cetin H, Cilek JE, Oz E, Aydin L, Deveci O, Yanikoglu A. Acaricidal activity of *Satureja thymbra* L. essential oil and its major components, carvacrol and gamma-terpinene against adult *Hyalomma marginatum* (Acari: Ixodidae). Vet Parasitol 2010, 170, 287–290.

[47] Garcia MV, Matias J, Barros JC, Lima DP, Lopes RS, Andreotti R. Chemical identification of *Tagetes minuta* Linnaeus (*Asteraceae*) essential oil and its acaricidal effect on ticks. Rev Bras Parasitol Vet 2012, 21, 405–411.

[48] Andreotti R, Garcia MV, Cunha RC, Barros JC. Protective action of *Tagetes minuta* (Asteraceae) essential oil in the control of *Rhipicephalus microplus* (Canestrini, 1887) (Acari: Ixodidae) in a cattle pen trial. Vet Parasitol 2013, 197, 341–345.

[49] Jaenson TGT, Palsson K, Borg-Karlson AK. Evaluation of extracts and oils of tick-repellent plants from Sweden. Med Vet Parasitol 2005, 19, 345–352.

[50] Tunon H, Thorsell W, Mikiver A, Malander I. Arthropod repellency, especially tick (*Ixodes ricinus*), exerted by extract from *Artemisia abrotanum* and essential oil from flowers of *Dianthus caryophyllum*. Fitoterapia 2006, 77, 257–261.

Deepali Koreti*, Anjali Kosre, Pramod Kumar Mahish,
Nagendra Kumar Chandrawanshi* and Shri Ram Kunjam
Chapter 11
Application of essential oils in alternative medicine

Abstract: Alternative medicine has gained much attention nowadays. It is also called a complementary therapy, and in this system, essential oils play a major role as a therapeutic agent for the treatment of various diseases. Whole plants as well as individual parts are the sources of essential oils, and various methods are used for essential oil extraction and purification. Essential oils are utilized in different way as alternative medicine such as aromatherapy, massage, and consumption. Essential oil can easily penetrate and show an effect on the affected area. This can be the most effective medicine for several problems like depression, headache, insomnia, indigestion, muscular pain, respiratory problems, swollen joints, and skin ailments. This chapter covers basic introduction of essential oils, its chemical constituents, sources, and extraction methods. It also discusses the use of essential oil as an alternative medicine and mechanism behind its works and its future aspects.

Keywords: Aroma, therapy, diseases, therapeutic, volatile

11.1 Introduction

Essential oils are aromatic chemicals abundantly found in oil glands or sacs located at various depths in fruit peels, particularly in the flavedo portion and cuticles [1]. Essential oils are fragrant, oily liquids extracted from various plant parts, such as

Acknowledgments: The authors are thankful to the Junior Research Fellowship (DBT/JRF/BET18/I/2018/AL/123), Regional Centre for Biotechnology, NCR Biotech Science Cluster 3rd Milestone, Faridabad-Gurugram Expressway, Faridabad 120001, for providing funding support. The authors are also thankful to the Head, School of Studies in Biotechnology, Pt. Ravishankar Shukla University, Raipur, for providing laboratory facilities and conducting the study.

*Corresponding author: Deepali Koreti, Nagendra Kumar Chandrawanshi, School of Studies in Biotechnology, Pt. Ravishankar Shukla University, Raipur, Chhattisgarh, India
e-mail: ranukoreti27@gmail.com
Anjali Kosre, School of Studies in Biotechnology, Pt. Ravishankar Shukla University, Raipur, Chhattisgarh, India
Pramod Kumar Mahish, Government Digvijay (Autonomous) Post Graduate College, Rajnandganv, Chhattisgarh
Shri Ram Kunjam, Government V.Y.T. PG Autonomous College, Durg, Chhattisgarh

https://doi.org/10.1515/9783110791600-011

leaves, bark, seeds, flowers, and peels. They are intricate substances generated from various plant metabolites and are characterized by their potent odor. They are transparent, have a lower density than water, and are soluble in lipid and organic solvents [2]. These essential oils' chemical composition varies greatly depending on their geographic origin, botanical parentage, genetics, bacterial endophytes, and extraction methods. They have been traditionally used for cleaning as an anti-inflammatory, calming, and stimulating agent, and they have the potential for and are currently being employed in clinical care [3]. Essential oils can are produced by using steam or hydro distillation. Around 3000–2500 BC, the first use of essential oils was documented. In the Middle Ages, they were created by the Arabs and utilized for their scents and medicinal qualities like antiseptic, antibacterial, virucidal, and fungicidal [1, 4]. The usage of essential oil extracts is also said to have co-occurred in China and India. It has been employed in localized anesthetic treatments, food preservation, analgesic, sedative, antibacterial, anti-inflammatory, and embalming processes. In nature, essential oils significantly impact plants by generally defending them against herbivores by lowering their hunger. Numerous insects are drawn to the aroma of plant essential oils, which aid in dispersing seeds and pollen. Plant organs such as buds, stems, roots, seeds, fruits, flowers, leaves, twigs, wood, or bark are a rich source of essential oils, which are then stored in canals, secretory cells, cavities, and epidermic layers [5]. The amount, quality, soil composition, climate, plant organ, age, and vegetative stage of plants are all factors that affect the products that are harvested [6]. It must be extracted under optimal circumstances from the same plant organ cultivated on the same soil and throughout the same season to maintain their constant composition [7]. Gas chromatography and mass spectrometric analysis are typically used for the quality assessment of major commercially available essential oils. Only 300 out of 3,000 essential oils are currently known for commercially significant uses in the agronomic, pharmaceutical, food preservation, cosmetic, perfume, and sanitary industries. Some essential oils, like D-limonene, D-carvone, and geranyl, are used in various items including perfumes, soaps, lotions, industrial solvents, fragrances for domestic cleaning products, and flavoring agents. Alternative medicine and its increasing demands generate suitable sources for alternative medicine, and this way, essential oils are one of the best alternative medicine sources. This chapter covers the basics of essential oil, its sources, extraction methods, and detailed applications of essential oils as employed for alternative medicine.

11.2 Extraction and production of essential oil

Plants are a good source of secondary metabolites that are used as protection from predators and to attract pollinators. For many thousands of years, plant oils and extracts have been used for a wide range of reasons. The secondary metabolism of aromatic and other plant varieties produces essential oils, which are a mixture of volatile components. Typically, volatile metabolites are separated from plant material using steam- or hydrodistillation techniques; the resultant fragrant mixture is known as an essential oil [2]. They can be produced commercially by steam distillation and hydro distillation, but they can also be obtained through expression, fermentation, effleurage, or extraction [3] and volatile terpenes and hydrocarbons make up the majority of the components found in essential oil. The goal of plant extraction is to remove specific components from plants. The necessary extract is subsequently obtained by solubilizing and containing these valuable plant components in the solvent. Essential oils have recently proven to be quite effective in helping the food industry manage foodborne bacteria; therefore plants are now receiving a lot of attention as potential alternatives to traditional antimicrobials. Numerous ways in which essential oils have been separated have enhanced their bioactive and medicinal properties. The best methods used in these extraction processes are freeze-drying, rotary evaporation, steam distillation, hydrolyzation, and GC chromatography assays, among others. Obtaining essential oils from various plants and specific compounds are shown in Table 11.1.

Table 11.1: Plant source, specific compounds and applications of essential oils.

S. no.	Plant source	Identified specific compound	Application	References
1	Clary sage (*Salvia sclarea* Linn.)	Linalool, linalyl acetate, alpha-terpineol, germacrene D, and geranyl	Tonic is used for womb and uterus-associated problems, and it also regulates the menstrual periods, ease tension, and muscle cramps	[8]
2	Eucalyptus (*Eucalyptus globulus* Labill)	Cineole (70–85%), aromadendrene, limonene, terpinene, cymene, phellandrene, and pinene	Regulate and activate the various systems like nervous system for neuralgia, headache and debility, antioxidant, anti-inflammatory, anti-proliferative, and antibacterial activities	[8]
3	Geranium (*Pelargonium graveolens* L. Herit)	Eugenol, geranic, citronellol, geraniol, linalol (linalool), citronellyl formate, citral, myrtenol, terpineol, methone, and sabinene	Dermatitis, eczema, aging skin, some fungal infections, along with anxiety and stress-related problems	[9, 10]

Table 11.1 (continued)

S. no.	Plant source	Identified specific compound	Application	References
4	Tea tree (*Melaleuca alternifolia* Cheel)	Terpinen-4-ol, an alcoholic terpene	Antiviral activity, antibacterial and antifungal, an immune booster, antiinflammatory, insecticidal, and immune-stimulant properties	[11, 12]
5	Lavender (*Lavandula officinalis* Chaix.)	Amphor, terpinen-4-ol, linalool, linalyl acetate, betaocimene, and 1,8-cineole	Supporting mental alertness and suppressing aggression and anxiety, antibacterial, and antifungal properties	[13]
6	Lemon (*Citrus limon* Linn.)	Terpenes, D-limonene, and limonene	Antiseptic, astringent, and detoxifying properties for blemishes associated with oily skin	[14]
7	Peppermint (*Mentha piperita* Linn.)	Carvacrol, menthol, carvone, methyl acetate, limonene, and menthone	Anti-inflammatory, analgesic, anti-infectious, antimicrobial, antiseptic, antispasmodic, astringent, digestive, carminative, fungicidal effects, nervine stimulant, vasoconstrictor, decongestant, and stomachic properties.	[15, 16]
8	Rosemary (*Rosmarinus officinalis* Linn.)	Resin, tannic acid, and volatile oil. The active constituents are bornyl acetate, borneol along with other esters, cineol, pinene, and camphene.	Liver and gallbladder tonic. It regularizes the blood pressure and retards the hardening of arteries, hysteria, and paralysis. encouraging hair growth.	[17]

11.3 Classification of essential oils

The essential oils present in various plants contain a wide range of compounds. Essential oils can be categorized in several ways depending on their various extraction techniques, chemical makeup, notes, scent, and so on. Detailed classification with examples is shown in Table 11.2.

Compounds responsible for essential oil characteristics are reported: hydrocarbons, esters, phenols, ketones, and so on; essential oils are classified according to the employed extraction method. Including steam distillation, in which plant material is put in a container and steam is poured through it to complete the process. The steam heats the oil and aromatic molecule-containing plant compartments. These

Table 11.2: Classification of essential oils.

	Names	Plant sources	References
Essential oils according to chemical compounds	Hydrocarbons	Citrus and pine, coriander, tea tree, and peppermint	[18]
	Terpenoids	Citronella, lemon balm, and lemon myrtle	[18]
	Cyclic aldehydes	Cumin, bitter almond, and cinnamon	[19]
	Carbonyl group is linked to two carbon atoms in ketone.	Pennyroyal, *Thuja*, Sage, and *Eucalyptus radiata*	[20]
	Phenol	Thyme and oregano	[21]
	Oxide	Eucalyptus, wormseed, and cajeput	[21]
	Sesquiterpenes and sesquiterpene lactone	German chamomile, yarrow, elecampane, and arnica	[22]
	Phenylpropanes	Aniseed, clove, tarragon, and myrtle leaf	[23]
	Ester	Lavender, wintergreen, and clary sage	[24]
Essential oils based on extraction	Steam-distilled	Aromatherapy oils	[25, 26]
	Expressed	Tangerines, grapefruits, lemons, oranges, and others have rinds	[26]
	Solvent-extracted	Including jasmine, rose, orange blossom (neroli), tuberose, and oak	[27]
Scent-based essential oil	Citrus	Bergamot, grapefruit, lemon, lime, orange, and tangerine [17]	[28]
	Herbaceous	Basil, chamomile, Melissa, Clary Sage, hyssop, marjoram, peppermint, and rosemary	[28]
	Medicinal/Camphorous	Cajeput, tea Tree, and other plants that have characteristics similar to borneol, mugwort, and rosemary with a fruity, dried plum-like undertone	[29]
	Woody	Cedar, cinnamon, cypress, juniper berries, pine, and sandalwood	[30]
	Earthy	Angelica, patchouli, vetiver, and valerian	[31]
	Spicy oils	Thyme, cloves, aniseed, black pepper, cardamom, cinnamon, coriander, cumin, ginger, and nutmeg	

molecules are liberated, rise with the steam, and move through a sealed system. The fragrant steam is then cooled and distilled with cold water after going through a cooling procedure. The essential oils condense and change into a liquid state during this process. Later, the liquid mixture is divided into two components: aromatic water or hydrosol and essential oils. The pressure of steam that is carried through plant material, the coolant used, the closed system temperature during oil production, and other factors are all considered during steam distillation. All of these elements, along with the distiller's ability, determine the quality and purity of the oil. Expressed method is also known as old-pressing or expressed processes. Fruit juice is used to extract oils using mechanical pressure. A separation method is used to separate oils from water since oils in their juicy form include much water. An issue with this method is that cold-pressed oils deteriorate more quickly than conventional oils. Therefore, these oils should be bought in small quantities and renewed as necessary. Solvent-extracted oil extraction is applied for those plant materials that cannot withstand cold pressing or heat (in the form of steam). The oil created when subjected to one of these methods could be tainted or of poor quality. Essential oils are extracted using ethanol, ether, methanol, hexane, alcohol, and petroleum. Essential oils are created by filtering and distilling the solvent mixture under low pressure. This technique has the drawback that the oils sometimes contain solvent remnants, which might result in allergic reactions in some people. Additionally, essential oils can be categorized according to their scent. Citrus, herbaceous, medicinal/camphorous, floral, resinous, woody, earthy, minty, and spicy oils can be found in this group of oils. Citrus oils come under this heading since they have a pronounced citrus flavor. Herbaceous oils derived from plants, most of which are valuable herbs. Camphoraceous oils are natural healing substances that are classified as essential oils. Floral oils include oils manufactured from floral components or that contain a plant's floral essence. Essential oils with a woody aroma derived from the bark and other woody portions of plants are referred to as "woody oils." These oils are made by plants including cedar, cinnamon, cypress, juniper berries, pine, and sandalwood. Essential oils that are either extracted from the roots or other earthy sections of plants or have a distinctly earthy fragrance. Spicy oils include oils derived from plants or spices.

11.4 Essential oil constituents

Every oil typically has more than 100 components, while the precise quantity varies depending on the source of oil and the implied extraction technique. However, the two chemical categories of terpenoids (monoterpenoids and sesquiterpenoids) and phenylpropanoids contain the most significant active substances obtained in essential oils. These two categories are produced by diverse metabolic routes and metabolism precursors. Essential oils, like all organic chemicals, are composed of hydrocarbon molecules and can be further divided into terpenes, alcohols, esters, aldehydes, ketones, and phenols, among other categories. There are also oxygenated molecules, phenols, alcohols, monoterpene alcohols, sesquiterpene alcohols, aldehydes, ketone esters, lactones, coumarins, ethers, and oxides that are present in essential oils. Some main constituents are described here in the following sections.

11.4.1 Terpenoids

Terpenes and terpenoids make up the bulk of the essential oils found in many different kinds of plants and flowers [32]. The monoterpenoid and sesquiterpenoid groups of terpenoids contain the majority of the significant components of essential oils of plants. Nearly all essential oils contain the molecules monoterpene and monoterpenoid, which have a structure of 10 carbon atoms, 2 double bonds, and 3 atoms. Geraniol, terpineol (lilacs), limonene (citrus fruits), myrcene (hops), linalool (lavender), and pinene (pine trees) are a few examples of monoterpenes and monoterpenoids [33]. Citrus oils do not survive very long because they are rich in monoterpene hydrocarbons, which react quickly to air and heat sources and are easily oxidized [34]. The most prevalent kind of functional group identified in essential oils is one that is oxygenated. Similar to terpenes, it is crucial to comprehend the various classes of oxygenated compounds that are present because each class provides diverse potential health advantages [34].

11.4.2 Esters

Esters are very prevalent chemicals that are formed by the process known as esterification, when an alcohol and an acid. They are present in many essential oils and are primarily present in oils used for the upper respiratory system and are beneficial for illnesses like dry asthma, colds, flu, and dry cough.

11.5 Essential oils used as alternative medicine

Complementary and alternative medicine are the natural products or formed by natural products and frequently used for therapeutic purposes. The use of alternative medicine is rising globally due to awareness of side effects of other drugs. Alternative medicine including wide variety of treatment including herbal preparations, aromatherapy, massage therapy, and foot reflexology [35, 36]. Aromatherapy is the practice or treatment that utilizes the natural oils extracted from bark, flowers, stems, roots, leaves, or other parts of a plant to enhance the psychological and physical well-being [37]. It is a form of supplementary medicine that seeks to alter a person's state of mind and mood by utilizing volatile oils and other aromatic molecules. The nature of volatile oils is hydrophobic. Different techniques, such as steam distillation, are used to extract essential oils. There is some research that suggests the medicinal potential of volatile oils. Volatile oils frequently enter the bloodstream through the skin, where they may aid in whole-body healing. There are several uses for essential oils, including the treatment of pain, mood enhancement, and improved cognitive performance. Essential oils have long been recognized for their medicinal value. For many years, they were employed as fragrances, tastes for food and drink, or to treat both the body and the psyche [38, 39]. They demonstrate their employment in numerous treatments in various forms in ancient civilizations like Chinese, Indian, and ancient Egyptian. The first civilization to adopt aromatherapy in folk medicine was the ancient Chinese, after which the ancient Egyptians invented an underdeveloped distillation machine that is still used today. Ancient Egyptians taught Greece a great deal, including how to use aromatic plants for their therapeutic and fragrant properties [3, 40]. Small aromatic molecules make up volatile oils, which are easily absorbed through the skin and respiratory system. Following their entry into the bloodstream, these medicinal substances disperse throughout the body, where they might develop their beneficial curative properties. They are too concentrated to be affected by modest amounts of volatile oil. Among one of the most widely used complementary therapies today, aromatherapy is a highly effective treatment for both acute and chronic illnesses. Our immune system is also strengthened by routine usage of aromatherapy and household items [41]. Numerous plant essential oils have been used as medicine for centuries and have shown to offer a number of health advantages, including impacts on acute, chronic, and chronic infectious disorders. It is generally known that plant essential oils and their individual components are used in medical preparations for the treatment of infectious diseases in humans. To prevent any adverse effects when they are administered, the choice of a suitable safe oil and the calculation of the best effective dose should be taken into account [42]. At ancient times, the essential oils have very importance role as a fragrance with a curative potential and relaxing for mind and spirit. These aroma molecules are organic chemicals from plants and it makes the surroundings free from disease, bacteria, virus, and fungus [36]. In aromatherapy oil is penetrated to reach the subcutaneous tissues that is one of the important characteristics in this therapy.

Nanotechnology recognizes as a new way for solving many of the problems in medical field and known for directed drug delivery. Nanoemulsions technology can be an attractive method for bioactive compounds delivery which is present in essential oils. Nanoemulsions are nano-sized emulsions (10 to 1,000 nm) and are utilized for increasing the delivery and surface area components. Nanoemulsions of essential oils are lipid droplets of nanometric size dispersed in an aqueous phase by using a suitable emulsifier at the oil/water interface and produced by high pressure homogenization. Nanoemulsification of essential oils makes essential oil's stable structural functional properties and improves absorption and bioavailability inside the body.

Nanoemulsion-based essential oil can be more effective as therapeutic agents, food, cosmetic, and pharmaceutical industries. There are many plant essential oils that are reported for extensive utilization in the nanoemulsions formation and treat many diseases [43, 44]. Many bioactive compounds of essential oil possess bioactivity like antioxidant, anti-inflammatory, antibiotics, anticancerous, and immunomodulatory [45, 46], and nanoemulsion-based delivery is the best option for efficient delivery and enhance bioavailability [47].

11.5.1 Mechanism of essential oil absorption

The basic mechanism behind the essential oil works is the absorption of essential oil by target area. The mechanism involves integration of essential oils into a biological signal of the receptor cells in the skin or nose when it is inhaled. Then the signal is transmitted to limbic and hypothalamus parts of the brain via olfactory bulb and signals cause brain to release neuro messengers (serotonin, endorphin, and noradrenalin) to link nervous and other body systems that assures a desired change – provide a feeling of relief [48]. *Pharmacokinetics* is the study of absorption and excretion of essential oils. Essential oils can be absorbed via olfaction through the external skin and through the "internal skin" lining of orifices (mouth, vagina, and anus) and via ingestion. Essential oil can be absorbed through the skin (both internal and external) and absorbed by ingestion. Essential oils contain many different chemical components, and it is these components that are absorbed by the body. Volatile oils first exert their effects by entering the body in one of the three ways: directly by ingestion, diffusion through the skin tissue, or direct absorption through inhalation. Essential oil components can be absorbed by the following routes.

11.5.1.1 The skin

The skin is the largest organ in the body and is a complex membrane, varying from less than a millimeter to approximately 3 mm thick. For many years the skin (stratum corneum) was thought to be a barrier that topically applied drugs, or essential oils, could not penetrate. Now we know that drugs can reach the basal epidermal cells. Scientist developed a model to predict the mass of a chemical absorbed into/through the skin from a cosmetic or dermatological formulation. Because the constituents of volatile oils are lipid-soluble, they can pass through epidermal membranes before being absorbed by the microcirculation and emptied into the systemic circulation, where they reach all target organs [49, 50]. Inflammatory illnesses are one example of this, which cause loss of vital functions and are characterized by pain, redness, and swelling. Tea tree oil has been demonstrated to accelerate the healing of chronic wounds by increasing monocytic differentiation in vitro and decreasing inflammation [51]. Aromatherapy utilizes certain essential oils for skin, body, face, and hair cosmetic products. These products are used for their various effects as cleansing, moisturizing, drying, and toning. A healthy skin can be obtained by the use of essential oils in facial products. In massage aromatherapy, almond, grape seed, and jojoba oil in pure vegetable oil have been shown to have wonderful effects and famous as healing touch of massage therapy [52, 53].

11.5.1.2 Consumption

Due to the potential toxicity of some oils, using essential oils orally requires caution. The remainder of the body may then absorb and supply ingested volatile oil molecules and/or their metabolites, which may then be transferred to various organs. Once ingested, volatile oils work through physiological processes to provide their medicinal effects. Roman chamomile, for instance, is frequently used to the lower abdomen to ease pain from physical disorders, menstrual cramps, and tension [54, 55].

11.5.1.3 Inhaled/breathing

The respiratory system is the route via which volatile oils enter the body. They are easily breathed through the upper respiratory tract and can quickly enter the lungs, where they can move to the blood stream because of their capacity to be volatile. The cutaneous channel is generally thought to be the second easiest entrance point after the respiratory tract [56]. Olfactory aromatherapy, which is a result of the inhalation of essential oils, has been shown to improve emotional wellness, tranquillity, relaxation, or bodily regeneration. Pleasurable scents that trigger odor memories are woven together with the decompression of stress. In olfactory aromatherapy, inhalation of essential oils has resulted in enhanced emotional wellness, relaxation, calmness, and rejuvenation of the human body by its pleasurable scents. And this way essential oils are complemented to medical treatment and can never be taken as a replacement for it [57].

11.6 Other applications

11.6.1 Essential oil in food sector

Essential oils are rich sources of biologically active chemicals with recognized antibacterial and antioxidant characteristics, which are attractive as additives in the food sector and have been classified as GRAS (generally recognized as safe) by the US Food and Drug Administration [58, 59]. However, due to their potent antibacterial and antioxidant capabilities, their use as a food additive has recently attracted increasing interest [60]. The fundamental strategy for ensuring food safety is to reduce the initial microbial load and/or to utilize active packaging to prevent the growth of any leftover bacteria during postprocess applications, such as manufacture and storage. Due to its numerous uses as a flavoring and scent component as well as an antibacterial agent, cinnamon essential oils have been referred to be the most relevant essential oils used in the food and cosmetic sectors [61]. Simionato et al. [62] claim that the incorporation of cinnamon oil into cyclodextrinnanosponges could be used as an antibacterial food packaging solution. Additionally, garlic essential oil nanophytosomes as a natural food preservative demonstrated its potential as a potential natural food preservative by successfully displaying suitable physicochemical properties, particularly in acidic food products, with its application in yoghurt as a food model [58, 63]. Studies have demonstrated that essential oils have potent antibacterial properties against foodborne microorganisms, which the food industry can utilize as a preservative or as an antimicrobial ingredient in food packaging [62]. How essential oils interact with pathogens on fruit and food surfaces. Active edible coatings with natural antioxidants may increase the stability of meat products and thus have application in the food sector. The loss of the produce's organoleptic quality and nutritional content is a result of the desired impact of microbial and enzyme inactivation. To preserve and enhance the storage quality and shelf life of goods, it is crucial to create efficient storage techniques and alternative technology. The use of essential oils and the parts of them as natural antibacterial agents to inhibit the growth of pathogenic and spoilage bacteria in food has enormous potential. Numerous plant essential oils have been found to have antibacterial activities [64, 65].

11.6.2 Role of essential oil in food safety and conservation

The consumption of food products with features of naturalness and little processing is becoming more and more popular today [66]. The World Health Organization estimated that there are around 600 million instances of foodborne illnesses each year (nearly 1 in 10 individuals worldwide) and 420,000 fatalities related to these illnesses. Food decaying is a metabolic process that results in changes in sensory properties that make food unsuitable or unpleasant for human consumption. Making sure that everyone on our world has access to enough food, both in terms of quantity and quality, is made possible by reducing postharvest losses [67]. Food safety has become a major concern in current scenario in terms of protect, secure, and safeguard food, which will ultimately ensure food security and the availability of fresh and healthy goods. Food preservatives are primarily used to protect and prolong the shelf life of food products, but they have some negative impact like allergies and attention-deficit hyperactivity disorder in children [68]. Extraction of specific compounds from plants is an alternative and good method for secure food in another form; in this sense essential oils extraction and utilization are the best way. They are currently gaining appeal as a natural, secure, and affordable therapy for a variety of health issues because of its antidepressant, stimulating, detoxifying, antibacterial, antiviral, and soothing characteristics.

11.7 Conclusion and future prospect

In conclusion, the essential oils are the natural and noninvasive gift of nature for humans. Essential oils are used in different sectors including uses as alternative medicine where essential oils are used for treatment of diseases and abnormality. Including aromatherapy, when regulates the physiological, spiritual, and psychological conditions. Pharmaceutical industries are working on for environmental-friendly, alternative, and natural medicine for disease. There are many possibilities of work optimization for enhancing the rate of reaction and bioavailability of drugs from the use of these essential oils. Essential oils can be a useful/best nonmedicinal option or can also be combined with conventional care for some health conditions, provided safety and quality issues are considered.

References

[1] Mahato N, Sharma K, Koteswararao R, Sinha M, Baral E. Citrus essential oils: Extraction, authentication and application in food preservation. Crit Rev Food Sci Nutr 2019, 59, 611–625.
[2] Herman RA, Ayepa E, Shittu S, Fometu SS, Wang J. Essential oils and their applications -Amini review. Adv Nutr Food Sci 2019, 4, 13–13.
[3] Bassole IHN, Juliani HR. Essential oils in combination and their antimicrobial properties Molecules 2012, 17, 3989–4006.
[4] Chouhan S, Sharma K, Guleria S. Antimicrobial activity of some essential oils-present status and future perspectives. Medicines 2017, 4, 58.
[5] Pichersky E, Noel JP, Dudareva N. Biosynthesis of plant volatiles: Nature's diversity andingenuity. Science 2006, 311, 808–811.

[6]   Angioni A, Barra A, Coroneo V, Dessi S, Cabras PJ. Chemical composition, seasonal variability, and antifungal activity of Lavandulastoechas L. ssp. stoechas essential oils from stem/leaves and flowers. Agric Food Chem 2006, 54, 4364–4370.

[7]   Masotti V, Juteau F, Bessiere JM, Viano J. Seasonal and phonological variation of the essential oil from the narrow endemic species *Artemisia molineri* and its biological activities. J Agric Food Chemistry 2003, 51, 7115–7121.

[8]   Sienkiewicz M, Głowacka A, Poznańska-kurowska K, Kaszuba A, Urbaniak A, Kowalczyk E. The effect of clary sage oil on staphylococci responsible for wound infections. Postepy Dermatol Alergol 2015, 32, 21–26.

[9]   Mulyaningsih S, Sporer F, Reichling J, Wink M. Antibacterial activity of essential oils from eucalyptus and of related components against multi-resistant bacterial pathogens. Pharm Biol 2011, 49, 893–899.

[10] Ghannadi A, Bagherinejad M, Abedi D, Jalali M, Absalan B, Sadeghi N. Antibacterial activity and composition of essential oils from Pelargonium graveolensL'Her and Vitexagnus-castus L. Iran J Microbiol 2012, 4, 171–176.

[11] Koh KJ, Pearce AL, Marshman G, Finlay-Jones JJ, Hart PH. Tea tree oil reduces histamineinduced skin inflammation. Br J Dermatol 2002, 147, 1212–1217.

[12] Pazyar N, Yaghoobi R, Bagherani N, Kazerouni A. A review of applications of tea tree oil in dermatology. Int J Dermatol 2013, 52, 784–790.

[13] Kim S, Kim HJ, Yeo JS, Hong SJ, Lee JM, Jeon Y. The effect of lavender oil on stress, bispectral index values, and needle insertion pain in volunteers. J Altern Complement Med 2011, 17, 823–826.

[14] YavariKia P, Safajou F, Shahnazi M, Nazemiyeh H. The effect of lemon inhalation aromatherapy on nausea and vomiting of pregnancy: A double-blinded, randomized, controlled clinical trial. Iran Red Crescent Med J 2014, 16, 14360.

[15] Tassou CC, Drosinos EH, Nychas GJ. Effects of essential oil from mint (Menthapiperita) on Salmonella enteritidis and Listeria monocytogenes in model food system at 4 degrees and 10 degrees C. J Appl Bacteriol 1995, 78, 593–600.

[16] Ravid U, Putievsky E, Katzir I. Enantiomeric distribution of piperitone in essential oils of some mentha spp., Calaminthaincana (sm.) heldr. and Artemisia indaica L. Flavour Fragr J 1994, 9, 85–87.

[17] Atsumi T, Tonosaki K. Smelling lavender and rosemary increases free radical scavengingactivity and decreases cortisol level in saliva. Psychiatry Res 2007, 150, 89–96.

[18] Schreiber WL, James NS, Manfred HV, Edward JS, Organoleptic uses of 1-(3, 3-dimethyl-2-norbornyl)-2-propanone in cationic, anionic and non-ionic detergents and soaps. Google Patents 1980.

[19] Marchese A, Barbieri R, Coppo E, Orhan IE, Daglia M. Antimicrobial activity of eugenol and essential oils containing eugenol: A mechanistic viewpoint. Crit Rev Microbiol 2017, 43, 668–689.

[20] Rolf D Inhalation antiviral patch. Google Patents 2004.

[21] Park JB. Identification and quantification of a major anti-oxidant and anti-inflammatory phenolic compound foundin basil, lemon thyme, mint, oregano, rosemary, sage, and thyme. Int J Food Sci Nut 2011, 62, 577–584.

[22] Lawless J. The Encyclopedia of essential oils: The complete guide to the use of aromatic oils in aromatherapy, herbalism, health, and well-being, Conari Press, 2013.

[23] Jones M. The complete guide to creating oils, soaps, creams, and herbal gels for your mind and Body: 101 Natural Body Care Recipes, Atlantic Publishing Company 2010.

[24] Ludwiczuk A, Skalicka-Wozniak K, Georgiev M. Terpenoids. In: Badal S, Delgoda R (eds), Pharmacognosy: Fundamentals, applications and Strategied. Elsevier, London, 2017, 233–266.

[25] Casselm E, Vargas RMF, Martinez N, Lorenzo D, Dellacassa E. Steam distillation modeling for essential oil extraction process. Ind Crops Prod 2009, 29, 171–176.

[26] Dima C, Dima S. Essential oils in foods: Extraction, stabilization, and toxicity. Curr Opin Food Sci 2015, 5, 29–35.

[27] Guan W, Li S, Yan R, Tang S, Quan C. Comparison of essential oils of clove buds extracted with supercritical carbon dioxide and other three traditional extraction methods. Food Chem 2007, 101, 1558–1564.

[28] Viuda-Martos M, Ruiz-Navajas Y, Fernandez-Lopez J, Perez-Alvarez J. Antifungal activity of lemon (Citrus lemon L.), mandarin (Citrus reticulata L.), grapefruit (Citrus paradisi L.) and orange (Citrus sinensis L.) essential oils. Food Control 2008, 19, 1130–1138.

[29] Yepez B, Espinosa M, Lopez S, Bolanos G. Producing antioxidant fractions from herbaceous matrices by supercritical fluid extraction. Fluid Phase Equilib 2002, 194, 879–884.

[30] Junming X, Jianchun J, Jie C, Yunjuan S. Biofuel production from catalytic cracking of woody oils. Bio Resour Technol 2010, 101, 5586–5591.

[31] Jirovetz L, Buchbauer G, Ngassoum MB, Geissler M. Aroma compound analysis of Piper nigrum and Piper guineense essential oils from Cameroon using solid-phase microextraction–gas chromatography, solid-phase microextraction–gaschromatography–mass spectrometry and olfactometry. J Chromato 2002, 976, 265–275.

[32] Thimmappa R, Geisler K, Louveau T, O Maille P, Osbourn A. Triterpene biosynthesis in plants. Annu Rev Plant Biol 2014, 65, 225–257.

[33] Breitmaier E. Terpenes: Flavors, fragrances, pharmaca, pheromones, John Wiley & Sons, 2006.

[34] Swamy MK, Mohanty SK, Sinniah UR, Maniyam A. Evaluation of patchouli (PogostemoncablinBenth.) cultivars for growth, yield and quality parameters. J Esse Oil Bear Plants 2015, 18, 826–832.

[35] Ahmad W, Ahmad A, Hassan YA, Sivapalan N, Daniel S, Anna RA, Al-Shurfa F, Albaharnah F, AlHayyan A. Awareness, knowledge, attitude, perception, and utilization of complementary and alternative medicines (CAMs) in the common population of Dammam, Saudi Arabia. J Phar Bioallied Sci. 2022, 14, 99-105

[36] Ali B, Al-Wabel NA, Shams S, Ahamad A, Khan SA, Anwar F. Essential oils used in aromatherapy: A systemic review. Asian Pac J Trop Biomed 2015, 8, 1–11.

[37] El-Anssary AAKE. Aromatherapy as Complementary Medicine. In: de Oliveira MA, Costa WAD, Silva SG (Eds.), Essential Oils – Bioactive Compounds, New Perspectives and Applications. IntechOpen, 2020.

[38] Baris O, Gulluce M, Sahin F, Ozer H, Kilic H, Ozkan H. Biological activities of the volatile oil and methanol extract of Achilleabiebersteini Afan. (Asteraceae). Turkish J Biol 2006, 30, 65–73.

[39] Shibamoto K, Mochizuki M, Kusuhara M. Aroma therapy in antiaging medicine. Anti-Aging Med 2010, 7, 55–59.

[40] Suaib L, Dwivedi GR, Darokar MP, Kaira A, Khanuja SPS. P

[41] Saeidi K, Moosavi M, Lorigooini Z, Maggi F. Chemical characterization of the essential oil compositions and antioxidant activity from Iranian populations of AchilleawilhelmsiiK. Koch Ind Crops Prod 2018, 112, 274–280.

[42] Elshafie HS, Camele I. An overview of the biological effects of some mediterranean essential oils on human health. Biomed Research Int 2017, 14, 9268468.

[43] Donsi F, Wang Y, Li J, Huang Q. Preparation of curcumin sub-micrometer dispersions by highpressure homogenization. J Agr Food Chem 2010, 58, 2848–2853.

[44] Koreti D, Kosre A, Kumar A, Chandrawanshi NK. Mushroom Bioactive Compounds: Potential Source for the Development of Antibacterial Nanoemulsion. In: Ramalingam K (Ed.), Handbook of Research on Nanoemulsion Applications in Agriculture, Food, Health, and Biomedical Sciences. IGI Global, 2022, 213–235.

[45] Chandrawanshi NK, Koreti D, Kosre A, Mahish PK. Mushroom-derived Bioactive-based Nanoemulsion: Current Status and Challenges for Cancer Therapy. In: Ramalingam K (Ed.), Handbook of Research on Nanoemulsion Applications in Agriculture, Food, Health, and Biomedical Sciences. IGI Global, 2022, 354–376.

[46] Kosre A, Koreti D, Chandrawanshi NK, Kumar A. Nanoemulsion Based on Mushroom Bioactive Compounds and Its Application in Food Preservation. In: Ramalingam K (Ed.), Handbook of Research on Nanoemulsion Applications in Agriculture, Food, Health, and Biomedical Sciences. IGI Global, 2022, 425–447.

[47] Wenzel E, Somoza V. Metabolism and bioavailability of trans-resveratrol. Mol Nutr Food Res 2005, 49, 472–481.

[48] Krishna A, Tiwari R, Kumar S. Aromatherapy-an alternative health care through essential oils. J Med Aromat Plant Sci 2000, 22, 798–804.

[49] Adorjan B, Buchbauer G. Biological properties of volatile oils: An updated review. Flav Frag J 2010, 25, 407–426.

[50] Baser KHC, Buchbauer G. Handbook of Volatile Oils: Science, Technology, and Applications, CRC Press, NW, 2010.

[51] Mart-nez-Perez EF, Juarez ZN, Hernandez LR, Bach H. Natural antispasmodics: Source, stereochemical configuration, and biological activity. Bio Med Research Int 2018, 2018, 32.

[52] Soden K, Vincent K, Craske S, Lucas C, Ashley S. A randomized controlled trial of aromatherapy massage in a hospice setting. Palliat Med 2004, 18, 87–92.

[53] Chang SY. Effects of aroma hand massage on pain, state anxiety and depression in hospice patients with terminal cancer. J Korean Aca Nurs 2008, 38, 493–502.

[54] Johnson AJ. Cognitive facilitation following intentional odor exposure. Sensors 2011, 11, 5469–5488.

[55] Wei A, Shibamoto T. Antioxidant/ lipoxygenase inhibitory activities and chemical compositions of selected volatile oil. J Agri Food Chem 2010, 58, 7218–7225.

[56] Moss M, Cook J, Wesnes K, Duckett P. Aromas of rosemary and lavender volatile oils several, ially affect cognition and mood in healthy adults. Int J Neuroscience 2003, 113, 15–38.

[57] Maeda K, Ito T, Shioda S. Medical aromatherapy practice in Japan. Essence 2012, 10, 14–16.

[58] Manso SD, Pezo Gomez-Lus R, Nerin C. Diminution of aflatoxin B1 production caused by an active packaging containing cinnamon essential oil. Food Control 2014, 45, 101–108.

[59] Wrona M, Bentayeb K, Nerin C. A novel active packaging for extending the shelf-life of fresh mushrooms (Agaricusbisporus). Food Control 2015, 54, 200–207.

[60] Aitboulahsen M, Zantar S, Laglaoui A, Chairi H, Arakrak A. Gelatin-based edible coating combined with menthapulegium essential oil as bioactive packaging for strawberries. J Food Qual 2018, 2018, 8408915.

[61] Haddi K, Faroni L, Oliveira EE. Cinnamon oil, in green pesticides handbook: Essential oils for pest control, Taylor and Francis, CRC Press, 2017, 117–150.

[62] Simionato I, Domingues FC, Nerin C, Silva F. Encapsulation of cinnamon oil in cyclodextrinnanosponges and their potential use for antimicrobial food packaging. Food Chem Toxi 2019, 132, 110647.

[63] Clemente I, Aznar M, Silva F, Nerin C. Antimicrobial properties and mode of action of mustard and cinnamon essential oils and their combination against foodborne bacteria. Inn Food Scie Eme Technol 2016, 36, 26–33.

[64] Rojas-Grau MA, Avena-Bustillos RJ, Olsen C, Friedman M, Henika PR. Effects of plant essential oils and oil compounds on mechanical, barrier and antimicrobial properties of alginate–apple puree edible films. J Food Eng 2007, 81, 634–641.

[65] Yemis GP, Candogan K. Antibacterial activity of soy edible coatings incorporated with thyme and oregano essential oils on beef against pathogenic bacteria. Food Sci Biotech 2017, 26, 1113–1121.

[66] Roman S, Sanchez-Siles LM, Siegrist M. The importance of food naturalness for consumers: Results of a systematic review. Trends Food Sci Techn 2017, 67, 44–57.

[67] Rawat S. Food Spoilage: Microorganisms and their prevention. Asian J Plant Sci Res 2015, 5, 47–56.

[68] Eigenmann PA, Haenggeli CA. Food colourings and preservatives- allergy and hyperactivity. Lancet 2004, 364, 823–824.

Vatsala Soni, Dipti Bharti, Meenakshi Bharadwaj, Vaishali Soni,
Richa Saxena and Charu Arora

Chapter 12
Toxicity of essential oils

Abstract: Natural essential oils (EOs) and their compounds have substantially shown beneficial effects, but even with related herbal remedies that have been used for centuries. Still, uncontrolled practice can be devastating and can have consequences. A 100- to 1,000-fold increase in the concentration of EO components (often <0.01% yield) compared to the concentration of EO components in whole plants indicates that EOs are not equivalent to whole plants. All aspects of safety are now under scrutiny and new policies will rapidly prevent the trade and use of numerous EOs and cosmetics as well as their use in food. Some EOs are noxious if used in especially high doses, particularly when in use orally or by children and infants. It has been found that there is occurrence of several noxious components in several EOs that intrinsically makes them toxic, even utilized at very low concentrations. Though, further comprehensive studies are necessary to make sure the drug safety and efficacy before its authorization. The aim of this chapter documented the efficient available data associated with different aspects of toxicity, efficacy, and shelter problems of EOs. Further research is needed to evaluate and predict the toxicological characteristics of EOs via in vitro and in vivo methods. This study seeks to provide insight and/or update on the current state of the field through multiple reviews of specific aspects of the toxicity of EOs and their compounds as well as government focus. However, further research is needed to confirm the drug's effectiveness and to evaluate the safety levels of the drugs proposed before approval.

Keywords: Herbal remedies, noxious components, toxicological properties

Vatsala Soni, PEC-Punjab Engineering College, Chandigarh 160012, India
Dipti Bharti, Department of Applied Science and Humanities, Darbhanga College of Engineering, Darbhanga, Bihar 846005, India
Meenakshi Bharadwaj, Department of Chemistry, IEC University, Baddi, Himachal Pradesh, India
Vaishali Soni, PGIMER-Post Graduate Government Institute of Medical Education and Research, Chandigarh 160012, India
Richa Saxena, Department of Biotechnology, Invertis University, Bareilly, Uttar Pradesh 243001, India
Charu Arora, Department of Chemistry, Guru Ghasidas University, Bilaspur, Chhattisgarh 495009, India

https://doi.org/10.1515/9783110791600-012

12.1 Introduction

Essential oils (EOs) are formed as bioactive compounds by aromatic medicinal plants, mostly belonging to several families, that is, Agavaceae, Anacardiaceae, Annonaceae, Apiaceae, Asteraceae, Burseraceae, Canellaceae, Cannabaceae, Cistaceae, Cupressaceae, Dipterocarpaceae, Ericaceae, Fabaceae, Geraniaceae, Illiciaceae, Lamiaceae, Lauraceae, Malvaceae, Myristicaceae, Myrtaceae, Oleaceae, Orchidaceae, Parmeliaceae, Pinaceae, Piperaceae, Poaceae, Rosaceae, Rubiaceae, Rutaceae, Santalaceae, Solanaceae, Styracaceae, Tiliaceae, Valerianaceae, Verbenaceae, Violaceae, and Zingiberaceae. EOs are natural volatile composite compounds illustrated via a strong odor [1, 2]. EOs are a composite combination of phenylpropanoid and terpenes complexes found in various classes of aromatic medicinal plants [3]. Several EOs like tea tree, lavender, sage, nutmeg, wintergreen, geranium, peppermint, fennel, thuja, lemon myrtle, woodworm, eucalyptus oil, and clove are common in nature. Usually, EOs come from different plant parts, for example, flowers (e.g., jasmine, rose, clove, mimosa, lavender, Clary sage, chamomile, and rosemary), fruits (orange, juniper, lemon, mandarin, and bergamot), rhizomes (e.g., orris, curcuma, calamus, and ginger), and oleoresin exudations or gums (e.g., benzoin, *Myroxylon balsamum*, balsam of Peru, storax, and myrrh), leaves (e.g., *Ocimum* spp., lemongrass, jamrosa, and mint), leaves and stems (e.g., petitgrain, patchouli, geranium, cinnamon, and verbena), bark (e.g., canella, cassia, cinnamon, laurel, and peppermint), wood (e.g., pine, sandal, and cedar), roots (e.g., *Angelica*, *Saussurea*, *Sassafras*, valerian, and vetiver), seeds (e.g., nutmeg, caraway, coriander, fennel, black pepper, and dill), and grass (e.g., lemongrass and palmarosa) [4].

EOs and their chemical compounds have various uses in flavors, fragrances, and perfumes, but they are also broadly used in traditional medicine. Salient features of EOs are color, aroma, volatility, and lipophilicity. As a result, more attention than ever has been directed to the biological activities of EOs [5]. EOs are commonly used for their antioxidant, antifungal, antibacterial, antiviral, antiinflammatory, anticancer, antihistaminic, antidiabetic properties, as well as for food preservation, feed additives, and ingredients in cosmetics. All these properties have remained largely unchanged to date, except that the mechanisms of antibacterial action are well known.

Since ancient times, all over the world, EOs have been used in lots of different traditional healing systems due to their pharmacological activities. EOs are recognized as one of the leading candidates for curing various diseases in the pharmaceutical industry.

A study by Stevanović et al. [6] showed that there are currently about 3,000 different EOs known, which are mainly composed of various volatile and nonvolatile oils such as terpenes, ketones, phenolics, epoxides, aldehydes, alcohols, acids, esters, sulfides, and amines. The beneficial compounds in many EOs, such as thymoquinone, eugenol, pepper family piperine, and curcuminoids, have promising therapeutic effects and offer extensive research potential for the pharmaceutical industry [7].

Several most active secondary metabolites which are found in EOs have mechanism to interrelate with numerous pharmacological objectives, for example, enzymes and receptors, which help in the improvement of effective drug opportunity for the industry. Some EOs of these families were used for their toxicological assessment. Acute toxicity to animals and humans at small levels is possible as a result of the reported changes in the composition of the components. Despite the need, no detailed toxicity studies have yet been reported to assess safety, especially in alternative animal models [8]. Inappropriate utilization of EOs can have unpleasant effects on humans, including headache, nausea, and skin irritation. In general, care should be taken in internal consumption of EOs or their use in food. EOs can cause functional damage to organs, for example, liver and stomach in animals and possibly in humans when it is used in nonrecommended doses [82]. Interestingly, lots of EOs are on the generally recognized as safe (GRAS) list of the US Food and Drug Administration (FDA) which allows the use of EOs in the formation of different products such as cosmetics, food, and feed. Although, current research has produced the conflicting results on EO toxicity in vivo and in vitro and also signifying that several EOs already show noxious effect at very low concentrations. In this regard, effects such as mucosal irritation, reproductive toxicity, respiratory damage, organ toxicity, and acute toxicity were discussed [9]. Toxicological effect of EO depends on the dose and the EO ingested [10].

Therefore, if EOs are to be used for therapeutic purposes or incorporated into cosmetics, feeds, or foods, their toxicity potential must first be investigated. However, EO toxicity testing is challenging due to the wide variation in active ingredient content, hydrophobicity, and viscosity. Furthermore, there is still a need for standardized and well-defined methods for reliable toxicological and molecular research as an alternative to animal testing [11–13].

Several toxicity end point investigations are mandatory for a precise prediction of the undesirable effects of constituents on living systems. At this time, many scientists reported a robust strategy for the prediction and assessment of the toxicological effects of EOs [14–16]. A study by Lanzerstorfer et al. [8] showed that eucalyptus, rosemary, and citrus EOs are commonly used as feed and food supplements in various cosmetics and products. Animal testing is the "gold standard" in the toxicology and safety evaluation of industrial chemicals, pesticides, cosmetics, and pharmaceuticals.

12.2 Essential oils' toxicities

EOs are readily available in supermarkets, pharmacies, or online and are used by large sections of the general public. Although the use of EOs is widespread and popular, the safety profile of EOs has not yet been entirely established [17]. Like many synthetic and natural medicines, the compounds in EOs can cause severe and even fatal poisoning if taken in excess. There are several reports of toxicity profiles of serious (nonfatal) and fatal consequences of EO consumption in children and adults [18]. Due to the chemical complexity of EOs, it is difficult to investigate the individual components responsible for specific adverse effects. However, some necessary steps have already been taken.

Eos are quickly absorbed orally and symptoms may develop as early as 30 min after ingestion or up to 4 h after exposures. The severity of EOs toxicity is dependent on the category of EOs and the amount of ingestions. Children are especially susceptible to toxic effects than adults and *Eucalyptus* oil as little dose as 2 mL (less than half a teaspoon) can cause significant poisoning in an infant [19, 20]. These cases are usually caused by accidental ingestion of babies, attempted abortions (in the past few years) [19].

Figure 12.1 illustrates some general aspects of the principles of toxicity as they apply to plants commonly used as EOs. These products are not regulated by the FDA, may be contaminated or adulterated with other chemicals, may not contain the specific herbs the buyer intended, or may contain unknown herbs. It may exist in possible concentrations.

Nomenclature is confusing, so "black cohosh" sold in different states and regions of the United States can have numerous natural herbs which have their own descriptions. In northern Appalachia cohosh refers to *Cimicifuga racemosa*, in southern Appalachia cohosh refers to *Caulophyllum thalictroides*, and in New England cohosh refers to Baneberry (*Actaea* sp.) [21]. It is very difficult to conclude which compound in EOs is useful, why, and which are naturally toxic. In addition, the strength of the Eos can vary and depend on the harvesting circumstances of the plant.

Some aromatic compounds, especially 1,8-cineole (found in *Eucalyptus* species), camphor (Borneon, as an isolated compound or in *Rosmarinus officinalis* CT camphor and *Lavandula latifolia*), and methyl salicylate (compounds derived as synthetic or found in *Gaultheria procumbens*) have specific toxic effects in much lower doses [23].

Several EOs and their compounds were investigated on laboratory animals for their potential toxic effect [24]. Table 12.1 shows examples of toxic EOs and their chemical compounds that are believed to be the main contributors to their toxicity.

The natural preservative form of EO is generally recognized and GRAS by the FDA, which allowing its use. Although these EOs must be used in high concentrations, toxicity issues cannot be ignored. There are currently available toxicity and safety assessments applied to various variables in EO. One of the most common methods used to evaluate the protection, efficacy, and toxicity of EO is LD50 or median lethal dose values which are determined in aminal model [25].

General Aspects of Essential Oils Toxicities and Other Natural Herbal Preparations

Product organization	Constituents	Conditions	Host Distinctiveness	Unidentified Toxicities
• Nomenclature: no international conventions, confusing synonyms • Quality control: no FDA regulation = buyer beware	• Natural toxicants: no adverse effect safety testing • Complex combinations of chemicals • combination with other herbs or medications regulation = buyer beware • Variable potencies: time of year, year to year, geography, soil, part of plant, maturity,	• Inappropriate use • Excessive use	• Placental transport/excretion into breast milk • Sensitization/anaphylaxis	• Mutagenesis • Carcinogenesis • Genotoxicity; • Embryotoxicity

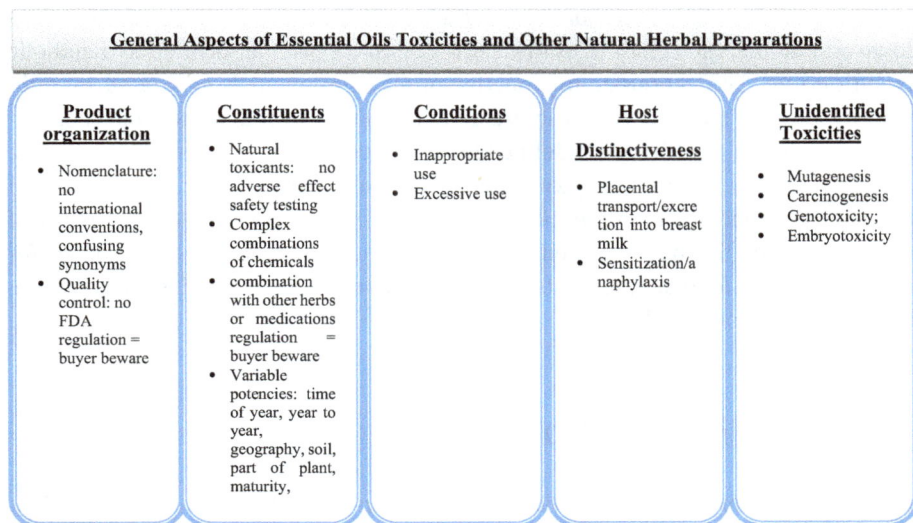

Figure 12.1: Classification of general aspects of essential oil toxicities [22].

Lis-Balchin [24] also noticed some exceptional EOs from *Chenopodium*, *Satureja hortensis* (savory), *Mentha pulegium* (pennyroyal), *Thuja*, and *Boldo* leaf who offered an LD50 between 0.1 and 1 g/kg in rats, indicating a noteworthy toxicity which advised essential safety measures for their use.

Tisser and and Young [26] evaluated the acute toxicity by LD50 test in rats, which discovered that some EOs have a LD50 of 1–20 g/kg, signifying a low toxicity. Lemon oil has an LD50 of above 5 g/kg in human. Consequently, for an adult of 70 kg, the lethal dose would be 350 g, hard to reach in normal conditions.

Most of the scientist investigated that the adequate range of EOs and their chemical constituents is 1–20 g/kg body weight with LD50 in animal models [25, 27]. In the recent years, these estimations are considered necessary because of the oestrogenic, carcinogenic, and abortifacient side effects [27].

12.3 Toxicological assessment of essential oils used as food supplements for safe oral ingestion

EOs and their main bioactive compounds are gradually more consumed as food supplements in food industry for improving the shelf life of consumable products. EOs safety should be guaranteed before being commercialized. In general, Most of the EOs did not demonstrate mutagenic activity and some genotoxic studies have been accounted on EOs in comparison to their main bioactive compounds. The genotoxic and mutagenic activities of these essences have been associated to metabolic activation [28]. The genotoxic prospective of all component is evaluated by published records or quantitative structure-activity relationship analyses and methodology provided in the ICHM7 guideline [29]. Safe dose for an EO is measured secure when the security margin ratio exposure for all compounds is all at least equal to 1 between the toxicological reference value and systemic exposure [30]. The study of Tamburlin et al. [30] have verified to be robust to found safe suggested doses for EOs employed as food supplementsand step-by-step evaluation is desired to assure their safe use in food packaging.

12.4 Acute intoxication

Most severe poisonings (intoxications) occur when these compounds are accidentally swallowed by children. Of particular concern are the epileptic effects of EOs such as camphor, eucalyptus, and thujone; severe hepatotoxicity of pennyroyal oil caused by pulegone and menthofuran and high concentration of clove oil caused by eucalyptol; salicylic acid poisoning by wintergreen oil; phototoxicity of bergamot oil, coumarin, and furocoumarin containing furocoumarin; uses for pregnant women because it passes easily and it may harm the fetus or baby especially for babies and toddlers because they are prone to severe poisoning [31, 32].

According to the study of Vostinaru et al. [7] severe intoxication belongings were accounted in the USA and Australia by tea tree oil, clove, cinnamon, eucalyptus, and wintergreen oils with the signofnausea, vomiting, polypnea, and convulsions.

Tisserand and Young [26] investigated that in the USA, 966 cases of acute intoxication were observed in children up to 6-year-old caused by tea tree oil intake. The current study of [33] identified the 1,387 reports of EOs intoxication among 2014 and 2018 in Australia. Infants and young children are at a special risk of acute poisoning due to their weight loss and immature enzyme systems capable of metabolizing EOs.

12.5 Dermatological toxicity

Nowadays, aromatherapy with some EOs and their ingredients can cause allergic dermatitis in rare cases because of their lipophilicity and ability to enter in the skin. Generally in aromatherapy, if essential oils diluted in carrier oil are applied directly on skin then they may cause skin irritation, sensitization, and photosensitization [59, 60]. The severity of skin reactions depends on the substance used (aldehyde and phenol), the vehicle used, the quality/mixture of the EOs, dilution, the process

Table 12.1: Examples of toxic essential oils and their chemical compounds.

S. no.	Source	Essential oil	Toxic chemical	Chemical constituent	Toxic effect	Effect on stored product pest	Reference
1	Nutmeg	Nutmeg oil	Myristacin, eugenol	Methoxysafrole	Hallucinations	*Tribolium castaneum* (Herbst) and *Sitophilus zeamais*	[22, 34]
2	Eucalyptus	Eucalyptus globulus oil	1,8-Cineole	Monoterpene	Seizures	*Rhyzopertha dominica*, *Sitophilus oryzae*, *Oryzaephilus surinamensis*, and *Tribolium castaneum*	[22]
3	Menthasp,	*Mentha spicata* essential oil	Menthol, menthone	Monoterpene	Ataxia, myalgia	*Tribolium castaneum*	[22, 35, 36]
4	Cinnamon	Cinnamon essential oil	Cinnamaldehyde	α,β-Unsaturated aldehyde	Dermatitis	*Tribolium castaneum* (Herbst) and *Sitophilus zeamais* Motsch.	[37, 38]
5	Pennyroyal	*Mentha pulegium* essential oil	Pulegone	Monoterpene	Hepatic necrosis	*Tribolium castaneum* and *Lasioderma serricorne*	[39–41]
6	Wormwood	*Artemisia absinthium* L. essential oil	α- or β-Thujone	Monoterpene ketone	Seizures, dementia, hallucinations, sleeplessness, and convulsions	*Callosobruchus maculatus* (F.), *Sitophilus oryzae* (L.), *Tribolium castaneum* (Herbst) (Curculionidae, Coleoptera)	[42–44]

(continued)

Table 12.1 (continued)

S. no.	Source	Essential oil	Toxic chemical	Chemical constituent	Toxic effect	Effect on stored product pest	Reference
7	Oregano and thyme	*Origanum vulgare* and *Thymus vulgaris* Essential oil	p-Cymene	1-Isopropyl-4-methylbenzene (monoterpene)	Decrease mitochondrial membrane potential, diarrhea, drowsiness, headache, nausea, vomiting, unconsciousness	*Tribolium castaneum*	[22, 45, 46]
8	Pine tree	Pine tree essential oils	α-Pinene	Monoterpene	Potential respiratory and skin irritation	*Acanthoscelides obtectus*	[22, 47, 48]
9	Almond	*Prunus amygdalus* essential oil	Benzaldehyde	Phenylmethanal	Irritation in the nose and throat which cause cough and shortness of breath	*Sitophilus oryzae*	[22, 46]
10	Lavender	*Lavandula angustifolia* and other species essential oil	Linalool	Terpene alcohol	Ataxia, a decrease in spontaneous motor activity, lateral recumbency, narcosis, and respiratory disturbances	*Oryzaephilus surinamensis*	[22, 49, 50]
11	Mexican tea plant	Chenopodium oil	Carvacrol	Monoterpenoid phenol	APAP-induced hepatotoxicity	*Oryzaephilus surinamensis*	[22, 51–53]
12	Tea tree oil	*Melaleuca alternifolia* essential oil	Terpinen 4-ol	Monoterpene alcohol	Skin corrosion/irritation	*Rhyzopertha dominica*	[22, 46, 54–56]
13	Camphor	*Cinnamomum camphora* Essential oil	D-Camphor	1,7,7-Trimethylbicyclo [2.2.1]heptan-2-one	Breathing, seizures, and death	*Tribolium castaneum* and *Lasioderma serricorne*	[46, 57, 58]

of administration, the anatomical site of exposure, the integrity of the skin, and age. Environmental circumstances such as ultraviolet (UV) light, ambient temperature, and humidity play a vital role in photosensitization, warm, and humid surroundings and these are more complimentary for augmented severity of adverse effects.

12.6 Skin irritation

Several researchers reported that some essential oils which are resultant from *Thymus vulgaris*, *Origanum vulgare*, *Cinnamomum zeylanicum*, *Cuminum cyminum*, *Tagetes minuta*, and *Syzygiumaromaticum* are usually known as common skin irritant oils [7, 25, 61–64]. The main skin irritant chemical compounds are carvacrol, thymol (thyme or savory, oregano), phenol, phenolic ethers (i.e., anethole (clove) and eugenol), or aromatic aldehydes (i.e., cinnamaldehyde (cinnamon)) [7].

12.7 Skin sensitization

Skin sensitization, as opposed to irritation, is the activation of the adaptive immune system to specific chemicals called sensitizers or haptens that alter skin proteins and can induce delayed allergic reactions mediated by T cells and this is a reaction and possibility exist [61, 65]. Some oils considered to be skin sensitizers include benzyl alcohol, cinnamyl alcohol, citral, eugenol, hydroxycitronellal, isoeugenol, benzyl salicylate, cinnamaldehyde, coumarin, geraniol, anisyl alcohol, and benzyl cinnamate [65–67]. To avoid sensitization, scientist recommends avoiding known skin sensitizers and avoiding long-term use of the same essential oil daily [68, 69].

12.8 Phototoxicity

Phototoxication or photosensitization is an effect between phototoxins from EOs that are applied to the skin in the presence of sunlight or UV rays. Interactions with light are either phototoxic or photoallergic. Generally, furanocoumarins (psoralens) are accountable for phytophototoxic effects in humans and mostly available in EOs derived from cumin, citrus species, marigold, and parsley leaf. Bergapten and psoralen are the most common compounds of furanocoumarins which are involved in photosensitization [25, 59, 70, 71].

12.9 Other physiological toxicities

EOs simply cross the blood–brain barrier and reach in to the central nervous system after being absorbed throughout the body. In the experimental study of Millet et al. [72], essential oils of *Hyssopus officinalis* and *Salvia officinalis* caused seizures at doses of 0.5 and 0.13 g/kg, respectively, after intraperitoneal administration. According to several studies, some essential oils from *Anethum graveolens*, *Cinnamonum-camphora*, *Cedrus* spp., *Salvia officinalis*, *Hyssopus officinalis*, *Mentha pulegium*, *Thuja plicata*, and *Eucalyptus* spp. created tonic–clonic convulsions, mostly in young children and particularly in individuals who has a history of epileptic syndromes [73, 74]. The investigation of Steinmetz et al. [75] reported that a powerful convulsive agent is pentylentetrazole resemble to some essential oil altering tissue gradients of K and Na and leading to enlarged cellular excitability in the brain.

Due to the neurotoxicity potential of EOs, the EMA (European Medicines Agency) has recognized that essential oils are used for the cure of respiratory diseases in seven European countries (Belgium, Italy, Luxembourg, Portugal, Spain, France, and Finland). The safety of suppositories containing terpenes was investigated. The report of EMA concluded that terpenes can cause seizures in children less than 30 months of age and recommended that terpenes be contraindicated in this particular patient population [76].

Over the past decade, evidence has accumulated that suggests the potential endocrine-disrupting effects of essential oils. Experimental data show that the effects of tea tree and lavender oils are caused by activation of the estrogen receptor (ER), and 50% of the potency of estradiol is due to folostrant, a pure antagonist of the ER receptor [77]. The most active EO chemical compounds with estrogenic activity were 4-terpineol, α-terpineol, and linalool. Further studies on more statistically significant populations are required to validate the importance of these findings [78].

The use of essential oils during pregnancy is a controversial topic and is still not fully understood. The main concern regarding the use of essential oils during pregnancy is related to the risk of the compounds crossing the placental barrier and directly affecting the conceptus as well as direct effects on abortion.

Some essential oils are abortifacient and may cause miscarriage or abortion. Essential oils such as Parsley oil rich in apiol (*Petroselinum sativum*), royal jelly oil rich in *Mentha pulegium*, Plectranthusamboinicus oil, Spanish sage oil (*Salvia lavandulifolia*), and Sabina oil (*Juniperus sabina*) should be enriched. Avoid savinyl acetate during pregnancy. Doses needed to induce abortion can also pose toxic risks to the mother, including kidney and liver damage and even death [24, 79–81].

12.10 Conclusion

Most of the time, essential oils are safe to utilize, except in cases of overdose, misuse, or hypersensitivity. Since the toxicity of essential oils is concentration-dependent, many unwanted side effects can be avoided by using low doses and limiting them to small areas of the skin. People with allergies should avoid essential oils because allergic reactions occur independently of the applied concentration and route of consumption. Proper storage of essential oils is essential, as oxidation products formed by exposure to light and/or air can play an important role in toxicity. To successfully exploit the diverse biological effects of essential oils, there is a need to raise awareness among healthcare professionals and the general public about the security profile of essential oils' uses.

References

[1]   Bakkali F, Averbeck S, Averbeck D, Idaomar M. Biological effects of essential oils–a review. Food Chem Toxicol 2008, 46, 446–475.

[2]   Jamalova DN, Gad HA, Akramov DK, Tojibaev KS, Musayeib NM, Ashour ML, Mamadalieva NZ. Discrimination of the Essential Oils Obtained from Four Apiaceae Species Using Multivariate Analysis Based on the Chemical Compositions and Their Biological Activity. Plants 2021, 10(8), 1529.

[3]   Ribeiro-Santos R, Andrade M, Sanches-Silva A, de Melo NR. Essential oils for food application: Natural substances with established biological activities. Food Bioproc Tech 2018, 11(1), 43–71.

[4]   Aziz ZA, Ahmad A, Setapar SH, Karakucuk A, Azim MM, Lokhat D, Rafatullah M, Ganash M, Kamal MA, Ashraf GM. Essential oils: Extraction techniques, pharmaceutical and therapeutic potential-a review. Curr Drug Metab 2018, 19(13), 1100–1110.

[5]   Unlu M, Ergene E, Unlu GV, Zeytinoglu HS, Vural N. Composition, antimicrobial activity and *in vitro* cytotoxicity of essential oil from *Cinnamomum zeylanicum* Blume (Lauraceae). Food Chem Toxicol 2010, 48(11), 3274–3280.

[6]   Stevanović ZD, Bošnjak-Neumüller J, Pajić-Lijaković I, Raj J, Vasiljević M. Essential oils as feed additives – Future perspectives. Molecules 2018, 23(7), 1717.

[7]   Vostinaru O, Heghes SC, Filip L. Safety profile of essential oils. Essential oils-bioactive compounds. New Perspect App 2020, 1–3.

[8]   Lanzerstorfer P, Sandner G, Pitsch J, Mascher B, Aumiller T, Weghuber J. Acute, reproductive, and developmental toxicity of essential oils assessed with *alternative in vitro* and *in vivo* systems. Arch Toxicol 2021, 95(2), 673–691.

[9]   Sandner G, Heckmann M, Weghuber J. Immunomodulatory activities of selected essential oils. Biomolecules 2020, 10(8), 1139.

[10] Horky P, Skalickova S, Smerkova K, Skladanka J Essential oils as a feed additives: Pharmacokinetics and potential toxicity in monogastric animals. Animals. 2019, (6):352.

[11] Dosoky NS, Setzer WN. Biological activities and safety of *Citrusspp*. Essential Oils. Int J Mol Sci 2018, 19(7), 1966.

[12] Mathlouthi N, Bouzaienne T, Oueslati I, Recoquillay F, Hamdi M, Urdaci M, Bergaoui R. Use of rosemary, oregano, and a commercial blend of essential oils in broiler chickens: *In vitro* antimicrobial activities and effects on growth performance. J Anim Sci 2012, 90(3), 813–823.

[13] Reyer H, Zentek J, Männer K, Youssef IM, Aumiller T, Weghuber J, Wimmers K, Mueller AS. Possible molecular mechanisms by which an essential oil blend from star anise, rosemary, thyme, and oregano and saponins increase the performance and ileal protein digestibility of growing broilers. J Agric Food Chem 2017, 65(32), 6821–6830.

[14] HesabiNameghi A, Edalatian O, Bakhshalinejad R. Effects of a blend of thyme, peppermint and eucalyptus essential oils on growth performance, serum lipid and hepatic enzyme indices, immune response and ileal morphology and microflora in broilers. J AnimPhysiolAnimNutr 2019, 103(5), 1388–1398.

[15] Kumar Tyagi A, Bukvicki D, Gottardi D, Tabanelli G, Montanari C, Malik A, Guerzoni ME. *Eucalyptus* essential oil as a natural food preservative: *In vivo* and *in vitro*antiyeast potential. Biomed Res Int 2014, 2014.

[16] Ozogul Y, Kuley E, Ucar Y, Ozogul F. Antimicrobial impacts of essential oils on food borne-pathogens. Recent Pat Food Nutr Agric 2015, 7(1), 53–61.

[17] Dima C, Dima S. Essential oils in foods: Extraction, stabilization, and toxicity. Curr Opin Food Sci 2015, 5, 29–35.

[18] Harnett JE, McIntyre E, Steel A, Foley H, Sibbritt D, Adams J. Use of complementary medicine products: A nationally representative cross-sectional survey of 2019 Australian adults. BMJ Open 2019, 9(7), e024198.

[19] Kumar KJ, Sonnathi S, Anitha C, Santhoshkumar M. *Eucalyptus* oil poisoning. Toxicol Int 2015, 22(1), 170–171.

[20] Webb NJ, Pitt WR. *Eucalyptus* oil poisoning in childhood: 41 cases in south-east Queensland. J Paediatr Child Health 1993, 29(5), 368–371.

[21] Gafner S Black Cohosh Adulteration Laboratory Guidance Document. American Botanical Council, Austin,TX. 2015.

[22] Woolf A. Essential oil poisoning. J Toxicol Clin Toxicol 1999, 37(6), 721–727.

[23] NDPSC. Compilation of Poisons Information Centre Reports Working Party on Essential Oils Toxicity monographs. 1998.

[24] Lis-Balchin M. Aromatherapy Science: A Guide for Healthcare Professionals, 1st edn, Pharmaceutical Press, London, 2005, 528. ISBN: 9780857111340.

[25] Kuttan R, Liju VB. Safety evaluation of essential oils. In: Hashemi SMB, Khaneghah AM, de Sant'ana AS (editors), Essential oils in Food Processing: Chemistry, Safety and Applications, 1st edn, John Wiley & Sons, UK. West Sussex, 2017, 339–358.

[26] Tisserand R, Young R. Essential Oil Safety, A Guide for Health Care Professionals, 2nd edn, Churchill Livingstone, London, 2014, 780.

[27] Falleh H, Jemaa MB, Saada M, Ksouri R. Essential oils: A promising eco-friendly food preservative. Food Chem 2020, 330, 127268.

[28] Llana-Ruiz-Cabello M, Pichardo S, Maisanaba S, Puerto M, Prieto AI, Gutierrez-Praena D, Jos A, Cameán AM. *In vitro* toxicological evaluation of essential oils and their main compounds used in active food packaging: A review. Food Chem Toxicol 2015, 81, 9–27.

[29] ICHM7(R1). Assessment and control of DNA reactive (mutagenic) impurities in pharmaceuticals to limit potential carcinogenic risk. International Council for Harmonisation of technical requirements for pharmaceuticals for human use. 2017.

[30] Tamburlin IS, Roux E, Feuillée M, Labbé J, Aussagues Y, El Fadle FE, Fraboul F, Bouvier G. Toxicological safety assessment of essential oils used as food supplements to establish safe oral recommended doses. Food Chem Toxicol 2021, 157, 112603.

[31] Akhila JS, Deepa S, Alwar MC. Acute toxicity studies and determination of median lethal dose. Curr Sci 2007, 93, 917–920.

[32] Lemmens-Gruber R. Adverse effects and intoxication with essential oils. In: Handbook of Essential Oils, CRC Press, 2020, 517–541.

[33] Lee KA, Harnett JE, Cairns R. Essential oil exposures in Australia: Analysis of cases reported to the NSW poisons information centre. Med J Aust 2019, 212(3), 132–133.

[34] Huang Y, Tan JM, Kini RM, Ho SH. Toxic and antifeedant action of nutmeg oil against Triboliumcastaneum (Herbst) and Sitophilus zeamaisMotsch. J Stored Prod Res 1997, 33(4), 289–298.

[35] Goudarzian A, Pirbalouti AG, Hossaynzadeh M. Menthol, balance of menthol/menthone, and essential oil contents of *Mentha piperita* L. under foliar-applied chitosan and inoculation of arbuscular mycorrhizal fungi. J Essent Oil-Bear Plants 2020, 23(5), 1012–1021.

[36] Snoussi M, Noumi E, Trabelsi N, Flamini G, Papetti A, De Feo V. *Mentha spicata* essential oil: Chemical composition, antioxidant and antibacterial activities against planktonic and biofilm cultures of *Vibrio spp.* strains. Molecules 2015, 20(8), 14402–14424.

[37] Qu S, Yang K, Chen L, Liu M, Geng Q, He X, Li Y, Liu Y, Tian J. Cinnamaldehyde, a promising natural preservative against *Aspergillus flavus*. Front Microbiol 2019, 10, 2895.

[38] Zhang Y, Liu X, Wang Y, Jiang P, Quek S. Antibacterial activity and mechanism of cinnamon essential oil against *Escherichia coli* and *Staphylococcus aureus*. Food Control 2016, 59, 282–289.

[39] Heydarzade A, Valizadegan O, Negahban M, Mehrkhou F. Efficacy of *Mentha spicata* and *Mentha pulegium* essential oil nanoformulation on mortality and physiology of *Triboliumcastaneum* (Col.: Tenebrionidae). J Crop Prot 2019, 8(4), 501–520.

[40] Jabba SV, Jordt SE. Estimating Fluid Consumption Volumes in Electronic Cigarette Use – Reply. JAMA Intern Med 2020, 180(3), 468–469.

[41] Salem N, Bachrouch O, Sriti J, Msaada K, Khammassi S, Hammami M, Selmi S, Boushih E, Koorani S, Abderraba M, Marzouk B. Fumigant and repellent potentials of *Ricinus communis* and *Mentha pulegium* essential oils against *Triboliumcastaneum* and *Lasiodermaserricorne*. Int J Food Prop 201, 20(3), S2899–913.

[42] Azab MM, Darwish AA, Mohamed RA, Mohamed HH. Efficacy of some plant oils against two stored product insects. Ann Agric Sci Moshtohor 2016, 54(1), 119–128.

[43] Pelkonen O, Abass K, Wiesner J. Thujone and thujone-containing herbal medicinal and botanical products: Toxicological assessment. RegulToxicolPharmacol 2013, 65(1), 100–107.

[44] Ueda J, Kato J. Isolation and identification of a senescence-promoting substance from wormwood (*Artemisia absinthium* L.). Plant Physiol 1980, 66(2), 246–249.

[45] Custódio JB, Ribeiro MV, Silva FS, Machado M, Sousa MC. The essential oils component p-cymene induces proton leak through Fo-ATP synthase and uncoupling of mitochondrial respiration. J Exp Pharmacol 2011, 3, 69.

[46] NCBI (National Center for Biotechnology Information). PubChem Compound Summary for CID 7463, P-Cymene. 2022. Available from https://pubchem.ncbi.nlm.nih.gov/compound/P-Cymene.

[47] NTP (National Toxicology Program) NTP Technical Report on the Toxicity Studies of α-Pinene (CASRN 80-56-8) Administered by Inhalation to F344/N Rats and B6C3F1/N Mice: Toxicity Report 81 [Internet]. Research Triangle Park (NC): National Toxicology Program; 2016. Available from: https://www.ncbi.nlm.nih.gov/books/NBK551127/

[48] Yang H, Woo J, Pae AN, Um MY, Cho NC, Park KD, Yoon M, Kim J, Lee CJ, Cho S. α-Pinene, a major constituent of pine tree oils, enhances non-rapid eye movement sleep in mice through GABAA-benzodiazepine receptors. Mol Pharmacol 2016, 90(5), 530–539.

[49] Pereira I, Severino P, Santos AC, Silva AM, Souto EB. Linalool bioactive properties and potential applicability in drug delivery systems. Colloids Surf B Biointerfaces 2018, 171, 566–578.

[50] Pokajewicz K, Białoń M, Svydenko L, Fedin R, Hudz N. Chemical composition of the essential oil of the new cultivars of Lavandula angustifolia Mill. Bred in Ukraine. Molecules 2021, 26(18), 5681.

[51] Monzote L, Stamberg W, Staniek K, Gille L. Toxic effects of carvacrol, caryophyllene oxide, and ascaridole from essential oil of Chenopodium ambrosioides on mitochondria. Toxicol Appl Pharmacol 2009, 240(3), 337–347.

[52] Palabiyik SS, Karakus E, Halici Z, Cadirci E, Bayir Y, Ayaz G, Cinar I. The protective effects of carvacrol and thymol against paracetamol–induced toxicity on human hepatocellular carcinoma cell lines (HepG2). Hum Exp Toxicol 2016, 35(12), 1252–1263.

[53] Suntres ZE, Coccimiglio J, Alipour M. The bioactivity and toxicological actions of carvacrol. Crit Rev Food Sci Nut 2015, 55(3), 304–318.

[54] Hart PH, Brand C, Carson CF, Riley TV, Prager RH, Finlay-Jones JJ. Terpinen-4-ol, the main component of the essential oil of Melaleuca alternifolia (tea tree oil), suppresses inflammatory mediator production by activated human monocytes. Inflamm Res 2000, 49(11), 619–626.

[55] Loughlin R, Gilmore BF, McCarron PA, Tunney MM. Comparison of the cidal activity of tea tree oil and terpinen-4-ol against clinical bacterial skin isolates and human fibroblast cells. Lett Appl Microbiol 2008, 46(4), 428–433.

[56] Shapira S, Pleban S, Kazanov D, Tirosh P, Arber N. Terpinen-4-ol: A novel and promising therapeutic agent for human gastrointestinal cancers. PLoS One 2016, 11(6), e0156540.

[57] Guo S, Geng Z, Zhang W, Liang J, Wang C, Deng Z, Du S. The chemical composition of essential oils from Cinnamomum camphora and their insecticidal activity against the stored product pests. Int J Mol Sci 2016, 17(11), 1836.

[58] Zuccarini P. Camphor: Risks and benefits of a widely used natural product. J Appl Sci Environ Manag 2009, 13(2).

[59] Burfield T. Safety of essential oils. Int J Aromather 2000, 10(1–2), 16–29.

[60] Michalak M. Aromatherapy and methods of applying essential oils. Arch Physiother 2018, 22, 25–31.

[61] Basketter D, Darlenski R, Fluhr JW. Skin irritation and sensitization: Mechanisms and new approaches for risk assessment. Skin Pharmacol Physiol 2008, 21(4), 191–202.

[62] Posadzki P, Alotaibi A, Ernst E. Adverse effects of aromatherapy: A systematic review of case reports and case series. Int J RiskSaf Med 2012, 24(3), 147–161.

[63] Price L. Power and hazards. In: Price S, Price L (editors), Aromatherapy for Health Professionals, 4th edn, Churchill Livingstone, London, 2011, 61–76. ISBN: 978070.

[64] Stea S, Beraudi A, De Pasquale D. Essential oils for complementary treatment of surgical patients: State of the art. Evid Complement Altern Med 2014, 2014.

[65] Vigan M. Essential oils: Renewal of interest and toxicity. Eur J Dermatol 2010, 20(6), 685–692.

[66] Buckle J. Clinical Aromatherapy-e-book: Essential Oils in Practice, Elsevier sci, 2014.

[67] Heisterberg MV, Menné T, Johansen JD. Contact allergy to the 26 specific fragrance ingredients to be declared on cosmetic products in accordance with the EU cosmetics directive. Contact Dermatitis 2011, 65(5), 266–275.

[68] De Groot AC, Schmidt E. Essential oils, part IV: Contact allergy. Dermatitis 2016, 27(4), 170–175.

[69] Uter W, Schmidt E, Geier J, Lessmann H, Schnuch A, Frosch P. Contact allergy to essential oils: Current patch test results (2000–2008) from the Information Network of Departments of Dermatology (IVDK). Contact Dermatitis 2010 Nov, 63(5), 277–283.

[70] Averbeck D, Averbeck S, Dubertret L, Young AR, Morliere P. Genotoxicity of bergapten and bergamot oil in Saccharomyces cerevisiae. J Photochem Photobiol B Biol 1990, 7(2–4), 209–229.

[71] Vangipuram R, Mask-Bull L, Kim SJ. Cutaneous implications of essential oils. World J Dermatol 2017 May 2, 6(2), 27–31.

[72] Millet Y, Jouglard J, Steinmetz MD, Tognetti P, Joanny P, Arditti J. Toxicity of some essential plant oils. Clinical and experimental study. ClincToxicol 1981, 18(12), 1485–1498.

[73] Burkhard PR, Burkhardt K, Haenggeli CA, Landis T. Plant-induced seizures: Reappearance of an old problem. J Neurol 1999, 46(8), 667–670.

[74] Halicioglu O, Astarcioglu G, Yaprak I, Aydinlioglu H. Toxicity of *Salvia officinalis* in a newborn and a child: An alarming report. Pediatr Neurol 2011, 45(4), 259–260.

[75] Steinmetz MD, Vial M, Millet Y. Actions de l'huileessentielle de romarin et de certains de sesconstituants (eucalyptol et camphre) sur le cortex cérébral de rat *in vitro*. J Toxicol Clin Exper 1987, 7(4), 259–271.

[76] European Medicines Agency. Assessment report for suppositories containing terpenic derivatives. 2012. EMA/67070/2012. Available from: https://www.ema.europa.eu/en/docu ments/referral/terpenic-derivatives-article-31referral-assessment-report_en.pdf.

[77] Henley DV, Lipson N, Korach KS, Bloch CA. Prepubertal gynecomastia linked to lavender and tea tree oils. N Engl J Med 2007, 356(5), 479–485.

[78] Ramsey JT, Li Y, Arao Y, Naidu A, Coons LA, Diaz A, Korach KS Lavender products associated with premature thelarche and prepubertal gynecomastia: Case reports and endocrine-disrupting chemical activities. J Clin Endocrinol Metab. 2019, (11):5393–5405.

[79] Black JM. Essential oils and miscarriage. Midwifery Today Int Midwife 2000, 56, 5–68.

[80] Dosoky NS, Setzer WN. Maternal reproductive toxicity of some essential oils and their constituents. Int J Mol Sci 2021, 22(5), 2380.

[81] Hollenbach CB, Bing RS, Stedile R, da Silva Mello FP, Schuch TL, Rodrigues MR, de Mello FB, de Mello JB. Reproductive toxicity assessment of *Origanum vulgare* essential oil on male Wistar rats. Acta Sci Vet 2015, 43(1), 1–7.

[82] Ngahang Kamte S.L., Ranjbarian F., Cianfaglione K., Sut S., Dall'Acqua S., Bruno M., Afshar F. H., Iannarelli R., Benelli G., Cappellacci L., et al. Identification of Highly Effective Antitrypanosomal Compounds in Essential Oils from the Apiaceae Family. Ecotoxicol. Environ. Saf. 2018;156:154–165. doi: 10.1016/j.ecoenv.2018.03.032.

Himanshu Pandey, Sushma Kholiya, RC Padalia, Priyanka Tiwari
and Ameeta Tiwari*

Chapter 13
Essential oil screening and bioactive potential of some selected trees from temperate zone of Kumaun Himalaya Uttarakhand

Abstract: Natural products are used by local and tribal communities to cure various diseases and alignment. According to the World Health Organization, traditional medicine is the primary line of treatment for numerous ailments for millions of people (80%) all over the world. Due to their active phytoconstituents like flavonoids, steroids, saponins, and tannins about 40% of currently approved pharmaceutical drugs come from natural sources. The essential oils obtained from various plant species have many medicinal and commercial applications. Because of the prominent use of medicinal and aromatic plants in fragrance, food, spices, and as pharmaceutical raw material the demand for these plants is rapidly increasing. The plant resources therefore have become important domain of intervention. This review includes ethnobotanical uses, essential oil constituents, and medicinal values of some plant species that are distributed in Kumaun Himalayan region. These species belong to five different families, namely, Myrtaceae (*Melaleuca bracteata, Melaleuca linariifolia, Melaleuca leucadendron, Callistemon citrinus, Eucalyptus globulus,* and *Eucalyptus citriodora*), Lauraceae (*Cinnamomum camphora* and *Cinnamomum tamala*), Rutaceae (*Murraya koenigii* and *Aegle marmelos*), Burseraceae (*Boswellia serrata*), and Cupressaceae (*Cryptomeria japonica*). According to previous research, plant extracts from various portions of these trees have been shown to have antimicrobial, antifungal, anti-inflammatory, antibacterial, and antioxidant activity.

Keywords: Essential oil, Myrtaceae, Lauraceae, Rutaceae, Burseraceae, Cupressaceae, pharmaceutical, phytoconstituents, ethnobotanical

*Corresponding author: Ameeta Tiwari, M.B.G.P.G College, Haldwani, Kumaun University, Nainital 263139, Uttarakhand, India, e-mail: Tiwari.ameeta@gmail.com
Himanshu Pandey, Sushma Kholiya, Priyanka Tiwari, M.B.G.P.G College Haldwani, Kumaun University, Nainital 263139, Uttarakhand, India
Sushma Kholiya, RC Padalia, CSIR-CIMAP Research Center, Pantnagar 263149, Uttarakhand, India

https://doi.org/10.1515/9783110791600-013

13.1 Introduction

The importance of medicinal and aromatic trees cannot be overstated. Despite their easy accessibility in their native habitat, they are frequently ignored. These trees play a crucial role in human existence. The locals and the impoverished who are directly dependent on these trees can earn a living from them in addition to providing food, fiber, fuelwood, medicine, fruit, and fodder. The primary sources of renewable organic resources are plants. They offer a wide range of complex chemical compounds with significant commercial potential as medicines, biochemicals, flavoring, fragrance, or coloring agents as well as unique and nutritionally dense proteins, lipids, and numerous effective agrochemicals [1]. However, the therapeutic value of these trees differs depending on the species and region. For instance, certain species (*Myrica*, mallotus, bamboos, etc.) are used for medicinal purposes locally and are rarely used for commercial purpose, while others (such as *Acacia*, *Aegle marmelos*, *Terminalia bellerica*, *T. chebula*, *Emblica officinalis*, *Commiphora mukul*, *Boswellia serrata*, *Pterocarpus marsupium*, and *Taxus baccata*) are commercially exploited. Contrarily, the therapeutic properties of species like Eucalyptus, sandalwood, pines, and deodar are of secondary importance and are largely employed for their wood. Similar to this, home grown trees like *Cinnamomum zeylanicum*, *C. tamala*, *M. koenigii*, and *Moringa oleifera* are used for flavoring food and for medicinal uses [2]. Local tribes are aware of the majority of aromatic and medicinal plants with beneficial effects, although they are yet mostly unexplored scientifically. However, governments aim to support indigenous forms of treatment rather than rely on imported medications, there are significant efforts to revitalize old cultures, and the usage of traditional medicine in emerging economies is growing as the population grows [1]. There are billions of dollars traded globally in plant-based medications. Essential oils are obtained by plant extracts. They are produced by steaming or pressing different plant components (flowers, bark, leaves, or fruits) to combine the fragrance-producing compounds. Essential oils are significant fragrant components of flocks and species, and from antiquity, people have used them in fragrances, food preservation, flavor, and healthcare. Essential oils' unique antibacterial, antifungal, and antiviral capabilities are clearly demonstrated by their antimicrobial activities, which show that they are more acceptable. Saturated and unsaturated hydrocarbons, alcohols, aldehydes, ethers, ketones, oxide, phenols, and terpenes are components of essential oils, which may produce distinctive odors [3]. They are transparent, odorless liquids with a high refractive index. These oils work on pressure points and rejuvenate since they are so strong and concentrated. Plants contain essential oils in a variety of places, including pockets and reservoirs, glandular hairs, specialized cells, and even intercellular spaces [4]. Aromatherapy's fundamental methods for treating mental and bodily balance involve inhalation and external application of these oils. These oils are used therapeutically because they are known to reduce stress, revitalize, and renew people in preparation for the next day's job [5].

13.2 Geographical distribution

The Myrtaceae family, which includes the species *M. bracteata, Melaleuca linariifolia, Melaleuca leucodendron*, and *Callistemon citrinus*, is widely distributed in tropical and subtropical regions and is particularly abundant in Australia, Indonesia, Papua New Guinea, tropical America, and south Asia [6]. The genera *Eucalyptus globulus* and *Eucalyptus citriodora* are native to Australia and Tasmania and are members of the Myrtaceae family. Most subtropical and warm temperate regions of Brazil, China, and India are also cultivating these plants [7]. Evergreen trees and shrubs are found in the genus *Cinnamomum*, which is a member of the Lauraceae family. It has over 250 species, most of which are found in Asia but also occur in South and Central America, Australia, and other tropical and subtropical climates. Along with the Seychelles, Madagascar, India, and Sri Lanka, these countries are the main producers of cinnamon [8]. *M. koenigii*, a deciduous to semievergreen aromatic tree found in India, Bangladesh, Nepal, Malaysia, Sri Lanka, and Burma, is a member of the Rutaceae family. It grows up to 1,650 m altitude and occurs in wild and cultivated forms [9]. Indian-born *A. marmelos*, which is a member of the Rutaceae family and belongs to the genus Aegle, is a plant that may be found in Bangladesh, Egypt, Malaysia, Myanmar, Pakistan, Sri Lanka, and Thailand [10]. With 17 genera and 600 species widespread across all tropical climates, the Burseraceae family has a significant presence in the plant world [11]. The *Boswellia* genus has roughly 25 species, the majority of which are found in the gulf countries (viz. Oman, Yemen, and Southern Saudi Arabia), East Africa (Somalia and Ethiopia), South Asia, and a profusion of dry highland regions of India, northern Africa, and the Middle East all have lush vegetation. A forest tree native to Japan known as *Cryptomeria japonica*, sometimes known as a Japanese cedar or sugi, is extensively dispersed throughout Asia's warm and cool temperate temperatures as well as the Azores Archipelago (Portugal). It is an essential timber tree for industry [12].

13.3 Plant morphology

Melaleuca plants can be either shrubs or trees. Every species in the genus is an evergreen, with small to big leaves. Sixty percent of recognized species have leaves that are between 10 and 30 millimeters long, while the remaining species have medium- to long-length leaves. The shape of a leaf can be elliptic, cordate, falcate, lanceolate, linear, oblong, ovate, obovate, triangular, or any combination of these. Leaves are cordate, attenuate, cuneate, truncate, or obtuse at the base. *Melaleuca* leaves typically have visible oil glands and may have a strong scent or none at all. Leaf venation can be subtle and can be pinnate, longitudinal, or longitudinal–pinnate [13]. *C. camphora* trees normally grow between 900 and 2,500 m above sea level and can grow up to many tens of meters (30–40 m) in height and 3 m in diameter. The bark is divided vertically and is either yellow or brown. Strong dormant buds encircled by expansive, silky-like recesses, and three to five distinct veins running alternately across the leaves [14]. *C. tamala* is a medium-sized monoecious evergreen tropical tree of the Lauraceae family [15]. This species grows to a height of 6–20 m and stem width of 150 cm. The short stalked, opposite, and alternately arranged leaves are lanceolate, globous, and three-nerved from the base. Mucilage is produced by its bark. The flowering season runs from March through May, and flowers are bisexual [16]. A fragrant deciduous to semievergreen tree, *M. koenigii* is a little spreading shrub that grows to a height of about 2.5 m. Its main stem is 16 cm in diameter and has a girth of 16 cm [9]. Its bark may be peeled off longitudinally to reveal the white wood beneath. Exstipulate, bipinnately compound leaves measure 30 cm in length and have 24 leaflets each. Bisexual, white, funnel-shaped, completely ripe fruits with a very shiny black surface make up the flowers [17]. The slow-growing *A. marmelos* tree can grow to a height of 25–30 feet. It has a soft, compact trunk and a few prickly limbs. The bark has a lot of sharp, straight spines and is gray or brownish in color. It has trifoliate leaves with a pointed tip and a round base. While the immature leaves are pale green, the adult leaves are a dark green color. The bisexual flowers have a greenish or yellowish color. Fresh leaves usually make it obvious. The bael fruit has a thick, 5–12 cm diameter outer shell that is resistant to damage [10]. The eucalyptus tree, which may reach heights of 40–70 m and has a straight, huge trunk that measures 0.6–2 m in diameter, is classified as an evergreen tall tree [18]. The leaves of this are petiolate, alternating, glossy, or waxy green and mainly hang downwards [19]. Numerous fluffy stamens, which can be white, cream, yellow, pink, or red, are present in the eucalyptus blooms. The operculum, also known as the cap above the stamens, is made of fused sepals, petals, or both. The forcing-off of the operculum on the proliferation of stamens is one of the distinctive characteristics of the genus *Eucalyptus*. The age of the plant and the depth of the furrowing affect the thickness, hardness, and color of the bark. The mature eucalyptus tree adds an annual covering of bark, which helps to increase the stems' diameter [20]. *B. serrata* is a deciduous tree that typically reaches a height of about ten feet (4–5 m). It has a 2.4 m circumference and is a medium to big branched tree (average 1.5 m). The thin bark's color changes from greenish gray, yellow, or reddish to ash, which can be removed with ease. The papery bark exudes translucent lumps, tears, or drops of sticky oleoresin that range in hue from white to yellow upon peeling or incision [21]. The gum has a harsh flavor and a balsamic fragrance. Odd pinnate leaves are 30–45 cm long, exstipulate, and have a variety of shapes. They are crowded near the ends of the branches. Leaflets are 8–15, 2.5–6.3 1.2–3.0 cm, elliptical or ovate-lanceolate, somewhat sessile, short-toothed, and primarily hairy. Bisexual, tiny, white flowers appear at the tips of the branches as axillary racemes or panicles. Trigonous, 1.25 cm long, cotyledous, and oblong in shape [22]. A very huge, conical, monoecious tree with a trunk diameter of up to 4 m, the *C. japonica*, can grow as high as 70 m (230 feet) (13 ft). It is a fast-growing tree that needs deep, moist soils with good drainage [23]. The bark peels off in vertical strips and is reddish-brown in color. The leaves are 0.5–1 cm (0.20–0.39 in) long, needle-like in structure, and smelly due to the presence of EO. The seed cones have 20–40 scales and are globular, measuring up to 1–2 cm (0.39–0.79 in) [12]. The images of selected species are mentioned in fig. 13.1.

13.4 Phytochemistry

Phenyl propanoids were found to be the main chemical constituents in tea tree oils of *M. bracteata* from diverse geographical locations. Methyl eugenol was said to make up 97% of the oil from Egypt, compared to barely 50% in Australian oil. Furthermore, Australian oil included elemicin, iso-elemicin, and methyl isoeugenol. Also, methyl eugenol (76%) is abundant in the EO of this species from Malaysia. According to reports, the Indian Bracteata oil contains phenylpropanoids including 89% methyl eugenol and 2–5% methyl cinnamate [24]. *Murraya koenigii* leaves from the southern part of Tamilnadu, India, were hydrodistilled to extract its essential oil. Linalool (32.83%), elemol (7.44%), geranyl acetate (6.18%), myrcene (6.12%), allo-ocimene (5.02), and α-terpinene (4.9%) were the major chemicals found in the oil [25].

The production and quality of *A. marmelos* leaves essential oil were closely correlated with the growing region and the harvesting season. According to Kathirvel et al. [26], the chemical composition of the essential oil of *A. marmelos* (L.) grown in the Western Ghats region contains *p*-mentha-1,4(8)-diene (33.2%), limonene (13.1%), *p*-cymen-α-ol (9.5%), γ-gurjunene (7.9%), and β-phellandrene (4.3%) as the major components, while from north India the major components of the essential oils were limonene (31.0–90.3%), α-phellandrene (<0.05–43.5%), (*E*)-β-ocimene (0.7–7.9%), α-pinene (<0.05–7.5%), (*E*)-caryophyllene (0.5–5.3%), β-elemene (<0.05–4.2%), and germacrene B (0.0–3.3%) [27].

1,8-Cineole, also called eucalyptol, is a major component of *E. globulus* oil. Different countries have different amounts of eucalyptol in their *E. globulus* leaf EO, with Tunisia having 53.7%, Germany having 14.5%, India having 33.6%–66.7%, Brazil having 71% [28], and Tanzania having 51.62% [29]. High concentrations of citronellal, citronellol, and citronellyl acetate can be found in *E. citriodora* oils from Australia, Morocco, or Burundi [30]. About 1.36% of the essential oil extracted from *E. citriodora* leaves in China is citronellal, which makes up the majority of the oil

Melaleuca
bracteata

Boswellia
serrata

Cinnamomum
camphora

Melaleuca
linariifolia

Eucalyptus
globulus

Melaleuca
leucodendron

Callistemon
citrinus

Cinnamomum
tamala

Aegle
marmelos

Murraya
koenigii

Figure 13.1: Images of selected tree species from Kumaun Himalaya Uttarakhand.

(57%) and is followed by citronellol (15.89%), citronellyl acetate (15.33%), and other chemicals [31].

Ent-kaurene and elemol are the two main components of *C. japonica* leaf essential oils in Nepal and East Asia, respectively. Sabinene and α-pinene, however, had a bigger weight percent than ent-kaurene in Corsica (France) [12]. The 1,8-cineole (77.4%) was found to be the main constituent in the essential oil analysis of fresh *M. linariifolia* leaves from North India. However, it was shown that methyl eugenol (86.8%) and (*E*)-methyl isoeugenol (1.4%) are the main components of the essential oil of *M. linariifolia* from Brazil [13].

Under varied climatic and ecological settings, the composition of the essential oils of distinct *Melaleuca* species revealed a wide range of variation. Australia and Brazil have reported methyl eugenol (99%) and (*E*)-methyl isoeugenol (88%)-rich chemotypes. Viridiflorol (28.2%) and 1,8-cineole (21.3%)-type compositions are reported from Cuba, while terpinolene (29.21%) and α-terpinene (22.55%) containing leaf oil composition of *M. leucadendra* are reported from Thailand. 1,8-Cineole (44.76–64.30%)-rich chemotype is reported from Egypt and Java, Indonesia. In the foothills of northern India, *Melaleuca leucadendra* L. was grown for its essential oils, which were primarily made up of oxygenated sesquiterpenes (81.23–93.50%) and sesquiterpene hydrocarbons (1.84–11.41%). (*E*)-Nerolidol (76.58–90.85%) was the primary component of essential oils from north India [32].

C. camphora can be divided into five chemotypes based on the primary constituents of its leaf oil: isoborneol, camphor, 1,8-cineole, linalool, and borneol types. The majority component of the essential oil recorded from Lucknow, India, was camphor, at about 74% [33]. An average of 49.8% 1,8-cineole was found in the leaf EOC of the cineole-type from eastern Australia by Stubbs et al. [34] Linalool (26.6%), 1,8-cineole (16.8%), α-terpineol (8.7%), isoborneol (8.1%), β-phellandrene (5.1%), and camphor (5.0%) were found to be the major components of the leaf EOC by Chen et al. [35] The oil of *C. tamala* was collected from the northern region (Chandigarh Botanical Garden, Chandigarh) and subjected to a GC–MS analysis. The results revealed 20 constituents, of which methyl eugenol (46.65%), eugenol (26.70%), *trans*-cinnamyl acetate (12.48%), and β-caryophyllene (6.26%) were found to be the main ones. Cinnamaldehyde (44.90%), *trans*-cinnamyl acetate (25.33%), and ascabin (15.25%) were identified to be the primary elements of the oil collected from the southern area according to the GC–MS study [36]. Most Callistemon species' essential oils contain a significant amount of 1,8-cineole. Oil from the northern plains only contained 38.1% of 1,8-cineole and 3.7% of α-pinene, whereas oil from the lower Himalayan region contained much larger quantities of 1,8-cineole (66.32%) and α-pinene (18.7%). This sharp discrepancy in the 1,8-cineole concentration of *C. citrinus* cultivated in the lower Himalayan region and the northern plains may be the result of agroclimatic factors [37].

According to Kasali et al. [38], *B. serrata* collected from Nigeria has a significant amount of α-pinene. According to Singh et al. [39] α-thujene (47%) was a prominent chemical in commercial samples, but α-pinene had not been found in the wild. Samples were taken from Jabalpur District, Madhya Pradesh, India. The main component of the essential oil of *B. serrata* found in several Indian sites is boswellic acid [40].

The major chemical constituents of these tree plants are compiled in Table 13.1.

Table 13.1: Major chemical constituents present in reviewed tree species.

S. no.	Tree name	Major chemical constituents	References
1	*Melaleuca bracteata*	Methyl eugenol, (*E*)-methyl cinnamate, methyl chavicol, elemicin	[6]
2	*Melaleuca linariifolia*	1,8-Cineole, α-terpineol, α-pinene, β-caryophyllene	[13]
3	*Melaleuca leucodendron*	(*E*)-Nerolidol, β-caryophyllene, viridiflorol, (*E*)-β-farnesene, β-humulene	[32]
4	*Cinnamommcamphora*	Camphor, camphene, α-pinene, β-pinene, 1,8-cineole	[14]
5	*Cinnamommtamala*	Eugenol, methyl eugenol, aromadendrene, β-caryophyllene, spathulenol, viridiflorene	[16]
6	*Callistemon citrinus*	1,8-Cineole, α-pinene, β-terpinene, p-cymene	[40]
7	*Murrayakoenigii*	α-Pinene, β-pinene, sabinene, (*E*)-caryophyllene, terpinen-4-ol, limonene	[9]
8	*Aegle marmelos*	Limonene, α-phellandrene, (*E*)-β-ocimene, α-pinene, (*E*)-caryophyllene, β-elemene, germacrene B, myrcene, α-humulene	[27]
9	*Eucalyptus globulus*	1,8-Cineole, α-pinene, limonene, α-terpineol, linalool, γ-terpinene	[7]
10	*Eucalyptus citriodora*	Citronellal, β-citronellol, geraniol, citronellyl acetate, δ-cadinene	[31, 41]
11	*Boswellia serrata*	Boswellic acids, serratol, α-pinene, *cis*-verbenol, limonene	[22]
12	*Cryptomeria japonica*	δ-Cadinene, epicubenol, α-muurolene, cubenol, β-eudesmol, torreyol, α-cubebene	[42]

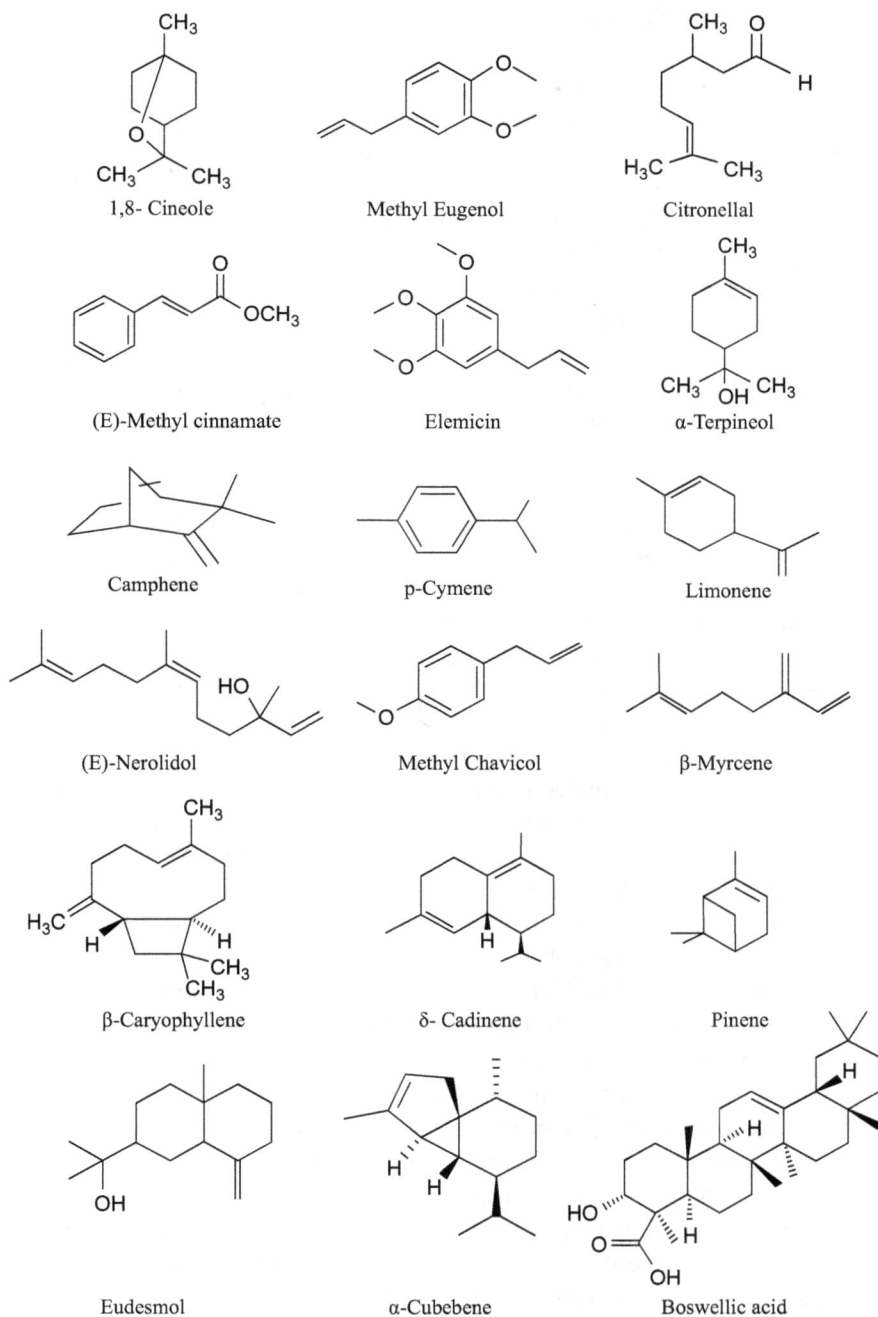

Figure 13.2: Structure of major chemical constituents present in reviewed tree species.

13.5 Ethnobotanical uses

Melaleucas were significant to the development of traditional aboriginal remedies. The inner bark and leaves of *M. Leucadendra* were used for medication (coughs and colds, aches and pains, cuts and sores, ringworm, vomiting and diarrhoea, and other malaises). Some melaleucas' bark was applied as a poultice to wounds and used to splint fractured bones; the bark's juice was thought to permeate the skin and promote healing [43]. *C. camphora* tree has historically been utilized for gynecological issues, metabolic and cardiac issues, bronchitis, colds, congestion, diarrhea, dysentery, edoema, influenza, and flatulence [8] *C. tamala* is used to treat spermatorhea, dry mouth, coryza, anorexia, bladder issues, and spermatorrhea in Ayurveda [44]. As a result of their fragrant, astringent, stimulant, and carminative properties, leaves and bark are used to treat a variety of conditions, including malaria, rheumatism, skin issues, cardiac issues, dental issues, diarrhoea, nausea, colic, ophthalmia, and vomiting. Children are given crushed seeds combined with honey to treat diarrhea and cough. To treat fever, anemia, and body odor, dried leaves and bark are utilized. Additionally, it is thought that this plant uplifts the spirit, revives, promotes mental clarity, and lessens exhaustion [16]. Traditional uses for *M. koenigii* leaves include antidiarrheal, antifungal, blood-purifying, anti-inflammatory, and antidepressant properties. Curry leaves can be used whole or in part. A great hair tonic for maintaining a normal hair tone and promoting hair development is made by boiling fresh curry leaves in a coconut oil mixture until they are reduced to a black residue. Treating snakebites with the bark is beneficial [25]. Vedic literature provides a detailed description of *A. marmelos* for the treatment of many different ailments. Bark is used to treat heart problems, intermittent fevers, and stomach problems. Leaves are used to treat a variety of ailments, including cholera, ulcers, dropsy, beriberi, heart problems, retching, stomach issues, and spinal and ocular complaints. Fruits are used to treat gonorrhoea, stomach problems, intestinal parasites, and epilepsy [11]. To manage diabetes mellitus, *Eucalyptus* tea is made from the leaves of the plant. Tea made from *Eucalyptus* leaves is also used to treat flu, sore throats, and colds. Chewing on leaves helps the gums grow stronger. Bark fine powder is employed as an insect dust [7, 19]. In traditional Ayurvedic and Unani texts, the gummy resin of this tree *B. serrata* is useful for treating arthritis. It is also mentioned as a potent treatment for diarrhea, dysentery, ringworm, boils, fevers (antipyretic), skin and blood diseases, cardiovascular diseases, mouth sores, bad throats, bronchitis, asthma, cough, vaginal discharges, hair loss, jaundice, hemorrhoids, syphilitic diseases, and irregular menstrual cycle. Additionally, it has diaphoretic, astringent, diuretic, and stimulating properties both internally and externally [11]. Sugi, also known as *C. japonica*, is grown in Japan as a decorative plant and produces high-quality wood products. Boiling sugi leaf extracts are used to treat gonorrhoea, eczema, tumors, and injuries [45].

13.6 Biological activities

Reported biological activities of selected plant species from temperate zone of Kumaun Himalayas along with their common name are listed in Table 13.2.

Table 13.2: Biological activities of selected tree species.

S. No.	Tree name	Family	Common name	Plant parts used	Biological activity	References *
1	*Melaleuca bracteata*	Myrtaceae	Black tea tree/river tea tree or mock olive	Leaves	Antibacterial, antimicrobial, antioxidant, and antiseptic	[6, 24]
2	*Melaleuca linariifolia*	Myrtaceae	Flax-leaved paper bark tree	Leaves	Antibacterial and larvicidal	[32]
3	*Melaleuca leucodendron*	Myrtaceae	White paperbark, long-leaved paperbark tree	Leaves, twig, and flowers	Antioxidant, antimicrobial, antibacterial, anti-inflammatory, antiproliferative, and nematicidal	[46]
4	*Cinnamomum camphora*	Lauraceae	Camphor tree	Leaves, stems, and fruits	Anti-inflammatory, anticancer, antifungal, antimicrobial, insecticidal, antiviral, antitussive, antibacterial, and antimutagenic	[47]
5	*Cinnamomum tamala*	Lauraceae	Tejpat, Indian bay leaf, or Malabar leaf	Leaves, bark, and fruits	Antibacterial, antidiarrhoeal, acaricidal, cytotoxic, antiulcer, antifungal, antidiabetic, analgesic, anti-inflammatory, antioxidant, antiviral, gastroprotective, and antimicrobial	[16]

Table 13.2 (continued)

S. No.	Tree name	Family	Common name	Plant parts used	Biological activity	References
6	*Callistemon citrinus*	Myrtaceae	Bottle brush	Leaves, flowers, and stem	Antioxidant, antimicrobial, antihyperglycemic, antibacterial, anti-inflammatory, hepatoprotective, antithrombin, antidiabetic, and hypolipidemic	[38, 48]
7	*Murraya koenigii*	Rutaceae	Curry patta	Leaves, root, bark, and fruit	Antidiabetic, antifungal, antineoplastic, antihelminthics, analgesics, antioxidative, hepatoprotective, antimicrobial, anticancer, neuroprotective, wound healing, and anti-inflammatory	[9, 25]
8	*Aegle marmelos*	Rutaceae	Bael	Leaves, fruits, stem, bark, and roots	Astringent, antidiarrheal, antidysenteric, demulcent, antipyretic, antiscourbutic, haemostatic, aphrodisiac antidote to snake venom. antimicrofilarial, radioprotective, analgesic, antihyperglycemic, antidyslipidemic, anticancer, antidiabetic, insecticidal, antigenotoxic, wound healing, antipyretic, antithyroid, immunomodulatory, and antioxidant	[10, 27]

Table 13.2 (continued)

S. No.	Tree name	Family	Common name	Plant parts used	Biological activity	References
9	*Eucalyptus globulus*	Myrtaceae	Blue gum	Seeds, leaves, and stem	Skin diseases, asthma, antimalarial, analgesic, antiseptic, antirheumatic, antibacterial, antifungal, antidiabetic, anti-inflammatory, antioxidant, antiviral, and antiacne	[49]
10	*Eucalyptus citriodora*	Myrtaceae	Lemon-scented gum	Leaves	Antimicrobial, antifungal, anticandidal, antibacterial, expectorant, cough stimulant activity, herbicidal, insecticidal, antihelmintic, antitumor, and antileech	[31, 50]
11	*Boswellia serrata*	Burseraceae	Salai guggul, Indian frankincense	Resin	Anti-inflammatory, antimicrobial, antiarthritic, anticancer, antihyperlipidaemic, antiasthmatic, analgesic, antirheumatic, antimicrobial, antidiarrheal, and antifungal	[22]
12	*Cryptomeria japonica*	Cupressaceae	Cedar	Leaves	Insecticidal, antibacterial, antifungal, cytotoxic, antiulcer, anxiolytic, and analgesic	[40]

13.7 Conclusion

The area of medicinal plants is increasing worldwide. However, its utilization to various sectors like to improve health, economic, food, and nutrition security is yet a dream to come true. In the modern world where the synthetic drugs are matter of concerns due to their side effects on the human health it becomes vital that cultivation of medicinal plants possessing the proper active constituent should be given prime importance. Medicinal and aromatic trees are being used in traditional medicinal practices. So, the promotion of these would give a boost to the economy of the state through the economic benefit that would accrue to the local rural economy in the hilly and tarai regions. To provide remunerative prices to the farmers for cultivating medicinal and aromatic trees market practices for medicinal and aromatic plants need to be organized systematically.

The present study reveals that the analysis of phytoconstituents and bioactive constituents of different species of aromatic trees of subtemperate zone of Kumaun Himalaya which belong to five different families viz. Myrtaceae (*M. bracteata*, *M. linariifolia*, *Melaleuca leucadendron*, *C. citrinus*, *E. globulus*, and *E. citriodora*), Lauraceae (*C. camphora* and *C. tamala*), Rutaceae (*M. koenigii* and *A. marmelos*), Burseraceae (*B. serrata*), and Cupressaceae (*C. japonica*). These tree extracts are rich in nutrients and are effective as antibacterial, antioxidant, antimicrobial, antiinflammatory, and antimicrobial.

References

[1] Garg GK, Gaur AK, Kumar A, Agrawal S. Biosources management of medicinal and aromatic plants using biotechnology as tool. Souvenir 2005, S-18, 105–115.
[2] Tiwari S, Kaushal R. Medicinal and aromatic trees: Status and Perspectives. Souvenir 2005, S-22, 141–143.
[3] Schiller C, Schiller D. 500 Formulas For Aromatherapy: Mixing Essential Oils For Every use, Sterling Publications, USA, 1994.
[4] Krishna A, Tiwari R, Kumar S. Aromatherapy- an alternative health care through essential oils. J Med Aromat Plant Sci 2000, 22(1B), 798–804.
[5] Perry N, Perry N. Aromatherapy in the management of psychiatric disorders clinical and neuropharmacological perspectives. CNS Drugs 2006, 20(04), 257–280.
[6] Goswami P, Verma SK, Chauhan A, Venkatesha KT, Verma RS, Singh VR, Darokar MP, Chanotiya CS, Padalia RS. Chemical composition and antibacterial activity of *Melaleuca bracteata* essential oil from India: a natural source of methyl eugenol. NPC Nat Prod Commun 2017, 12(6), 965–968.
[7] Surbhi, Kumar A, Singh S, Kumari P, Rasane P. Eucalyptus: phytochemical composition, extraction methods and food and medicinal applications. Advances in traditional medicines 2021, https://doi.org/10.1007/s13596-021-00582-7.
[8] Kumar S, Kumari R. *Cinnamomum*: Review article of essential oil compounds, ethnobotany, antifungal and antibacterial effects. Open Access J Sci 2019, 3(1), 13–16.

[9]   Verma RS, Padalia RC, Arya CA. Aroma profiles of the curry leaf, *Murrayakoenigii (L.)Spreng.* chemotypes: Variability in north India during the year. Ind Crops Prod 2012, 36, 343–349.

[10]   Savita SAP, Singh AP. *Aegle marmelos (L.)* (Bael): A Systematic Review. J Drug Deliv Ther 2021, 11(3-S), 131–136.

[11]   Siddiqui MZ. *Boswellia serrata*, a potential anti-inflammatory agent: An overview. Indian J Pharm Sci 2011, 73(3), 255–261.

[12]   Lima A, Arruda F, Medeiros J, Baptista J, Madruga J, Lima E. Variations in essential oil chemical composition and biological activities of *cryptomeria Japonica* (Thunb. ex L.f.) D. Don from different geographical origins – A critical review. Appl Sci 2021, 11, 11097.

[13]   Padalia RC, Verma RS, Chauhan A, Goswami P, Verma SK, Darokar MP. Chemical composition of *Melaleuca linarrifolia* Sm. from India: A potential source of 1,8-cineole. Ind Crops Prod 2015, 63(2015), 264–268.

[14]   Lee SH, Kim DS, Park SH, Park H. Phytochemistry and applications of *cinnamomum camphora* essential oils. Molecules 2022, 27, 1–10.

[15]   Sharma G, Nautiyal AR. *Cinnamomum tamala*: A valuable tree from Himalayas. Int J Med Aromat Plants 2011, 1(1), 1–4.

[16]   Abha KM, Shukla G, Chakravarty S. The spice Tree of India: *Cinnamomum tamala*. J Tree Sci 2021, 40, 92–100.

[17]   Parmar C, Kaushal MK. Murraya koenigii. In: C. Parmar and M.K. Kaushal (Eds), Wild Fruits, Kalyani Publishers, New Delhi, India, 1982, pp. 45–48.

[18]   Hardel DK, Laxmidhar S. A review on phytochemical and pharmacological of *Eucalyptus globulus*: A multipurpose tree. Ijrap 2011, 2(5), 1527–1530.

[19]   Dixit A, Rohilla A, Singh V. *Eucalyptus globulus*: A new perspective in therapeutics. Int J Pharm Chem Sci 2012, 1(4), 2020–2025.

[20]   Vecchio M, Loganes C, Minto C. Benefcial and healthy properties of *Eucalyptus* plants: A great potential use. Open Agri J 2016, 10, 52–57.

[21]   Senghani MK, Patel PM. Pharmacognostic and phytochemical study of Oleo gum resin from *Boswellia serrata*. Res J PharmacogPhytochem 2013, 5, 244–245.

[22]   Iram F, Khan SA, Husain A. Phytochemistry and potential therapeutic actions of Boswellic acids: A mini review. Asian Pac J Trop Biomed 2017, 7(6), 513–523.

[23]   Nagakura J, Shigenaga H, Akama A, Takahashi M. Growth and transpiration of Japanese cedar (*Cryptomeria japonica*) and Hinoki cypress (*Chamaecyparisobtusa*) seedlings in response to soil water content. Tree Physiol 2004, 24, 1203–1208.

[24]   Yasin M, Younis A, Javed T, Akram A, Ahsan M, Shabbir R, Ali MM, Tahir A, El-Ballat EM, Sheteiwy MS, Sammour RH, Hano C, Alhumaydhi FA, El-Esawi MA. River tea tree oil: Composition, antimicrobial and antioxidant activities, and potential applications in agriculture. Plants (Basel) 2021, 10(10), 2105.

[25]   Balakrishnan R, Vijayraja D, Jo SH, Ganesan P, Su-Kim I, Choi DK. Medicinal profile, phytochemistry, and pharmacological activities of *murrayakoenigii* and its primary bioactive compounds. Antioxidants (Basel) 2020, 9(2), 101.

[26]   Poonkodi K, Vimaladevi K, Suganthi M, Gayathri N. Essential Oil Composition and Biological Activities of Aegle marmelos (L.) Correa Grown in Western Ghats Region-South India, Journal of Essential Oil Bearing Plants, 2019, 22(4), 1013–1021.

[27]   Verma RS, Padalia RC, Chauhan A. Essential oil composition of *Aegle marmelos (L.)* Correa: Chemotypic and seasonal variations. J Sci Food Agric 2014, 94(9), 1904–1913.

[28]   Boukhatem MN, Boumaiza A, Nada HG, Rajabi M, Mousa SA. *Eucalyptus globulus* essential oil as a natural food preservative: antioxidant, antibacterial and antifungal properties in vitro and in a real food matrix (orangina fruit juice). Applied Sciences 2020, 10(16), 5581.

[29] Almas I, Innocent E, Machumi F, Kisinza W. Chemical composition of essential oils from *eucalyptus globulus* and *eucalyptus maculata* grown in tanzania. Sci Afr 2021, 12, e00758.

[30] Chalchat JC, Muhayimana A, Habimana JB, Chabard JL. Aromatic plants of Rawanda II. Chemical composition of essential oils of ten *Eucalyptus* species growing Ruhande Arboretum, Butare Rwanda. J Essent Oil Res 1997, 9(2), 159–165.

[31] Luqman S, Dwivedi GR, Darokar MP, Kalra A, Khanuja SP. Antimicrobial activity of *Eucalyptus citriodora* essential oil. Int J Essent Oil Ther 2008, 2, 69–75.

[32] Padalia RC, Verma RS, Chauhan A, Chanotiya CS. The essential oil composition of *Melaleuca leucodendra L.* grown in India: A novel source of (E)-nerolidol. Ind Crops Prod 2015, 69, 224–227.

[33] Pragadheesh V, Saroj A, Yadav A, Chanotiya C, Alam M, Samad A. Chemical characterization and antifungal activity of *Cinnamomum camphora* essential oil. Ind Crops Prod 2013, 49, 628–633.

[34] Stubbs, B, Specht A, Brushett, D. The Essential Oil of *Cinnamomum camphora* (L.) Nees and Eberm. —Variation in Oil Composition Throughout the Tree in Two Chemotypes from Eastern Australia. Journal of Essential Oil Research 2004, 16, 200–205.

[35] Chen J, Tang C, Zhang R, Ye S, Zhao Z, Huang Y, Xu X, Lan W, Yang D. Metabolomics analysis to evaluate the antibacterial activity of the essential oil from the leaves of *Cinnamomum camphora* (Linn.) Presl. J. Ethnopharmacol 2020, 253, 1–35.

[36] Kumar S, Sharma S, Vasudeva N. Chemical compositions of *Cinnamomum tamala* oil from two different regions of India. Asian Pac J Trop Dis 2012, 2, S761–S764.

[37] Srivastava SK, Ahmad A, Jain N, Aggarwal KK, Syamasunder KV. Essential oil composition of *Callistemon Citrinus* leaves from the lower region of Himalayas. J Essent Oil Res 2001, 13, 359–361.

[38] Kasali AK, Adio AM, Oyedeji AO, Eshilokun AO, Adefenwa M. Volatile constituents of *Boswellia serrata* Roxb (Bursearaceae) bark. Flavour Frag J 2002, 17, 462–464.

[39] Singh B, Kumar R, Bhandari S, Pathania S, Lal B. Volatile constituents of natural *Boswellia serrate* oleo-gum-resin and commercial samples. Flavour Frag J 2007, 22, 145–14.

[40] Mande P, Sekar N. Comparison of chemical composition, antioxidant and antibacterial activity of *Callistemon citrinus* skeels(bottlebrush) essential oil obtained by conventional and microwave-assisted hydrodistillation. J Microwave Power Electromagn Energy 2020, 54(3), 1–15.

[41] Costa AV, Pinheiro PF, de Queiroz RVM, Marins AK, Valbon WR, Pratissoli D. Chemical composition of essential oil from *eucalyptus citriodora* leaves and insecticidal activity against myzuspersicae and frankliniellaschultzei. Teop 2015, 18(2), 374–381.

[42] Mizushina Y, Kuriyama IC. *Cedar (Cryptomeria japonica)* Oils. Essential Oils in Food Preservation, Flavor and Safety, Elsevier (Eds. Victor R. Preedy) London, 2016, 317–324.

[43] Williams C. Medicinal plants in Australia, Volume 2: Gums, resins, tannin and essential oils, Rosenberg Publishing, Sydney, 2011.

[44] Kapoor IPS, Singh B, Singh G. Essential oil and oleoresins of *Cinnamomum tamala (tejpat)* as natural food preservatives for pineapple fruit juice. J Food Process Preserv 2008, 32(5), 719–728.

[45] http://medicinalplants.us/cryptomeria-japonica-don-japanese-cedar.

[46] Monzote L, Scherbakov AM, Scull R, Satyal P, Cos P, Shchekotikhin AE, Gille L, Setzer WN. Essential oil from *Melaleuca leucadendron*: antibacterial, anti kinetoplastid, antiproliferative and cytotoxic assessment. Molecules 2020, 25, 1–13.

[47] Malabadi RB, Kolkar KP, Meti NT, Chalannavar RK. Camphor tree, *Cinnamomum camphora* (L.); Ethnobotany and pharmacological updates. Biomedicine 2021, 41(2), 181–184.

[48] Kumar R, Gupta A, Singh AK, Bishayee A, Pandey AK. The antioxidant and antihyperglycemic activities of Bottlebrush plant (*Callistemon lanceolatus*) stemextracts. Medicines 2020, 7(11), 1–16.

[49] Sharma S, Sushma N. A Review on *Eucalyptus Globulus* – An Authentic Herb. J Pharm Res Int 2021, 33(53B), 107–114.

[50] Dhakad AK, Pandey VV, Beg S, Rawat JM, Singh A. Biological, medicinal and toxicological significance of *Eucalyptus* leaf essential oil: A review. J Sci Food Agric 2018, 98(3), 833–848.

Index

https://doi.org/10.1515/9783110791600-014